Why **Less is More** for Children Learning Math: How Parents Can Help Their Child Succeed by Concentrating on Essential Topics

Herbert S. Gaskill, Ph.D.

September 8, 2015

Contents

This book is dedicated to the children who need to learn arithmetic and to the mentors willing to help.

Preface

During the period I was Head of the Mathematics and Statistics Department at Memorial University of Newfoundland, the problem of how to deal with entering students who lacked basic math skills reached unacceptable levels. Unfortunately, in spite of changes in Newfoundland K-12 math curriculum, this problem still exists and was the essential reason I started working on an arithmetic book for parents in 2013. The book was based on the premise, derived from 30 years of teaching, that there was a short list of topics that children needed to learn to succeed in the rest of their math schooling. Thanks to advice from a colleague in the US, I became aware of the Common Core State Standards in Mathematics (CCSS-M) which were, according to their stated guiding principles, aimed at producing a focused and coherent curriculum. The standards articulated by the CCSS-M for K-6 appeared to achieve my goals and I adopted the CCSS-M as a guide with the end result being *Parents' Guide to Common Core Arithmetic* jointly authored with my wife. The story could have ended there.

However, in response to the *Parents' Guide* some questions about K-6 curricula arose which caused me to undertake a serious review of the underlying rationale for the CCSS-M. The principal study on which the CCSS-M is based is an analysis of The International Math and Science Study data from the late 1990s. The lead researcher was Wm. Schmidt of the Education Policy Center at Michigan State University. The analysis performed by the Schmidt group showed that successful countries teach a narrowly focused coherent curriculum. The sad part, as far as I am concerned is that the CCSS-M did not, in my view, carefully follow the prescription for achieving focus set out by the Schmidt group in respect to K-6.[1] As a result, the K-6 Standards, although an improvement, still include too many topics.

The purpose of this book is to remedy that by carefully following Schmidt's prescription which produces a narrow list of essential topics on which parents can focus to ensure their child is able to succeed no matter what math curricula their child's school is using.

The book is also designed to be used as a guide for developing a Schmidt type curriculum for K-6. Such curricula would improve learning outcomes on both sides of the US/Canada border.

Finally, many, if not most, of us believe math is hard. This is a tragedy and it doesn't have to be so. We can make math hard by choosing to teach the poorly designed curricula used in the under-performing countries identified by Schmidt. Or, we can make it straightforward and fun by teaching narrowly focused and coherent curricula in which most children succeed as used in the high-performing countries identified by Schmidt.

---

[1]Links to all supporting documents can be found at: www.mun.ca/math/people/ppl-faculty/retired/

Acknowledgements

The essential difficulty in writing a book like this is in making sure it is accessible to its intended audience. To the degree that this book achieves that goal is entirely due to the efforts of my partner Catherine Gaskill. She has been patient and painstaking in her approach and I am truly grateful for help on this project.

I would also like to thank the current head of the Department of Mathematics and Statistics at Memorial, Chris Radford who has supported all my projects.

Herbert Gaskill
Memorial University of Newfoundland
August, 2015

Chapter 1

Why Less Really is More in K-6 Math

On the face of it, the assertion of the title *Less is More for Children Learning Math* sounds absurd, particularly in the context of math. Nevertheless, this assertion when applied to K-6 math curricula is true and the reader needs to understand why.

1.1   The Concept of an A+ Curriculum

We all know what it means to get an A+ in school, although for most of us, receiving that grade on anything was probably not a common experience. So a K-6 math curriculum that could be described as being **A+** would certainly have to have the following properties:

- almost all children would succeed in mastering the learning tasks set by an A+ math curriculum;

- children learning from an A+ math curriculum would perform better than children learning under all alternative curricula;

- children learning from an A+ math curriculum would be able to continue learning math topics to mastery at higher levels.

The notion of an A+ math curriculum with these properties sounds too good to be true. Moreover, if such a curriculum did exist, it would seem that the children studying under it must do nothing but math all day. In actuality A+ math curricula do exist. By focussing on a narrow list of key topics, these curricula produce successful learning outcomes at high levels for almost all children.

1.2 The Research Identifying A+ Curricula

The concept of an A+ math curriculum was developed by researchers working at the Education Policy Center (EPC) at Michigan State University.[1] The particular body of work that identified the criteria for an A+ curriculum was derived from an analysis of the data set that was collected in 1997 during the Third International Math and Science Study (TIMSS). This analysis was performed by a group led by Wm. Schmidt and published in 2002.[2]

The Schmidt group found that the education systems in various countries have a similar hierarchical structure. At the top is a department of education whose jurisdiction may be nation-wide in a small country like Singapore, or state- or province-wide in a large countries like the U.S and Canada. At the bottom are classroom teachers.

The principal activity of a department of education is to specify what is to be taught to the nation's children. This is accomplished through promulgating an **intended curriculum**[3] consisting of lists of topics to be taught, grades in which these topics are to be taught, and the amount of time allocated to each topic. Of course, for various reasons what is actually taught in a particular classroom may differ from the intended curriculum and for this reason ACC refers to what is actually taught as the **enacted curriculum**. Among the important findings of this TIMSS study were[4]

1. in all countries, what students learn from the enacted curricula is what is in the intended curriculum;

2. some intended curricula are much more effective than others in respect to measurable student learning outcomes;

3. effective intended curricula have common features in respect to the number of topics presented each year, the order and duration of presentation of individual topics, and so forth;

4. successful intended curricula are narrowly focused and coherent;

5. North American intended curricula extant in 2002 tended to be ineffective in respect to learning outcomes precisely because they lacked focus and coherence; hence, they were termed *a mile-wide and an inch-deep*.

To understand the import of the findings for parents and children we need to dig a little deeper.

---

[1]For more information on the work of the EPC visit: http://education.msu.edu/epc/.

[2]*A Coherent Curriculum: The Case of Mathematics*, Wm. Schmidt, H. Houang and L. Cogan, **American Educator**, Summer 2002, available on-line. Hereafter, ACC.

[3]See ACC.

[4]See ACC.

1.3 When Students Fail, Who or What Gets the Blame?

When a country's rankings sink on an international study, assigning blame becomes an important activity. Since the delivery system for the intended curricula is the teacher, it is easy to understand why, as noted by Schmidt *et al.*, there is a tendency to blame the teachers.[5] That such a response is typical may be seen from the following quote excerpted from an editorial on the math curriculum in the Province of Newfoundland and Labrador:

> *The fact that marks have dropped almost universally in tandem with this new math is explained away as a matter of teacher training alone, not curriculum.*[6]

Now consider the problem of poor performance in math from the perspective of a parent. The critical fact to remember is that what a child learns has two components: the teacher (as presenter of the enacted curriculum) and the intended curriculum (as a list of topics promulgated by the central authority). What most of us who were not trained as curriculum professionals assume is that the intended curriculum must be good, since it was prepared by our best trained educators. Thus, if a particular child is failing, it is very easy for a parent to blame the teacher for not getting the job done. Simultaneously, from the teacher's perspective, it is very easy to blame the child, or the home environment, etc. The point is, the curriculum is almost never seen as the fundamental cause of failure.

Now consider the first conclusion of ACC to the effect that universally teachers are teaching children what is in the intended curriculum set out by your state/province's department of education, and that a given teacher is likely to be doing a good job of teaching that curriculum as measured by tests. That's the upside.

The downside is that some curricula are much less effective in producing positive student learning outcomes, so that even though a child has a great teacher, the constraints imposed by an ineffective intended curriculum lead to reduced learning outcomes.[7] As Schmidt *et al.* observed (ACC, p. 3), teachers using typical North American curricula are instructed:

---

[5]ACC, p. 3.

[6]*I used to be good at math. Now I find it hard.*, by Peter Jackson, **The Telegram**, April 29, 2015.

[7]The EPC surveyed teachers on exactly this point and found the principal determinant of what that teacher did was what was in the intended curriculum. More information can be found at the EPC website in Working Paper 33.

Teach everything you can. Don't worry about depth. Your goal is to teach 35 things briefly, not 10 things well.[8]

In short, the curriculum set out by your state/province's department of education really does matter when it comes to your child's long-term success because that is what your child's teacher is going to teach. The question for parents is whether an ineffective math curriculum really will have a measurable negative impact on their child's future success. The fact that at every post-secondary institution in North America students are regularly denied access to programs because of missing math skills attests to the negative impacts. Moreover there are still entry-level jobs that require an ability to correctly perform arithmetic computations as a condition of employment.[9] Being denied access to a program or losing a job because of poor math skills is devastating to a young adult.[10]

The other side of this is whether there really is a curriculum under which essentially all students succeed?

1.4 The Effect of an A+ Curriculum on Student Success

We need to make explicit what the difference is in terms of student outcomes between a coherent curriculum and a typical mile-wide North American curriculum. Arriving at such an assessment starts with asking students from different countries having different math curricula to perform the same, or equivalent, mathematical tasks. This is easy to do in a discipline like mathematics because computations don't change when borders between countries are crossed. Let's look at how a data set like TIMSS is analyzed.

Within each country individual students are rank ordered by percentile. Thus, if a child is ranked in the 100 th percentile in the US, it means no US student answered more questions correctly than that child. If a child was ranked in the 75 th percentile in Canada, it means 25 % of Canadian students answered more questions correctly,

---

[8]Unfortunately, in the opinion of this writer, the Common Core State Standards in Math (CCSS-M) also suffer from too many topics in K-6.

[9]I recently had breakfast at a Waffle House. The waitress was great at her job until it came time to add up the bill. Her computation contained $12.00 + $1.38 = $12.38. I suggested she check her arithmetic which she did and fixed it. She thanked me and assured me that an error like that could result in her dismissal.

[10]This remark is obvious on its face. But as head of a math department having to witness the pain experienced by such students provides an entirely different perspective, particularly when it does not have to happen.

while 74 % answered less correctly. Finally, if a child was ranked in the 25 th percentile it means that 75 % of students answered more questions correctly. Percentile rankings provide no information as to how many actual questions were correctly answered by any student. They only indicate how a particular test score by one child ranks relative to the test scores generated by all children in the given country. Percentile data in one form or another is generally what is available on public websites such as the one maintained by the National Center for Educational Statistics in the US.[11]

Countries can also be rank ordered based on the relative performance of that country's students. The simplest way to do this is to average the scores of all students within a country and then list the averages in descending order to obtain a ranking in comparison to other countries. As an example, the US was ranked 25 th on PISA[12] (UT 2013).[13] This ranking is not a percentile, nor does it tell us anything about actual student performance. It simply means that the students of 24 other countries were judged to perform better on this series of tests, perhaps by the method outlined above. However, within each country there are still percentile/raw-score tables which inform researchers exactly how many correct answers were required to achieve a given percentile ranking in that country. Because the same test instrument is used by each country, the table can be used to determine exactly what an individual raw score in one country means as a percentile in any other country. Thus for example, the researchers at EPC found that the raw score of a US student in the 75 th percentile on TIMSS was ranked **below** the 25 th percentile in Singapore (ACC)! Another way to think of this is that more than 75 % of students being taught under Singapore's curriculum achieve at levels that only 25 % of US students achieve at using a US curricula! In essence, the overwhelming majority of students in Singapore are achieving results every North American parent would be proud of. This is the functional difference in individual student outcomes between learning from a coherent curriculum, such as Singapore's and the typical mile-wide curricula extant in North America prior to 2012.[14] This difference is the reason why parents need to understand exactly what constitutes an A+ curriculum.

---

[11]For example, data from the National Assessment of Educational Progress (NAEP) is kept here: http://nationsreportcard.gov/math_2011/.

[12]PISA is the Program for International Student Assessment run by the Organization for Economic Co-operation and Development (OECD).

[13]Uri Treisman, *Iris M Carl Equity Address: Keeping Our Eyes on the Prize*, NCTM, Denver, April 19, 2013. The address in a variety of formats can be found at: http://www.nctm.org

[14]As I have already noted, the CCSS-M in K-6 still contain far too many topics, although they are certainly better.

1.5 A Model Coherent (A+) K-6 Curriculum

The following table is derived from the effective curricula identified by Schmidt **et al.** in their 2002 paper (ACC) and which they term A+ and illustrates the shared features that make all effective curricula effective. The table faithfully reproduces the data reported in ACC for A+ curricula but has been truncated to omit the data from Grades 7 & 8 because our focus is elementary school. As well, the table has been rearranged so that like topics are presented as a group, e.g., numbers and operations on numbers are presented together.

TOPIC & GRADE:	1	2	3	4	5	6
Whole Number Meaning	●	●	●	○	○	
Whole Number Operations	●	●	●	○		
Common Fractions			□	●	●	○
Decimal Fractions				○	●	○
Relationship of Common & Decimal Fractions				○	●	○
Percentages					○	○
Negative Numbers, Integers & Their Properties						□
Rounding & Significant Figures				○	○	
Estimating Computations				○	○	○
Estimating Quantity & Size				□	□	
TOPIC & GRADE:	**1**	**2**	**3**	**4**	**5**	**6**
Equations & Formulas			□	○	○	○
Properties of Whole Number Operations				□	○	
Properties of Common & Decimal Fractions					○	○
Proportionality Concepts					○	○
Proportionality Problems					○	○
TOPIC & GRADE:	**1**	**2**	**3**	**4**	**5**	**6**
Measurement Units	□	●	●	●	●	●
2-D Geometry: Basics			□	○	○	○
Polygons & Circles				○	○	○
Perimeter, Area & Volume				○	○	○
2-D Coordinate Geometry					○	○
Geometry: Transformations						○
TOPIC & GRADE:	**1**	**2**	**3**	**4**	**5**	**6**
Data Representation & Analysis			□	□	○	○

This table is adapted from the A+ curricula identified by ACC. Topics have been reorganized to reflect domains identified in the CCSS-M and only grades 1-6 are shown. Topics identified with a • are in the intended curricula of all the A+ countries in the grade shown; ○ identify 80% of A+ countries; □ identify 67% of A+ countries. Topics not on this list are **not** in the intended curricula of A+ countries in Grades 1-6!

Solid circles (•) indicate that the topic at the left was taught in the grade at the top by 100% of high-performing countries, i.e., countries that met Schmidt's criteria for being A+ (ACC). A blank indicates the topic was taught by fewer than two thirds of the A+ countries; most likely, the topic was not taught. A careful read of the Schmidt study leads to the conclusion that what is not taught in a particular grade is as important as what is. In other words, **more is not better**. Rather, fewer, carefully selected topics in the right order need to be taught in a concentrated fashion so that students learn to mastery and retain what they have learned over the long-term. This is the essence of **less is more** when it comes to K-6 math.

Focus for a moment on the first group of topics in the table related to numbers and operations. Notice that within this group each consecutive line of the table represents more difficulty and/or sophistication. For example, the first line of the table is concerned with the meaning of whole numbers and reflects the fact that the computational procedures of arithmetic depend on the place-value system of notation. The intention is for children to learn to count and through this process learn about the Arabic (place-value) notation system (see Chapter 7). On the next line the topic is whole number operations. The first operation studied here is single-digit addition for which there is a finger counting procedure that is based on counting. More advanced topics like two-column addition are founded on the simpler topic combined with a deeper knowledge of place-value. A+ countries expect children to have mastered operations with whole numbers by the end of Grade 4.[15] Indeed, 20% of the A+ curricula complete the learning of operations by the end of Grade 3. To ensure children succeed at this, A+ curricula limit the topics taught in Grades 1 and 2 to whole numbers, addition and subtraction and making measurements, as with a ruler.

In A+ countries there is little or no study of common fractions before Grade 3, and a full third of the A+ curricula delay any discussion of fractions until Grade 4. In making this decision, A+ countries recognize the children must master operations with whole numbers first because operations with fractions utilize whole-number operations. Again, introducing more topics in earlier grades does not produce better results. It simply detracts from mastering the important concepts and procedures associated with whole number arithmetic.

Another feature of the A+ curricula is the avoidance of topics that derive from

---

[15]My memory of my own public schooling is that we had finished long division by the end of Grade 4.

algebra and data analysis in the early grades. Rather, the entire focus in Grades 1-3 is on number, operations with numbers and measurements related to the geometry of the line. The reader may wonder, if students are dealing with fewer topics, why don't they learn less? The answer is that students learning from an A+ curriculum are expected to have a more extensive knowledge of what they do learn. For example, the place-value system of notation (§7.1.2) for whole numbers connects to our system of computations (§8.6) and A+ curricula expect that students will come to terms with these relationships and be able to execute the procedures fluidly.[16]

In summary, the best way to help your child is to provide learning support for the limited number of topics identified in the A+ curriculum. Thus, for example, in Grade 1 the issue for your child is coming to terms with the number system, the meaning of addition and subtraction, and how to add and subtract counting numbers less than 20. So the purpose of this book is to enable parents to focus on ensuring that your child masters the critical computational skills that are essential to continued success. In the long run, ensuring your child masters the short list of topics identified on the A+ list is what will provide the best foundation for success in the workplace and/or post-secondary.[17]

---

[16]A detailed analysis of the difference between an A+ curriculum and North American curricula was performed by Liping Ma: *A Critique of the Structure of U.S. Elementary School Mathematics*, found at www.ams.org/notices/201310/fea-ma.pdf This paper is essential reading for anyone who wants to understand how North American curricula in K-6 math went seriously astray.

[17]Recently, a friend asked my advice about a child who was having trouble with the math taught at the start of Grade 4. I told him to make sure his child learned her addition and times tables well. At the end of the year his child got the prize as the best math student in her class. The point is: small amounts of knowledge learned really well go a long way in arithmetic. There is also huge gratification to a child who knows they know the answer.

Chapter 2

The Nature of Whole Number Arithmetic

A familiar complaint of today's parents is:

Math has changed so much that I am unable to help my child.

This complaint would not be so meaningful if today's children were generally successful with the material being taught in our schools. That our children are having difficulty dealing with today's math curricula is evidenced by the number of services that provide tutoring in math.[1] The point is that parents are frustrated by a perceived inability to help their children. And children are frustrated by a curriculum that makes learning difficult for all but a very few, and even those few do not perform at the levels achieved by most students in A+ countries. Clearly we ought to be able to do better on both scores!

At the societal level, doing better would require completely revising the curricula. The US has tried to do this with the Common Core State Standards (CCSS) initiative which began in 2009[2] and the process was still on-going in 2014. In consequence, expecting an immediate societal response that will help children and their parents is simply not possible.

The purpose of this book is to make it possible for parents to help their children in spite of the nature of modern K-6 math curricula. The book accomplishes this by providing a complete explanation of the A+ portions of the curriculum and instructions on how to communicate those explanations to children. I can appreciate

---

[1]A Google search under the phrase *math tutor* turned up more that 26 pages of listings. Exactly how many I don't know because I stopped looking at 26.

[2]See   http://www.corestandards.org/about-the-standards/development-process/#timeline   for the exact timeline.

that there are many who might like to do this but suffer from the *I've never been good at math.* syndrome. The fact is that if you can count, you can master both whole number arithmetic and the arithmetic of common fractions. Not only that, but you can **understand** it at the level intended by those who drafted the CCSS-M. I know this is hard to believe for some, but remember, in A+ countries, essentially all students achieve at high levels, so why not all of us?

2.1   The Nature of Arithmetic and Mathematics

The purpose of mathematics in general and arithmetic in particular is to answer questions. So the first thing to consider when confronted with a piece of arithmetic is:

What question is being answered by this computation?

For example, consider $5 + 4$. The implicit question is:

> Given a collection of five objects and a second collection of four objects, if the two collections are combined into a single collection, how many objects will be in the combined collection?

It is essential that every child know that providing answers to questions like this one is the primary purpose of the addition operation applied to counting numbers.

The next important fact to recognize is that knowledge of what question is being posed is fairly useless without a procedure for finding the answer. The key fact that makes the remainder of this book possible is that for every arithmetical question there is a straightforward procedure that provides the answer. Indeed, for every question that is posed in math and/or physics up until at least the end of first year university, there is a procedure that yields the answer. Thus, the second critical fact that every child must know is:

Every question I will be asked is solved by using a procedure.

Consider again finding $5 + 4$. The first procedure a child should learn for solving this problem is:

> count out five objects to form a collection and then count out four different objects to form a second collection, then combine the two collections and count the result.

This process provides a concrete method for solving all addition problems and the process tells the child the question that is being addressed.[3] Later the child learns a second, more effective procedure, namely, count-on starting at the larger number, and finally, the child will find the sum by recall. But always there is a beginning method that guarantees a successful conclusion.

2.2 Rote Learning vs Understanding in K-6 Math Curricula

Rote learning has a bad name. In part this was due to the fact that in the period prior to the introduction of the post-Sputnik New Math curricula, the focus of K-6 school math curricula was on a seemingly endless repetition of computations in which almost no theory was taught. So we need to ask the question: Does rote learning serve a useful purpose?

2.2.1 The Two Functions of Rote Learning

The most important fact about rote learning is that there are some things that can only be learned by rote. The most important examples are the meanings attached to symbols. For example, the symbols

$$dog \quad \text{and} \quad 5$$

have specific meanings.[4] We cannot sensibly question the meanings assigned to these symbols. They are definitions and simply are. So to ask for an explanation as to why when someone says *dog* we think of a four-legged pet like a spaniel is ridiculous. Similarly, to ask a child to explain why the numeral 5 is associated with a diagram like

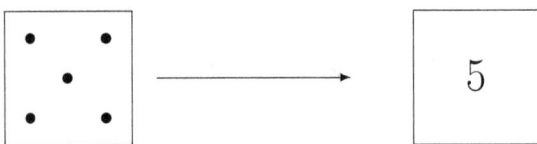

---

[3]Clearly this procedure is not efficient for large numbers, but for the moment we are not concerned with efficiency issues.

[4]A frustrated parent called a local radio station to complain that her daughter had been asked: $2 \times 2 = ?$ Explain. Since any explanation comes down to saying this is a definition, it is no wonder the parent was frustrated.

is absurd. These are definitions and definitions have no explanation. But they must be learned exactly by rote. For this reason rote learning has an important place in mathematics education.

The second function of rote learning relates to the transition from external problem solving to internal problem solving. The process of transition between the two methods has been extensively studied by cognitive psychologists. A brief summary of recent work follows.

Cognitive psychologists identify two kinds of problem-solving. The first is **procedure-based** which simply means the problem solver uses an external procedure to solve the problem, e.g., as described above in respect to $5 + 4$. Since, arithmetic is completely procedure-based, learning these procedures is the first step in a successful math education.

The second kind is **memory-based** problem-solving which means the problem solver recalls the solution from memory. Obviously, an internal-to-the-brain memory-based problem-solving process is much a more efficient means of problem solving.

Until recently, cognitive psychologists knew that children learning math transition from procedure-based to memory-based problem-solving, but they did not understand how. That changed when the most recent research on learning found that the transition to memory-based problem-solving involved changes in a child's brain and these changes continue to enhance the individual's skill at problem-solving as they become adults.[5] We repeat: these changes in a child's brain **last a lifetime** and continue to enhance problem-solving skills. This is why the effect of rote learning of math is so important. And, as we all know, rote learning is accomplished through **practice**.

While memory-based problem solving is the ideal, actual arithmetic problems often require a combination of internal and external procedures. For example consider, 357×924. To perform this computation a child would use recall to produce the set up on paper, use recall to find products like 7×4, use recall of the process to properly place partial results on paper, and so forth. The product example also illustrates the hierarchical nature of arithmetic. New procedures are developed using old procedures as a base which is why it is essential that children learn the procedures for solving problems to the point of mastery. Then they able to successfully incorporate old procedures into new ones. As well, they will always have the most basic procedures, e.g., the procedure for single digit addition, as a fall-back and a means to check their memories.

Successfully transitioning to memory-based knowledge of the facts and procedures of arithmetic is critical because unlearning a wrong memory-based skill is next to

---

[5]See http://www.nature.com/neuro/journal/vaop/ncurrent/full/nn.3788.html and http://www.medicaldaily.com/math-skills-childhood-can-permanently-affect-brain-formation-later-life-298516

impossible at a later point in their education.[6]

So here are two simple things that every parent/mentor can do that are guaranteed to help a child succeed. First, ensure the child knows the addition table from memory by the start of Grade 2, and second, ensure the child knows the multiplication table from memory by the end of Grade 3. Of course there are many other things parents can do, but these two are guaranteed to have immediate substantive benefit to your child's success in math.

Skills requiring rote learning are only effectively acquired by practice. Our presentation will focus on achieving fluidity with computational procedures because these same procedures apply at all levels. For example, the simple procedure we will give for adding common fractions applies equally well to algebraic fractions. This is another reason why these skills have to **last a lifetime**: they will be used over and over due to the hierarchical nature of mathematics.

2.2.2   What is Meant by 'Understanding' in K-6 Curricula

The notion that North American curricula were deficient in that students did not *understand* became prevalent in the 1960's with the advent of the **New Math**. There is substantial literature on this subject but we focus on the discussion in a paper by Z. Usiskin.[7] What is clear from Usiskin's paper is that even today, there is no universal agreement on what it means to **understand a given mathematical idea**. Thus, until we can all agree on what exactly it means to understand and how to demonstrate that understanding, educators will have difficulty assessing the **understanding component** of math curricula, particularly at the primary and elementary level.

With the above caveat in mind, we note that in the 2012 paper, Z. Usiskin[8] suggests that there are four independent components to mathematical understanding:

1. procedural understanding — how to correctly perform a computation;

2. use-application understanding — being able to recognize when a computation should be applied;

3. representational understanding — how to pictorially, or otherwise, represent a computation;

---

[6]I make this last statement based on working with post-secondary students who had weak math kills. Their problems were almost always associated with wrongly learned methods for performing arithmetic computations.

[7]See Z. Usiskin, (2012) http://www.icme12.org/upload/submission/1881_F.pdf (hereafter, ZU 2012)

[8]See ZU 2012.

4. proof understanding — how to derive the formula underlying a computation.

The meaning of **procedural understanding** is the most transparent. We have a procedure and to find out if a child meets the procedural understanding criteria, we would ask the child to execute the relevant computation. For example: Find 26×42 without using a calculator.

The meaning of **use-application understanding** is also fairly transparent. A child is presented with a problem which can be solved by performing a computation. Consider the following:

> A child has a marble collection containing 13 marbles. If a friend gives this child 8 marbles, how many total marbles will the child have?

To solve this problem the child has to know that the correct tool to apply is addition and the required computation is: $13 + 8$. Clearly, this is a higher level of knowledge and illustrates one aspect of the hierarchical nature of mathematics. A child may be unable to solve this problem for two fundamentally different reasons. First, the child may be able to execute $13 + 8$ but may not know that this is the required computation because he/she does not know what question addition is supposed to answer. Alternatively, the child may know that finding $13 + 8$ will answer the question, but not know how to perform the computation. Readers should know that this type of understanding is a major focus of all modern curricula.

Representational understanding is also a major focus of modern curricula. A typical question designed to test this form of understanding is the following:

> Find $10 + 7$ and explain your thinking by drawing a picture.[9]

What would be expected is a diagram something like the following:

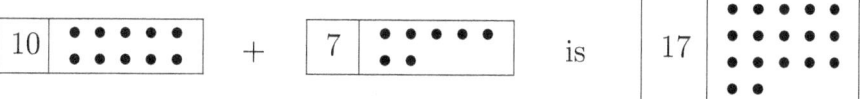

Asking students to explain their work with diagrams like this is ubiquitous in K-2. Producing such a diagram demands that the child know the meaning of each counting number on the left and that they can represent this counting number visually as in §2.2.1. In addition, they must know the operational meaning of $+$ and be able to represent that visually. The diagram shown also makes use of $=$ which represents an additional level of abstraction. Many of the websites containing worksheets have

---

[9]A+ curricula avoid the early introduction of algebra, i.e., prior to Grade 3. Thus, we do not use 'equation' as a descriptor. The CCSS-M does not adhere to this admonition.

questions of this type on which children can get practice. We also note that producing a solution is procedural.[10]

In an A+ curriculum we would expect students in Grade 3 to respond to the following more abstract form of the same question:

Formulate an equation that expresses the sum of 10 and 7 as a quantity to be found.

by writing

$$10 + 7 = \boxed{?}$$

and to understand the meaning of the equality symbol and the $\boxed{?}$ as being an unknown.

The last type is proof understanding. We consider the question of whether proof understanding is, or ought to be, a component of K-6 math curricula.[11] Usiskin equates proof understanding with the knowledge of how to derive a formula (ZU, 2012). Certainly the ability to derive theorems and formulae is basic to understanding. But the ability to reproduce a derivation is not equivalent to *understanding that derivation* as the phrase is used by mathematicians. That type of understanding implies complete knowledge of the ideas and concepts on which the derivation is based, an ability to explain each step of the derivation and a further ability to apply the ideas to discover related theorems and construct proofs for same. In short, proof understanding is highly complex and requires what mathematicians refer to as **mathematical sophistication.** The complexity described suggests that working with proofs would appear to have little place in K-6 and runs counter to the idea that mathematics is hierarchical and is best learned from the particular to the general. The designers of the A+ curricula were clear on this when they made the choice to avoid all algebra before grade 3. Thus, their answer was that proof understanding ought not to be within the province of K-6, a decision with which I completely agree. For this reason, **proof understanding can, and will, be ignored**. We will revisit proof understanding briefly when we discuss discovery in the next section.

To summarize, of the types understanding identified by Usiskin, only three are important for children in K-6. These are procedural, use-application and representational. All questions designed to test one of these three types of understanding are amenable to solution using a recalled procedure because the required information on

---

[10]Step 1: The child must diagram each counting number on the left. Step 2: The child produces a diagram on the right containing one item for each item appearing in their diagrams on the left. Step 3: The plus sign and the equality relation have to be introduced.

[11]My comments are based on more that 30 years of teaching university math courses in which understanding proofs was a significant course requirement.

which answers are based is the factual knowledge of what defines particular operations on numbers. The question above about $10+7$ illustrates this perfectly. For this reason, none of these questions should ever require guessing on the part of a child.

2.3   Discovery Learning in K-6 Math Curricula

Many modern math curricula contain a component of 'discovery'. The idea is that students will discover the principles for themselves with minimal guidance.[12] They are expected to achieve this by completing exercises that lead them to the principle. One of the greatest thinkers ever was Albert Einstein and this is what he said about the task of finding the principles:

> *... But the former task, namely to establish these principles which can serve as the basis for his deductions, is one of a completely different kind. Here there is no learnable, systematically applicable method which would lead to the objective. ...*, p. 24[13]

The procedures and principles on which our mathematics is based took tens of thousands of years to develop. It is a fact that these procedures and principles can be mastered by most children. However asking children to develop these, even in the presence of a qualified mentor, is problematic for the reason identified by Einstein. To do so is wasteful of a child's time, a source of major discouragement for many children and destructive for many in the form of lost potential. The question is what to do about it. The answer provided by this book is to focus on the topics in the A+ curriculum and ignore the rest. Ensuring your child learns these topics to mastery will enable your child to succeed in K-6 and master algebra which is the key to post-secondary and full participation in our technologically based society. That is the goal.

---

[12]An analysis of this instructional paradigm by the eminent cognitive scientists Paul Kirschner, John Sweller and Richard Clark can be found at:
projects.ict.usc.edu/itw/vtt/Constructivism_Kirschner_Sweller_Clark_EP_06.pdf
This paper explains why instructional methods based that expect novice learners to teach themselves do not work and may be counter productive. The paper is informed by the neuroscience of the brain and just what it means for learning to occur. It is essential reading for all who want to know about this method of instruction.

[13]**Einstein's Unification**, J. Van Dongen, Cambridge University Press, 2010, 212pp.

Chapter 3

How to Use This Book to Help Your Child

3.1 Learning the 3 R's

In the primary and elementary grades children learn three essential life skills: reading, writing and arithmetic. This learning is of a profoundly different nature than much of that which occurs in most other subjects later in their schooling. The difference is that the 3 R's all involve

doing[1]

which means being able to perform a **new behavior**.

Consider what it means to have learned to read. After a successful conclusion, we expect a child to be able to read and understand written **material they have never seen before**, even if the material contains new words. This is what it means to **do reading**.

Similarly, for a child who has learned how to add whole numbers, we expect the child to correctly find a sum, **even if they have never seen that particular sum before**.[2] This is what it means to be able to **do addition**. And, we could make the same observation about writing.

In short, children who have mastered the 3 R's are able to continually perform new behaviors *on the fly*. This is remarkable because the number of possible sums is infinite.

---

[1]Learning a new language would be a further example of a subject that requires **doing**.

[2]Fluidly execute the standard algorithm in the words of the CCSS-M.

Learning new behaviors continues to be an essential component of math classes in university.[3] However, this component is most intense in primary and elementary school. What is important to realize here is that successfully acquiring new behaviors requires practice. In this sense, learning the behaviors required to do reading, writing and arithmetic are just like learning the skills required to succeed at sports or playing a musical instrument. They require practice and helping children get the required practice is an area where parents can provide valuable support.

To succeed in K-6 children must have feedback and respond to that feedback in effective ways. This process is essential to a child's success and because it is labor intensive, it will require co-operation between parents and teachers. So this is what we consider next.

3.1.1 The Concept of Formative Assessment

The importance of assessment/feedback as a component of the K-6 math curricula is described in a position paper on formative assessment at the National Council of Teachers of Mathematics (NCTM)[4] website.[5] In particular, the paper recommends:

1. The provision of effective feedback to students

2. The active involvement of students in their own learning

3. The adjustment of teaching, taking into account the results of the assessment

4. The recognition of the profound influence that assessment has on the motivation and self-esteem of students, both of which are crucial influences on learning

5. The need for students to be able to assess themselves and understand how to improve

(quoted from NCTM-FA).

The purpose of formative assessment is that students would immediately know whether they have successfully acquired a body of material and, if not, that information would be available in a timely fashion enabling an immediate response to correct deficits. Without doubt, using assessments to correct deficits in the manner described is a desirable outcome for all students, parents and teachers, since it would prevent

---

[3]For example in a calculus class a new behavior might take the form of learning to differentiate a new class of functions.

[4]The NCTM is the primary professional organization for teachers of K-12 mathematics in the US.

[5]See **Formative Assessment** A position of the NCTM, found at: www.nctm.org/ uploaded-Files/About_NCTM/Position_Statements/Formative%20Assessment1.pdf (NCTM-FA).

the kind of knowledge deficits evidenced by the standardized math tests in Grade 4 that found less than 40% of students proficient in math.[6]

3.1.2 How the Process Works

The implied theory of formative assessment as described in the five points involves three steps: teachers/mentors will identify difficulties, communicate those difficulties in a suitable manner to learners, and the learners will take responsibility for fixing the problem. Consider the following example from Grade 2.

Step 1

A child computes five sums of two digit numbers and gets the first two questions right and the last three wrong. Specifically, the child gets this, and similar problems right:

$$\begin{array}{r} 23 \\ +45 \\ \hline \end{array}$$

but gets this and similar problems wrong:

$$\begin{array}{r} 27 \\ +45 \\ \hline \end{array}$$

The fact that the child gets the first and similar problems correct suggests two things to the teacher: the child knows all single digit sums less than 10, and the child knows the basics of the standard algorithm for addition (see §8.6).

The fact that the child gets problems of the second type wrong suggests a problem either with single digit sums more than 10, or with the carrying procedure. To determine which, the teacher considers the answers given. For example, here are three possible wrong answers: 73, 62 and 61. Compared with the correct answer, 72, the first answer is wrong in the *ones* place but correct in the *tens* place. Thus, the child is able to correctly execute carrying; but the error suggests the student believes $7 + 5 = 13$. The second answer, 62, suggests that the student has an incomplete knowledge of place value and carrying while the third answer suggests the student has problems with single digit sums that have a two digit answer and also has difficulty carrying. Confirming the exact source of each error may require asking the child:

Exactly how did you arrive at this answer?

---

[6]These deficits were identified as part of the 2011 National Assessment of Educational Progress. See http://nationsreportcard.gov/math.2011/summary.aspx and click on Grade 4 in the **Proficient** paragraph.

At the conclusion of evaluating the child's responses, the teacher/mentor should have a fair idea of what the child knows, what the child doesn't know and where more practice is needed.

Step 2

Communicating results to children is complicated because these results are loaded with success/failure connotations.[7] Point 4 above recognizes that discussing results with children is delicate. We all need to recognize that the purpose of evaluation is to determine those areas in which a child needs more practice and to act on test results accordingly. Testing is not about finding fault or failure. It is about identifying future learning actions.

Step 3

The intended response to test results is that after the error is explained, the child will undertake sufficient additional practice in the identified area to correct the difficulty. This step is critical because once a wrong fact becomes fixed in memory, e.g. $7+5 = 13$, it becomes almost impossible to unlearn.[8]

For a successful response to be effected, all parties, teachers, parents and students, must believe that essentially every child has the intellectual capacity to learn arithmetic to the standard expected by the A+ curriculum. Without such a belief, the non-believer will subvert the learning process. The only party that can conceivably be excused is the child who is likely to be frustrated by the need for additional work. The point is that all must believe that success can be achieved.

The reason I say that all children can achieve at the required level is that the skills of arithmetic that the A+ curriculum expects children to master are algorithmic. By this I mean there is a fixed solution procedure which may also require knowledge of a small data base, e.g., the addition table in the example above. And even the items in that data base can be found by a procedure, namely **counting-on**, so if a required fact is not remembered, it can be constructed on the spot by a child who knows the counting-on procedure. For this reason, achieving mastery is one of practicing procedures and/or recalling facts.

Once we all agree that the issue is one of practice, we have to consider whether it is reasonable to expect a child in Grade 2 to act responsibly in this matter. My answer is that ensuring an individual child in elementary school undertakes the practice requires

---

[7]Test results are, and have been, wrongly used to make judgements about student abilities, teacher abilities, school quality, and the list goes on. We will keep our focus on the child.

[8]This is why having a simple procedure available for checking results as part of the recall process is an essential component of a child's math toolbox (see §8.6).

adult supervision[9] and in providing this supervision, teachers and parents should be allies.[10]

3.1.3   Meeting the Standard

As is evident from the last section, ensuring children achieve mastery of the A+ topics will be **labor intensive**. Further, it seems unlikely that governments will provide the necessary resources in the form of money and qualified personnel to ensure all children meet the goal. Indeed you can already find evidence of this on Internet news sites.[11]

Given these realities, it seems plausible that even if the A+ curriculum were universally adopted, the educational system may continue to fail for many children due to lack of resources. The remedy for this circumstance is that responsibility for ensuring a child's success is taken on by that child's parents. What is truly encouraging in this respect is that survey data by the EPC of parents in CCSS adopting states indicate that parents are willing to undertake this responsibility.[12] This book exists to help parents who want to act as mentors to help their children. It does this by providing the detailed knowledge of procedures mentors will need in a framework they can understand. The book has been written with the specific intent that it should be accessible to all parents, even those who consider themselves weak at math.

3.2   What a Mentor Needs to Know

An adult who contemplates undertaking this responsibility may be reluctant on the grounds that it is absurd to expect the average person to be able to help a child through the entirety of the school math curriculum. Moreover, recent studies seem

---

[9]In my experience, there are many members of extended families who are willing to take this responsibility on.

[10]I recall an article in *Scientific American* on why the children of immigrants tended to out-perform local children in their school work. The answer was that such children did homework around the kitchen table after supper as a regular practice. In other words, there was parental supervision on a regular basis. This practice was dropped in the 2nd generation and as a result the grandchildren of immigrants performed like local North American children.

[11]A search of Huffington Post has more than 50 pages of articles on Common Core. Some focus on resource/training requirements and whether such resources/training will be available in particular states. See, for example, http://www.huffingtonpost.com/stephen-chiger/to-improve-teaching-get-s_b_3655190.htm

[12]See the EPC Working Paper 34, on-line.

to confirm this idea.[13] These studies suggest that when a child enters middle school, parental efforts can become counter productive and actually result in lowering their child's test scores. An identifiable cause is that

> parents may have forgotten, or never truly understood, the material their children learn in school (see Goldstein, p. 85).

However, based on more than forty years of teaching in universities, the only portion of the curriculum you need to concern yourself with to ensure your child will succeed is primary and elementary.

> During primary and elementary children are at an age most amenable to help from parents and if a child completes this portion of the curriculum with no deficits, the child will do just fine on the rest of the school math curriculum.

So in terms of math content, this book confines itself to topics from arithmetic identified in the A+ curriculum. As already noted, all of math at this level is procedural and the book presents procedures in an easily understood manner. For this reason, with the aid of this book, even parents who consider themselves weak at math will be able use this book to help their children.

In respect to helping your child, you need to know various things:

1. Is my child performing up to standard on a grade-by-grade basis?

2. You need knowledge of the underlying content to be able to intervene and help your child where necessary.

3. You need to know where to find additional practice materials to support your intervention.

Information regarding each of these points is included.

3.3 The Content

The remainder of this book is about numbers, their representation and the operations on numbers. These topics are found in the first two sections of the A+ curriculum table on p. 6.[14] Topics related to proportionality are not treated and geometry is

---

[13]... *And Don't Help Your Kids With Their Homework*, Dana Goldstein, **The Atlantic**, April, 2014, p. 85

[14]They comprise the **number strand** in the CCSS-M.

treated only in respect to length and area. Data analysis is not treated and to the degree that it should contribute topics in a K-6 curriculum, it should be in the context of science.

We focus on topics related to arithmetic because the factual evidence based on testing is that students who master this limited set of topics out perform students taught under any expanded curriculum. Moreover, in the experience of the author, students who can do and understand these arithmetic topics have no difficulty getting through courses in calculus and statistics of the type required for a career in business, the sciences, engineering or economics. It's also the case that if a student can't do arithmetic they are not only blocked from the above, but also from many industrial trades having a substantial math component. Finally, the focus of the book is on topics children learn in primary and elementary because the test data tells us that this is where deficits begin to accumulate.

3.3.1 Using the Content

In previous sections we have made the point that the computations of arithmetic can all be performed using straight-forward mechanical procedures. For example, I think most would accept that the most feared topic in elementary school arithmetic is **fractions**. In §14.3.3 a three-step procedure is presented for adding fractions that for common fractions involves only computations with whole numbers and in all cases always produces a correct answer, even for algebraic fractions. Once a child has learned this procedure, that child is set for life when it comes to adding fractions. At a minimum, every parent can familiarize themselves with this procedure and make sure that their child knows how to use it.

At this point you may be saying something like: If it's that simple, why is there a problem with math? The key here is that the nature of mathematical knowledge is hierarchical — new procedures use and incorporate previous procedures. Thus, unless a child has mastered the previous procedures, that child will be unable to effectively learn the dependent procedure. As test results show, a mile-wide, inch-deep curriculum makes this type of mastery learning impossible for all but a few.

The following point by point outline presents the key procedures of arithmetic in a fashion that illustrates their hierarchical relationship. The only knowledge that is not procedural is the definitions of abstract notions and symbols related to number. These items must be mastered in the order shown:

1. the concept of counting number as measure of the relative size of a collection and conservation of counting numbers — Chapters 4-7;

    (a) using pairing to show which collection has more;

23

(b) the process of counting by rote;

(c) meanings of the numerals;

(d) 1 is the least counting number;

(e) for every counting number there is a next counting number;

(f) the process of counting-on beginning at a specific number;

(g) the place-value naming system for counting numbers;

(h) ordering of counting numbers using numerals;

2. addition of counting numbers — Chapter 8;

(a) understanding addition as finding total number in combined collections;

(b) procedure for single-digit addition based on counting-on — addition table;

(c) representing single digit sums visually (see diagram for $10 + 7$ in Chapter 2);

(d) procedure for two two-digit sums using addition table and counting-on;

(e) procedure for up to four two-digit sums using addition table;

(f) procedure for two multi-digit sums using addition table;

3. the equality symbol, $=$ — Chapter 8;

(a) essential properties of the equality relation are discussed in relation to addition;

4. subtraction of counting numbers — Chapter 9;

(a) understand subtraction as removing members from a collection;

(b) procedure for subtraction as take-away;

(c) visually represent subtraction process;

(d) procedure for subtracting two-digit numbers using standard method;

(e) borrowing;

5. length as a numerical attribute — Chapter 10;

(a) units of length;

(b) making measurements;

(c) the half-line as a way of representing counting numbers;

(d) using lengths to model the addition of counting numbers;

6. multiplication — Chapter 11;

   (a) defining multiplication of counting numbers as repetitive addition;

   (b) visually representing products of two using area;

   (c) commutative law from multiplication from area;

   (d) distributive law from multiplication from area;

   (e) visually representing products of three using volume;

   (f) multiplication of single-digit numbers as repetitive addition — multiplication table;

   (g) multiplication of two-digit numbers by single-digit numbers using standard procedure;

   (h) multiplication of multi-digit numbers by single-digit numbers using standard procedure;

   (i) multiplication of two-digit numbers by two-digit numbers using standard procedure;

   (j) multiplication of multi-digit numbers by two-digit numbers using standard procedure;

7. division — Chapter 12;

   (a) defining division as splitting counting numbers into equal groups;

   (b) concepts of multiples and factors;

   (c) skip-counting to generate multiples;

   (d) standard procedure for division;

8. concept of fractions — Chapter 13;

   (a) making measurements that are not counting numbers;

   (b) concept of unit fractions and dividing the unit interval into equal parts;

   (c) naming unit fractions;

   (d) the Fundamental Equation governing the behavior of unit fractions;

   (e) repetitive addition and common fractions;

   (f) notation for common fractions and the Notation Equation;

9. basic computations with fractions — Chapter 14;

    (a) procedure for multiplication of fractions;

    (b) procedure for addition of fractions with same denominator;

    (c) procedure for addition of fractions with different denominator;

    (d) three-step procedure for adding fractions that always works;

10. advanced computations with fractions — Chapter 15;

    (a) subtraction of common fractions;

    (b) dividing of common fractions;

    (c) visual representation of multiplication of common fractions as scaling;

    (d) putting fractions in lowest terms;

    (e) mixed numbers;

11. ordering fractions — Chapter 16;

    (a) using computational procedures to order common fractions;

    (b) placing common fractions on the line;

12. common fractions and decimals — Chapter 17;

    (a) understanding decimal notation;

    (b) the notion of decimal fraction;

    (c) placing decimals on the line;

13. operations with decimals — Chapter 18;

14. negative whole numbers —Chapter 19;

    (a) the notion of additive inverse;

    (b) computational definition of negative numbers;

    (c) algebraic properties of the integers;

    (d) essential properties of arithmetic;

    (e) the whole number line;

15. the arithmetic of real numbers — Appendix A;

    (a) real numbers from geometry;

(b) the Fundamental Equation and notion of multiplicative inverse;

(c) the rules of arithmetic;

16. ordering the real numbers — Appendix B;

    (a) arithmetic and $<$;
    (b) procedure for finding length;
    (c) absolute value;
    (d) addition and the real line;

17. the arithmetic of exponents — Appendix C.

The most important thing to take from this outline is how each procedure builds on and/or incorporates previous procedures. This is why mastery of prerequisite procedures is essential both for you as mentor and for your child who is learning. It is also why children need to focus their attention on learning the specific procedures that are supported by and mesh with our system of numeration. In my experience, mastery of the topics outlined above and the laws on which they are based will enable children to succeed in higher level math courses precisely because knowing how to do the underlying computations will not be an issue. Finally, thorough knowledge of these procedures is the foundation on which **mathematical understanding** rests.

   In summary, the purpose of this book is to provide mentors with the mathematical information they need to ensure the children in their care master arithmetic. There are five aspects to this:

1. applying the A+ Table to identify topics critical to a child's success on a grade-by-grade basis;

2. providing the procedural knowledge on which computations are based so mentors are able to help children master the computations of arithmetic;

3. providing theoretical knowledge so that mentors can explain the **why** of computations;

4. providing detailed information as to what mentors should expect a child to know on completion of each grade levels;

5. suggesting specific activities that help and encourage children engaged in the learning process so as to ensure their child achieves appropriate levels of mastery.

3.3.2 Tracking Tests

Testing has something of an evil name, certainly among students. The fact is, there is only one way to determine whether a child knows something and that is to ask them.

In the opinion of the author, the only legitimate purpose of testing in mathematics is to:

determine whether a student has learned the material that has been taught.

This may seem obvious, but often times test questions seem to have another purpose; for example, finding out if a student is clever. The results of tests should be used only to inform parents and teachers whether a child has assimilated the mathematical knowledge required to proceed.

Here is what we recommend in respect to testing. Each year your child should have a notebook devoted solely to tests in arithmetic. As each new test is added to the collection, you and your child should go over the test question by question. Going over all questions provides an opportunity to hand out praise. Going over questions that are wrong provides the opportunity to find out if your child knows why the answer was wrong and whether there are issues in respect to the question that need to be addressed. We stress: **going over a test is not an adversarial process**. It is about helping your child learn what is needed.

There may be rare situations in which neither you nor your child knows what is wrong and what the correct solution is. In that case, ask the teacher for the correct solution with an explanation.

Finally, this notebook will provide pointers as to what sorts of additional practice your child needs to master a topic. The next section shows where you can get help with this.

3.4 Math Websites for Mentors and Teachers

There are any number of websites that have been developed to support K-6 math and offer material suitable for children focused on A+ topics. Some sites offer worksheets that are available for download at no cost. We provide a sample list of such sites with descriptive material taken from the site. We also provide a list of subscription sites.

The website

http://www.achievethecore.org/parent-community-common-core/parent-resources/

has more information on the Common Core movement, including detailed information on expectations.

The website

$$\text{http://www.pta.org/advocacy/content.cfm?ItemNumber=3552}$$

contains numerous items on the Common Core movement including a number of math videos on specific topics, for example, *coherence*. Parental guides from PTA can be found at http://pta.org/parents/content.cfm?ItemNumber=2910

The website

$$\text{http://everydaymath.uchicago.edu/parents/}$$

has information for parents on curriculum topics at all grade levels. Of particular interest to parents is the material on alternative algorithms for performing standard computations. Short videos demonstrate the use of these methods, for example, the partial sums method, for performing computations. It also provides access to on-line learning games such as **Bunny Count** and **Connect the Dots**. We will refer to this site as EDM.

3.4.1 Free Websites Containing Worksheets

The website

$$\text{http://www.math-aids.com/}$$

contains the following content description:

> The website contains over 72 different math topics with over 847 unique worksheets. These math worksheets may be customized to fit your needs and may be printed immediately or saved for later use. These math worksheets are randomly created by our math worksheets, so you have an endless supply of quality math worksheets at your disposal. These high quality math worksheets are delivered in a PDF format and include the answer keys. Our math worksheets are free to download, easy to use, and very flexible. These math worksheets are a great resource for Kindergarten through 12th grade. A detailed description is provided in each math worksheets section.

The home page at this site has a list of topics on the left. Clicking on a topic produces a worksheet page that begins with a description of the types of worksheet. The *Kindergarten* button produces worksheets on a variety of entry-level topics. Across the top is a list of buttons, one of which is a site map which is very useful. There is also a button leading to links to other resources, most of which are commercial. We will refer to this site as M-A.

The website

http://www.superkids.com/aweb/tools/math/

has an interactive engine that will produce worksheets on a fairly complete list of computational topics from arithmetic. It also covers pre-algebra and exponents. Access is free. The following is from the site description:

> Have you ever wondered where to find math drill worksheets? Make your own here at SuperKids for free! Simply select the type of problem, the maximum and minimum numbers to be used in the problems, then click on the button! A worksheet will be created to your specifications, ready to be printed for use. We will refer to this site as SKids.

The website

http://www.softschools.com/

provides math worksheets and interactive games on an extensive list of topics. There is a lot of interactive stuff on counting, so it's good for Pre-K. The following is from their site description

> SoftSchools.com provides free math worksheets, free math games, grammar quizzes and free phonics worksheets and games. Worksheets and games are organized by grades and topics. These printable math and phonics worksheets are auto generated. There are many counting games at Pre-K level. We will refer to this site as SS.

The website

http://www.kidzone.ws/

has **free** printable worksheets. Unfortunately, the items covered are limited in scope to counting, the basic operations, and word problems related to these computations. The grade levels covered are Pre-K to Grade 5. The site does not have material on fractions.

The website

http://www.pbs.org/parents/education/math/

has lots of general education stuff. There are on-line games for children and their parents. It is probably more valuable to new parents and parents of the very young.

3.4.2 Subscription Websites Containing Worksheets

The website

http://www.adaptedmind.com/Math-Worksheets.html?type=hs

has worksheets organized by grade level. The list of topics is extensive and parents will have to pick and choose. There is useful material on place value in topics for Grades 1 and 2. Not free: $10/mo.

The website

http://edhelper.com/math.htm

offers extensive list of worksheets and activities covering K-12 at a cost of $19.99 per/year for a limited subscription and $39.99 for a complete subscription.

The website

http://themathworksheetsite.com/

has worksheets available on all aspects of arithmetic. A subscription is $25.

The website

http://ca.ixl.com/

contains practice problems that are interactive. It is comprehensive and geared to the Canadian curriculum. For this reason parents will need to be choosey about which activities they ask their child to undertake. As long as parents focus on curriculum elements from the CCSS-M, there's lots of great interactive material to provide practice for your child. Unfortunately, access is not free, but twenty free worksheets can be had from http://www.math-drills.com/ which appears to be a subsidiary of IXL. Other subsidiary sites are

http://www.mathsisfun.com/worksheets/
http://www.dadsworksheets.com/
http://www.mathworksheetsland.com/

Chapter 4

Collections and Numbers

4.1 What Numbers Are and Why They Exist

People have always been interested in answering the questions:

How much? How many?

as they pertain to collections of things in the world. The most primitive cultures, mathematically speaking, answered these questions with either: one, two, or many. Indeed, there are still some cultures in the world today whose mathematics is limited to one, two, or many. Over time, limiting the answers to one, two, and many, proved inadequate and more complete systems of numbers for answering these questions were developed.

Historically, the first numerical questions posed by our ancestors were probably applied to collections. For example, a wife might want to know how many birds were brought home for supper, or how many clams had been collected at the beach, or in another vein, how many days until the next full moon.[1] Such questions involve identifying groups of objects and assembling them into collections. So in the first instance, our focus will be on **counting numbers** that arise as a response to the question:

How many objects are in this collection?

The answer to this question is a **fact about the real world** that is obtained by **counting** the number of objects in the collection. In this sense, the answer is the result of an experiment and there is a universally held belief that

---

[1]The earliest known potentially mathematical object is a baboon's leg bone with 29 notches in it dated from 35,000 years ago. (see Wikipedia: Prehistorical mathematics)

any two competent counters counting the same collection will get the same result.

Every time anyone counts a group of objects as part of their normal activities, that person is performing a counting experiment that witnesses the truth of this universally held belief.

The number of objects in a collection is referred to as an **attribute** of the collection. Numbers that capture this attribute of a collection are called **counting numbers**, or **cardinal numbers**.

Obviously, there are other kinds of numbers that answer questions about objects, for example, the **length** of a soccer pitch, the **area** of a plot of land, or the **weight** of a cow, etc. All of these are **numerical attributes** of real-world objects. While we will deal with such numbers later, as stated, our initial focus will be on the counting numbers associated with collections because that is where a child's mathematical education begins.

Before continuing our investigation of what numbers are, we will concern ourselves with the fundamental question:

Why should numbers exist at all?

4.2   Conservation

The most important and deepest laws in the natural world are conservation laws. Conservation laws have application in every area of science and indeed, one such law is at the heart of arithmetic. To see what it is, consider the following **thought experiment**.

> Suppose you have a pile of buttons. Count them, place them all in a jar; seal the jar. Place the jar on the shelf. Now suppose you come back at a later time and observe that your seal is intact. How many buttons will be in the jar? More specifically, do you need to count the buttons in the jar to determine your answer?

Obviously, you do not need to recount the buttons. Without counting you know that the number of buttons remains the same as when you placed them in the jar. If the seal is not broken, no buttons can be removed, and none can be added.

Consider still another experiment:

> Unseal the jar of buttons and carefully pour the buttons into another empty jar making sure all the buttons are transferred. How many buttons are in the new jar when you are finished? Do you need to count?

Again, we do not need to recount the buttons so long as we are sure that all the buttons were transferred to the new jar and none were added.

Consider one last experiment:

> Suppose we have some empty jars. Unseal the jar containing buttons and distribute some of the buttons among the empty jars, possibly retaining some in the original jar. How many buttons are in **all** the jars? Do you need to count?

In all of these cases there is no need to count the buttons. We started with a fixed number of buttons, and unless there is a source of new buttons or a sink for the old ones, the number of buttons is unchanging in time.

These ideas are illustrated in the following diagram:

> A drawing showing a box containing six buttons on the left. These buttons have been distributed between the two boxes on the right. There is no need to count to know the number of buttons in the two boxes is the same as the number we started with, because buttons are conserved.

In summary:

> **Conservation Principle (CP)**. The number of things in an isolated collection is unchanging in time.

The effect of this fact about the world is that the cardinal (counting) number of a collection is a **stable** attribute of that collection. This observation about the world is referred to as **Conservation of Number**. The reason underlying the principle is that buttons, and physical things more generally, are not spontaneously created or destroyed by the world. This observation may seem obvious, or, trivial. It is neither and as we will see is the source of both the Commutative and Associative Laws.

The age at which children come to terms with this principle in its various forms was studied by a famous cognitive psychologist, Jean Piaget.[2] His results are complex

---

[2]Cognitivist (learning theory) is the theory that humans generate knowledge and meaning through sequential development of an individual's cognitive abilities, such as the mental processes of recognize, recall, analyze, reflect, apply, create, understand, and evaluate. (From Wikipedia.)

and show children's understanding of this conservation principle develop and change over time as the brain matures.[3]

To see that conservation is absolutely essential to all our reasoning, ask yourself how the world would work if the principle weren't true. Among other things, this principle is what enables you and your banker to agree on the contents of your bank account.

We prefaced this discussion with the question:

Why can there exist numbers at all?

Now we have our answer.

Numbers can exist because we can form collections having constant membership, that is, that have a numerical attribute attached to them that remains constant over time.

This fact that isolated collections have constant membership permits us to have standard collections against which we can test any others. In this respect, such test collections would be like the cylinder of platinum and iridium in Paris which is the standard kilogram.

4.3   What are Collections?

Since our focus will be to develop numbers as a response to: How many are in a collection?, it is important that we know what collections are. To be precise:

a **collection** is any identifiable group of objects in the real world.

This definition gives rise to the first observation we will use about collections: **collections always contain at least one object**.

Since it is always useful to have a concrete idea in the back of one's mind, we could think of a standard collection as some buttons in a jar. Thus, the collection consists of all the buttons in the jar. The jar, or container, is not part of the collection, only the buttons. The buttons themselves, are referred to as **members**, or **elements** of the collection.

Consider two jars of buttons, that is, two distinct collections. In the real world, we know the buttons in the first jar are different objects from the buttons in the second. This assertion merely reflects the fact that it is not possible for a physical object to

---

[3]For example, Piaget found that realizing that the total things in a collection are conserved when the collection is subdivided, required substantial learning. In another example, Piaget found that young children believed that the contents of a tall thin container was more than those same contents transferred to a flat, wide container, even though they watched the transfer!

be in two places at once. This is the second observational fact we will take about collections, namely: **a single object can not belong to two different collections simultaneously.**[4]

In our development, we will identify the numerical properties of collections. In doing this, we will depend on the fact that every collection considered, or group of collections considered, can, in principle, exist in the real world. As such, the same object cannot belong to two collections simultaneously.

4.3.1 Sets vs Collections

Mathematics is about abstractions. As a professor in one of my classes observed more than fifty years ago as he put some vectors on a blackboard: "These are not vectors, they are squiggles on a blackboard. Vectors do not exist in the real world."

"What does this have to do with us?" you may ask.

Almost certainly, your child's curriculum uses the word **set** in respect to collections. For this reason we need to go over how mathematicians use the word.[5]

Sets are the things mathematicians use to **model** collections.

We need to provide some explanation of what we mean by *model*. So let's start with a physical models. Perhaps when you were small you built a model airplane. It was a device that captured some of the properties of the actual object, and that could be used to study that object. Model cars, model trains and the like are ubiquitous. Dolls are also models, although we don't usually think of them that way. The point is that every child has played with physical models of various kinds.

Engineers build more realistic models to study properties like loads on a dam or lift on an airplane wing. Such models must accurately reflect key properties of the object under study, or when the actual structure is built, it could fail with catastrophic consequences.

Not all models are physical. Some are entirely abstract. If you have a bank account, there is an abstract model that you regularly use. It is the record of checkbook transactions. The squiggles recorded there are not money, which is real, nor are they numbers, which are entirely abstract. But these squiggles provide a means of keeping track of how much money is left in your account. In this sense they are a model. And

---

[4]To the truly mathematically inclined reader, I am aware of counterexamples. But for purposes of clarity and simplicity, we should think of two collections as being two urns containing some buttons.

[5]When sets were introduced into the school curriculum as part of something called the **New Math** back in the 1960's, all but a few found it intimidating. Readers of the *Peanuts* comic strip from the 1960's which is being recycled in papers today, may recall seeing individual strips complaining about aspects of the New Math. Such strips were indicative of the degree of discomfort that greeted the New Math in the general population.

correctly used, your bank book provides a perfect means of predicting what is in your account at any moment in time.

As stated above, sets are the abstractions that mathematicians study to model the behavior of collections. Because they are abstract, rather than try to say what sets are, mathematicians concern themselves with rules that describe how sets behave and how sets can be constructed. For the most part, these rules reflect what we know about real-world collections, but there are some differences.

Much of the nomenclature regarding sets is identical to that for collections. Thus, things in sets are called **members**, or **elements**, these words being used in exactly the same way as for real-world collections. Thus, if we have a set A and something in the set called b, we would say b is a member of A, or b is an element of A. The standard mathematical notation for b is a member of A is

$$b \in A.$$

where the symbol $\in$ is read *is a member of*. We will see this notation in Chapter 10 and remind the reader of this discussion at that time.

But there is a fundamental difference between *collection*, as we are using the term, and *set* as used by a mathematician. It is this. If I put a button in a collection, you cannot put that same button in a different collection. On the other hand, if I put 1 in a set, there is nothing to stop you from putting 1 in a different set, even while I still have 1 in my set. Moreover, the 1 in my set is exactly the same as the 1 in your set. They are indistinguishable.

This difference between collections and sets has consequences. For us, the most important consequence is the following. Suppose you have a collection with one button, and I have a collection with one button. If we combine our collections, the new collection will have two buttons. Alternatively, suppose you have a set having a single member, 1. And I also have a set having a single member, namely, 1. If we combine our sets, that is put the single member of your set and the single member of my set into a new set, that set will contain only the single member 1.

It is clear that sets have an added level of complexity over collections. But not to worry, as we will be working with collections. Indeed, in what follows we will spend a lot of time thinking about jars of buttons, and if you wanted, you could actually physically reproduce any part of the discussion with your own jars of buttons.[6]

Although the terms *collection* and *set* are often used interchangeably, we will continue to apply the noun *collection* only to groups of real-world objects. Thus, any

---

[6]In my lifetime I have met and worked with some really fine mathematicians. All thought in the most concrete terms possible, and were not afraid to say so. Moreover, all believed that the most effective strategy for doing research begins with understanding the work that went before.

collection we will speak of could, in principle, physically exist. *Set* will be reserved for groups of abstract objects which do not exist in the world.

We adopt this approach for the expressed purpose of emphasizing that the generation of the ideas being discussed in respect to arithmetic arises from considerations about the real world. Since collections are things all of us deal with everyday, the rules collections obey are so ingrained in our thinking, we don't even know they are there. But we can identify these rules and use them as a basis for arithmetic.

Chapter 5

A Child's First Perception of Numbers: Pairing

Our presentation will follow the outline set by the A+ curriculum. The table presented in §1.5 begins in Grade 1. But children begin to learn about numbers long before that.[1] So that's where we begin.

5.1   The Notion of Pairing

Consider the thought experiment:

> Suppose you and your very young child are driving by a field that contains two cows and three horses. You ask your child: Are there more cows or more horses? What I want you to consider is what your child needs to know to answer this question. Specifically does your child need to know the names *two* and *three* to draw a conclusion?

To make this more concrete, consider the schematic field shown below.

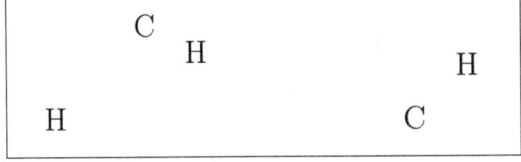

> A schematic field outlined by the frame. Horses are denoted by an H and cows by a C.

---

[1] I have observed mothers working on counting with children as young as 14 months.

A child that knows nothing about numbers could answer the question about **more** or **less** based on proximity by noticing that two cows are feeding close to two horses, while the third horse is far away. So by **pairing**, as explicitly shown below, all the cows with some of the horses, and noticing there is an unpaired horse, the child can draw the correct conclusion that there are more horses without ever knowing about numbers.

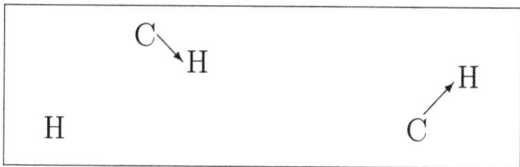

A schematic field outlined by the frame. One horse is not paired with a cow, so there are more horses than cows.

Thought Experiment. Alternatively, suppose you and your very young child come to another field that has three cows and three horses. You pose the same question, more or less?

In this case the diagram reveals an exact pairing between the cows and the horses which creates a problem because there are neither more cows than horses, nor more horses than cows. A child would most likely say: **the same**. But the immediate question then is: What do we mean by **same** in this context? Clearly, it cannot mean cows are the same as horses. So when the word 'same' is used it has to be referring to something else. It is exactly at this point that the notion of an abstract attribute of collections arises which is independent of whether the collection contains cows, or horses, or something else.

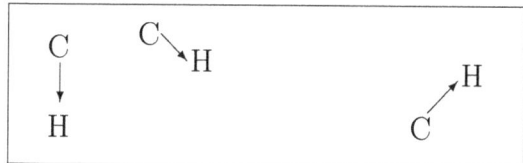

A schematic field outlined by the frame. There is an exact pairing between cows and horses, hence an equal number of cows and horses.

An exact pairing between the elements of two collections is called a **one-to-one correspondence**. We will continue to use **exact pairing** to describe this situation because we think pairing is more easily understood as a concept.

Here is an example of how the notion of exact pairing can be useful:

Suppose you are setting the dinner table for eight — the cousins are coming over — and your young child, who can't count, wants to help. You could lay out the place mats and tell the child to put one fork, one knife and one spoon at each place mat.

Again, pairing can substitute for knowledge of number. But doing the pairing leads to the notion of number as an attribute of collections.

5.1.1 Why Pairing Has Limited Utility

The pairing process requires proximity in space and time as the next thought experiment shows.

Suppose you and your very young child are driving by a field that contains four cows. The child says, "Look! Cows." Later, you come to another field that has five horses. The child says, "Look! Horses." And now you say, "Are there more horses in this field, or cows in the last field?"

We illustrate this situation below:

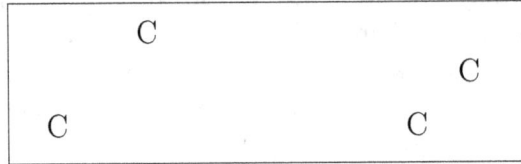

Schematic fields with four cows (above) and five horses (below).

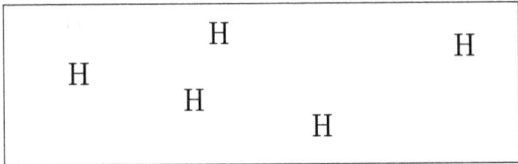

Because the fields, and hence the animals, are separated in space and time, it is no longer possible for a child who knows nothing about numbers and counting to determine whether there are more cows, or more horses using a direct pairing based on direct proximity of the animals. Thus, the requirement of the pairing process, that collections be proximate in space and time, is a severe limitation.

41

5.2   Counting Numbers as a Solution

To answer the question: Which has more? for collections separated in space and time, the child has to deal at a much higher level. What is required is the notion that there is something fundamental — a property — about all collections for which there is an exact pairing with a collection containing four cows. It is the abstract notion of *fourness*. Moreover, this abstract property of fourness must be recognized as being **stable**, that is, **conserved**. Hence, to operate at this level the child must also believe that were you to return to the field with the cows, there would still be four. In other words, the child must come to terms with conservation at some level.

In a separate vein, the child must also understand that when we say a collection has the numerical attribute of *fourness*, this tells us nothing about the physical nature of the members of that collection. It only answers the question;

How many are in the collection?

Because it is an abstract idea, the attribute of fourness may be equally well applied to cows, or horses, or spoons, or whatever. Similarly, all collections containing five items have a common attribute that gives rise to the notion of *fiveness*. And so on for other numbers that answer the question: How many?

The realization that every collection has an attribute which answers the question, *How many are in the collection?*, gives rise to the abstract notion of **counting numbers (cardinal number)**.

> **Counting Numbers** are the numbers we use to name the attribute of collections of things that tells us: *How many items are in the collection.*

We stress that **cardinal number** is just another name for counting number.

5.2.1   Equality Between Counting Numbers

Once we have identified counting numbers as being things we want to use, we need a fixed and robust procedure for determining when two counting (cardinal) numbers are equal. Based on the discussion above:

> **Equality Principle.** Two collections will be assigned the same counting number exactly if there is an **exact pairing (one-to-one correspondence)** of all the objects in one collection with all the objects in another, with none left over in either collection.

We illustrate the equality principle using two collections having four members each:

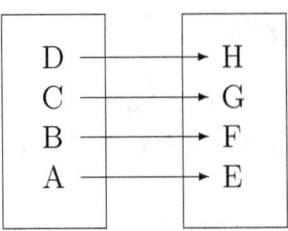

A diagram illustrating an exact pairing between two collections. So, as a consequence of the Equality Principle, both must be assigned the same counting number as the answer to: *How many?* Note that the decision as to whether the two collections have the same counting number does not depend on what that number actually is. Nor does it depend on what the elements of the collections are. It only depends on whether there is an exact pairing between the two collections.

Thus, the procedure for determining when collections have the same cardinal number is derived from our understanding of the nature of collections in the real world. We emphasize that the determination does not depend on the nature of what is actually in the collection; nor does it require a notation for the counting number that captures the cardinality attribute because the pairing process does not mention number. As such it captures our real-world experience. I want to emphasize this point strongly. What we are doing is creating an abstract model of certain aspects of the real world. This model must exactly replicate the aspects it seeks to capture. Thus, if we have two collections for which there is an exact pairing between the objects, they must be assigned the same counting (cardinal) number.

As we have already noted, the exact pairing process has limited utility due to the proximity requirement. Thus, for counting numbers to be the solution, there must be another process for determining equality between collections. There is, but to describe that process we need to explore the deeper properties of real-world collections.

Chapter 6

Counting Numbers in Pre-K

Counting is the foundation on which our system of arithmetic is built. For this reason it is essential that every child master the counting process for numbers less than 10 during the years prior to entering Kindergarten.

6.1   The Two Uses of Counting Numbers

There are two distinct ways in which we use counting numbers. The first is to record the number of objects in a collection found in the world. It seems most likely that this use is the one that led to the recognition and generation of counting numbers. The second use is in constructing collections whose size has been specified by a given counting number. These two uses of counting numbers are illustrated below:

The first use: count the members on the left to get the numeral on the right.

The second use: given the numeral on the left, produce the collection on the right.

As the reader can see, both of these uses require a way to represent numbers, in other words, a **system of numeration**. As well, both uses require a procedure for getting from the left side of the diagram to the right. In both cases, the procedure is the same, namely, **counting**, which is why mastering counting is so important. However, any form of counting requires a system of numeration, so we briefly consider that topic first.

6.2   Systems of Numeration

In the scenarios developed in Chapter 5, we found it was possible to compare the size of small collections with no direct knowledge of numbers so long as the collections were not separated in time and or space.

The simplest solution to the problem of separation is the use of a **tally stick** as a means of recording numbers. For example, consider a scout for a tribe of early hominids who wants to convey information to the chief on the number of interlopers crossing the boundary into their territory. By observation the scout could make a tally stick that contains one mark for each interloper and send this to the chief.

A schematic of an early hominid tally stick that simply matches marks to members of the collection of interest.

Or, consider the child with cows in one field and horses in another trying to judge whether there are more cows or horses. Again a tally stick with one mark for each cow in the first field solves the problem. The child can use the tally stick as a standard against which to compare the number of animals in any other field.

As well, the example of §5.1.1 of a mother putting eight place mats on a table and then asking the young child to put one knife, one fork and one spoon at each place, in essence uses the tally stick idea to solve the problem of constructing a collection of a fixed size. Thus, the tally stick idea plays a role in a child's development through the pairing process.

While tally sticks will correctly transmit counting numbers, they are not a good solution because they are cumbersome, even for relatively small counting numbers like 50. What is required is an efficient representation of each counting number that can be accurately transmitted through time and space, and from one person to the next. The solution that enables men to transmit a culture of ideas through space and time is **language**. And so it is with numbers. What is required is a language for identifying numbers. At a minimum, this language must contain a name for every counting number, a tall order given there are an infinite number of counting numbers.

To solve the problem of language, we have to agree on a name and notation — a way of writing the number — for each number we assign to collections. In addition,

the name of each number must be unique.

Coming up with a system for naming the counting numbers was a challenge because, as we have seen, there are an unlimited number of numbers, and the name and notation for each individual number must be **universally recognized**.

To see why having to name each member in an unlimited collection causes difficulties, consider using tally markings based on groups of five (shown below) which we have used to denote the number eighteen:

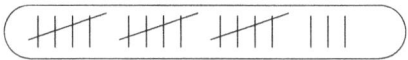

A schematic for a more sophisticated tally stick based on groups of five.

This system of notation does produce a unique representation for each counting number. In addition, any notation constructed in this system will, almost certainly, be universally recognized and correctly interpreted. But the notations are unwieldy to say the least. For example, the notation for one hundred, which is not a particularly large number, requires twenty groups of five.

But a system of notations based on tally markings has a greater failing than merely being unwieldy.

To see why, consider again the problem of the scout. Suppose the scout observes the number of interlopers shown on the tally stick represented above. How can he communicate this to the chief? If the only system available to the scout for communicating numbers is the tally stick, he must show the tally stick directly to the chief to transfer the information. This is a severe limitation on the communication process. Thus, to realize the full utility of numbers, we must have not only a notation for each number, but also a verbal name.

One way around the unwieldy problem associated with tally marks is to introduce a unique symbol for each counting number. Presumably, a verbal expression would correspond to each symbol. The problem with this approach is that such a system would require an unlimited number of different symbols. Moreover, as ever larger numbers are needed, new symbols would have to be created. How would these new symbols become universally accepted in a commercially active world that demands universal acceptance of mathematical symbols at all times in order to function? This is simply not possible if new symbols have to be made up each time a larger number is needed.

Solving the number notation/naming problem was a major step forward in human intellectual development. Aside from tally sticks, several candidates were tried and discarded before the Arabic system came into use. We will study this system of numeration in detail because it is the heart of our computational system. Mastery of this system of numeration will be one of two great intellectual steps a child will take during the study of arithmetic. The other is learning to count and from these two, all else will follow. The two are intimately connected and mastery of the numeration system begins with learning the meaning of the symbols.

6.3   The Single Digit Numerals for Counting Numbers

In the Arabic system of numeration for counting numbers with which we are all familiar, each number has been given a **name** and a **numeral**. The single digit numerals in increasing order are:

one or 1; *two* or 2; *three* or 3; *four* or 4; *five* or 5; *six* or 6; *seven* or 7; *eight* or 8; *nine* or 9.

These numerals are special because they are single digits and do not involve the notion of **place**. For this reason, this list is the starting point for a child learning about numbers and counting.

To be clear, names are the words we use in language. Numerals are the symbols (notations) we use in computations. Neither the numeral nor the name is the counting number they represent anymore than your name is you. Thus, the name *one* is the name of the number for which we employ the symbol (numeral) 1. The number denoted by this name and numeral is an abstraction and does not exist other than in our minds. That being said, in common usage we speak of *the number* 1 instead of the *number denoted by the numeral* 1.

As discussed in Chapter 5, each single digit represents a counting number. The counting number, in turn represents the size attribute attached to a particular collection. Further, two collections are assigned the same counting number if and only if there is an exact pairing of members between the two collections. Thus, for numbers to have a meaning, we must have a standard collection attached to each numeral that specifies the size attribute. The following diagrams show the numerals for the first nine counting numbers together with a standard collection containing the requisite number of items specified by the counting number that the numeral names.

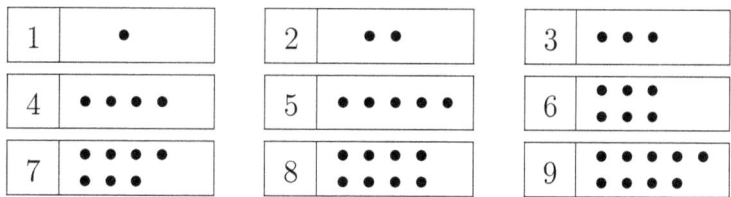

Single Digit Definitions. Schematics showing the nine single digit numerals for the first nine counting numbers together with the defining standard collection. The numerals are given in increasing order and each successive numeral/collection contains exactly one more item than its predecessor.

It is critical to understand that each of the individual diagrams constitutes a **definition**. For example, consider the following diagram in respect to 4 :

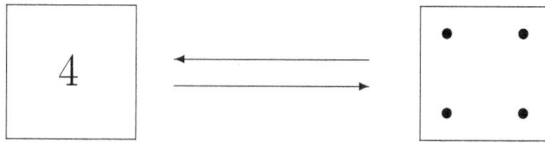

A diagram showing a box containing the numeral 4 on the left and a standard collection containing four items on the right **defines** the meaning of the numeral.

Children must understand that because the meanings assigned to numerals like 4 are definitions, they are not subject to question. They simply are and as such, the relationship between a numeral and its defining collection is a cultural fact.

The second critical fact is that the collection defining the first numeral in the list, namely 1, is special because removal of any member destroys the collection. This property is not shared by any other collection and for this reason the property characterizes smallest collections. Such collections are said to consist of a **single** member and have cardinal (counting) number 1.[1]

The third thing to notice is that the numerals are listed in a specific order determined by the **more** concept and reflected in the following list:

$$1, \quad 2, \quad 3, \quad 4, \quad 5, \quad 6, \quad 7, \quad 8, \quad 9.$$

Consider any given numeral in the list ≤ 8 and the numeral adjacent to it on its right. Looking at the definitions above, observe the defining collection for the numeral on

---

[1] We will us cardinal as an alternative adjective to identify the counting numbers. They are also referred to as **natural** numbers. In speaking to your child you may want to use counting number.

the right has exactly one more member than the defining collection for the numeral on the left. This fact is illustrated for 2 and 3 in the next diagram:

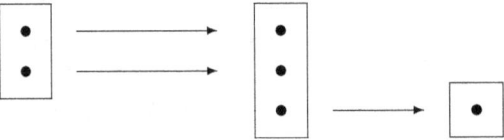

Any pairing between a collection having 2 members and one having 3 members will leave 1 member of the collection having 3 members unpaired.

The collection at the left is associated with 2, the one in the center with 3 and the one on the right with 1. If we construct a pairing between the collection for 2 with the collection for 3, some member will always be left over. Thus, 3 is associated with collections having more members than the collections 2 is associated with, so 3 is more than 2. How many more? There is a single element that is unpaired. How do we know this? Because removal of the unpaired element makes the pairing exact. So the collection consisting of left-over members has cardinal number 1 as can be seen from the diagram. Taken together, these facts tell us there is no counting number that is less than 3 but more than 2. This is the reason why we can say 3 is the **next** counting number after 2.

A similar relationship exists between each numeral/collection and the one to its right, e.g., 7 and 8, or 6 and 7. The counting number that is one larger than a given counting number is referred to as its **successor**. Thus, 7 is the successor of 6, and so forth. Knowledge of the natural order of counting numbers based on successor as it relates to collections is critical information that children should acquire as part of their pre-K experience.

6.4 Applying the Numeral Definitions

Children must be able to apply the numeral definitions in the two distinct ways identified above. Thus, consider the numeral 5. First, a child must know that given any collection for which there is an exact pairing with the standard collection for 5 must be assigned 5 as the numeral that denotes its cardinal number, e.g. the toes on one foot. Second, it means that when asked to construct a collection having five members, a child will produce a collection for which there is an exact pairing with the standard collection in the defining diagram for 5. The process by which these identifications and constructions are accomplished is counting.

6.4.1 The Counting Process

Counting is second nature. We all learned how to do it so early in life that we've all forgotten exactly what we learned. But counting is so critical to arithmetic that it needs to be carefully considered. We do this in the context of the first nine counting numbers.

As a rote process, counting up to nine is simply the ability to list the numerals for the first nine counting numbers in increasing order:

$$1, \; 2, \; 3, \; 4, \; 5, \; 6, \; 7, \; 8, \; 9.$$

Learning this order is simply a matter of repeating the list in the prescribed order enough times that it becomes a matter of recall. Learning this list by rote is certainly where the child starts. One way this can happen is with a set of blocks or plastic symbols that list the alphabet and the numerals. So the above list is concrete, think of the list as being a collection of nine plastic numerals as shown:

$$\boxed{1 \quad 2 \quad 3 \quad 4 \quad 5 \quad 6 \quad 7 \quad 8 \quad 9}$$

The single digit numerals as a collection of plastic symbols.

A child should be able to use this collection as a basis for counting. Consider counting the collection shown below:

To physically count this collection, a child constructs a pairing between the collection of numerals and the collection to be counted. The essential piece of the construction process that makes it counting is that the numerals are paired one-at-a-time in increasing order with members of the collection being counted. A partial pairing is shown next:

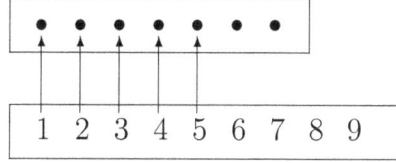

Partial counting of a collection having seven members. The count is finished when 6 and 7 are paired with the remaining unpaired members.

The one-at-a-time process occurs in increasing order as follows: first 1 is paired, then 2 is paired, then 3, and so forth. The process stops when all the objects to be counted have been used and each object in the collection is paired with exactly one numeral. The last numeral paired with an object names the cardinal number of the collection being counted. In the example given, that numeral is 7.

6.4.2   Counting Beyond 9

Counting beyond nine requires a discussion of the Arabic system of notation and the role of **place**. Coming to terms with the Arabic system and place is the critical next step in a child's learning. Because this system is central to all the algorithms of arithmetic, we give this discussion its own chapter.

6.5   What Your Child Needs to Know

The information in this section, and ones like it, has been constructed by identifying topics in both the A+ curriculum and the explicit list of goals in the CCSS-M for K-12. Although there are no goals in the A+ and CCSS-M for Pre-K, this situation was remedied by the educators at the University of Chicago who constructed a list of Pre-K topics relevant to the CCSS-M.[2] The list below is therefore based on topics from the Chicago list that are prerequisite to A+ topics.

6.5.1   Goals for Pre-K

On entering Kindergarten it is expected your child can:

1. Verbally count in sequence by rote to 10 and beyond.

2. Develop flexibility in counting, including counting-on and counting backward from a given number.

3. Count objects in a collection with one-to-one correspondence and know the last counting word tells "how many."

4. Develop an awareness of numbers and their uses; associate number names, quantities, and written numerals; recognize and use different ways to represent numbers (for example, groups of objects or dots).

---

[2]See grade-level-goals in parents' section of http://everydaymath.uchicago.edu based on the CCSS-M.

5. Compare and order groups of objects using words such as more, fewer, less, same.

6. Solve and create simple number stories using concrete modeling; explore part-whole relationships (for example, 5 is made of 2 and 3).

Let's go through these goals to make clear the A+ intention and at the same time identify where you can find support materials.

Goal 1 says that your child should be able to count to 10 and beyond. What is being discussed is **rote counting**, which is clear at the website. This means your child could repeat in order the numbers up to 15, but would not necessarily be able to count the buttons in a jar containing 15 buttons and report that fact. It is expected that the tasks set out under this goal can be executed by rote. If you want to operationally see what's required, try the **Connect the Dots** game with numbers at the EDM website. The system of notation we use for counting numbers is discussed in Chapter 5.

Goal 2 says that your child should be able to count a small group of objects in the manner that an adult would do it. This process works by assigning counting numbers in order with the last numeral assigned giving the cardinal number of the collection. Using this process represents a deeper level of knowledge. Again, these ideas are covered in Chapter 5. If you want to operationally see what's required, try the medium level **Bunny Count** game with numbers at the EDM website.

Goal 3 says that your child should be able to directly associate numerals and collections having the cardinal number named by that numeral. For example, 3 and a jar with three buttons (see Chapter 5). Clearly this represents a still deeper level of knowledge. If you want to operationally see what's required, try the **Bunny Count** game with numbers at the EDM website.

Goal 4 says that your child should be able to compare and order groups of objects. This begins with the processes discussed in the present chapter. If you want to operationally see what's required, try the **Bunny Count** game with characters at the EDM website.

Goal 5 says that your child should be able to solve simple number stories involving counting. For example:

> Tom stacks three blocks. Mom asks: If you put two more blocks on the stack, how many blocks will be in the stack?

The intention of such stories is to develop a sense of the operation of addition. Addition is discussed in Chapter 8.

Returning to the issue of what and when in respect to Pre-K, encouraging a child to explore and experiment can begin very early. For example, I have observed such an interaction counting a small collection of objects between a mother and her 14 month-old child in which the child was both interested and engaged. In another context, a friend describes their 18 month-old grand child negotiating the number of episodes of a video she could watch before bedtime. The parent would offer 1, the child would respond with 5; the negotiated settlement was usually 2. The point is number concepts are being absorbed by children at a very early age, 2^+ years according to research data.[3] Obviously, the younger the child, the simpler the questions that should be posed. So for example, for a 2-year old: Are there more forks (2 or 3), than spoons (1 or 2), would be a good place to start. Or more forks (2 or 3) than fingers? The point is that this is about finding ways to encourage your child to think about the size of collections and counting.

Obviously, children are individuals, and what is appropriate for one, may not be for another, even in the same family.[4] Thus, you will have to gauge the response of your child to determine how to proceed. A good rule is that learning is like play; it should be fun.

In respect to providing guidance, research shows that young children respond very differently when provided with instruction, namely, they tend to focus on the solution provided by the instructor, as opposed to exploring the situation as widely as possible.[5] So an alternative to instructing is to simply pose the problem and let the child work it out. At the point the child has a potential solution, get the child to tell you how the solution was arrived at. When you understand the child's thinking, you can suggest new problems.

There is one very important thing that should be understood. Because a young child is trying to find a solution to a problem for which **no solution is known to them**, all solutions they might propose have value. The process is one of **trial and error**. It is a fact that most trials end in error.[6] There is no fault in this. There is only the learning that arises from eliminating an incorrect solution. Trial and error is the process by which we humans managed to create culture. However, learning from those with experience is how we expand thousands of years of culture. Balancing these paradigms is the key to successful learning.

---

[3]See *Scientific Thinking in Young Children: Theoretical Advances, Empirical Research, and Policy Implications*, Science, **337**, 1623-1627 (28 September 2012).

[4]I have a colleague who has twin sons, who, according to his descriptions, are completely different in respect to their mathematical behavior and interest. While both are competent, one is clearly deeply interested in numbers, and has been so since the age of two. The other is much more interested in artistic activities.

[5]See *Scientific Thinking in Young Children: Theoretical Advances, Empirical Research, and Policy Implications*, Science, **337**, 1623-1627 (28 September 2012).

[6]See **Adapt: Why Success Always Starts With Failure**, Tim Harford, Farrar, Strauss and Giroux, 2011.

Lastly, with respect to Pre-K, there are a number of sites like EDM that offer interactive games and/or interactive worksheets. Particularly with small children you could jointly play some of the games and do some of the worksheets. Building counting skills is essential. **Counting and conservation** are the basis for all of what we do. They are the foundation of every principle that is presented in this book. All the key laws are experimentally verifiable by a counting process that depends on conservation. It is by doing the activities and worksheets that your child comes to terms with these things.

Chapter 7

The Arabic System of Numeration

In the last chapter we discussed counting and numerals for the first nine counting numbers. An important fact was that for each counting number there was a unique **next** counting number called the **successor** of that counting number (see §6.3). It is clear that every counting number will have a successor. Therefore, there are an infinite number of counting numbers and hence we will need an infinite number of names and numerals.

The diagrams defining the first nine numerals as discussed in Chapter 6:

1	•		2	• •		3	• • •
4	• • • •		5	• • • • •		6	••• •••
7	•••• •••		8	•••• ••••		9	••••• ••••

The standard collection associated with the successor of the counting number 9 is:

We have used a '?' to denote the numeral for the cardinal number of this collection because all symbols for counting numbers have been used.

The central question is how should we create a numeral for the cardinal number of this collection? Because we have used all previous single digits, providing a notation for this cardinal number will require at least one new symbol.

7.1 The Roles of Zero

There is one special number having a single digit numeral which has not been listed because it is not associated with a real-world collection. Recall that collections, as we are using the term, **have to have members**. It took human beings a long time to realize that the result of removing the last member from a collection resulted in something that needed a numerical description. What we are saying is,

nothing also needs a number.

The idea that nothing also needs to be counted is really abstract, much more abstract than counting numbers themselves. The evidence for this is that it took human beings an extra 2000 years to discover that nothing needed a number even though they were using 0 in its other role as a **place-holder** in the number system.[1]

The use of zero as a place-holder will be discussed at length when we consider the Arabic System of numeration in detail. So we leave the second use for the time being and return to considering zero as a number.

If you ask yourself, How do I think about nothing?, you begin to see the problem with arriving at zero as a number that represents a quantity. What seems most direct is the notion of **empty set**, that is, a set with nothing in it. While a collection in the real world ceases to exist when we take out the last element, in our minds we can imagine the **empty set** as being what is left. And if we are asked how many elements are left, we would say: None. Assigning this a number leads directly to the idea that a set with nothing in it has **zero** members. The numeral for zero is: 0 , and a graphic descriptor analogous to the ones above is:

$$\boxed{0 \mid }$$

A schematic of an empty set on the right and the numeral of its counting number on the left.

The numeration system we now use is the **Arabic System**. It uses ten digits, which in order starting from zero are:

0, 1, 2, 3, 4, 5, 6, 7, 8, 9

Corresponding to each symbol is a verbal name, which for completeness in corresponding order is:

---

[1] According to Wikipedia, zero as a number was identified in India by the 9th century AD. The difficulty of recognizing zero as a number is evident from the fact that the requirement for a zero-like place holder was known to Babylonians 2,000 years earlier!

zero, one, two, three, four, five, six, seven, eight, nine.

The symbols 0 – 9 are the only symbols that occur in the Arabic notation for any counting number. Since there are an unlimited number of numbers, these symbols may have to be used more than once in the expression for a particular number. How this is done is one of the really clever features of Arabic notation. Other systems, for example Roman numerals, also use the same symbol multiple times, but not nearly as effectively, and not in a way that connects to the system of computation. The connection to computations is, perhaps, the key reason why Arabic notation became universally accepted. Understanding the Arabic System of notation, how it supports the computations of arithmetic and developing fluidity with its computational schemes is an essential goal of the A+ curriculum.

7.1.1 A Concrete Realization of the Arabic System

To understand arithmetic and to be able to help your child, you need a thorough understanding of the Arabic numeration system. For this reason, we develop an example, based on a mythical button supplier that provides a physically based realization of the Arabic System of numeration. We will use this realization as the starting point for all our explanations of how the computational procedures work. These explanations will ultimately trace all computations back to counting! So it is important that you thoroughly understand this example. It is also the case that many of the manipulatives used in pre-K–Grade 1 to explain the Arabic system to children incorporate many of the same ideas.[2]

> **Button Dealer's System.** The button supplier has a very large supply of buttons. Customers show up, tell the supplier how many buttons they need, and he fills their order. To do this efficiently, the button supplier keeps his buttons in jars labeled with one of the numerals 1, 10, 100, 1000 and 10000. Each jar contains the number of buttons identified by its numeral. So a jar with 100 on the front contains one hundred buttons. The jars are stored on shelves according to the following system he has devised:
>
> on the first shelf, jars labeled with: 1;
> on the second shelf, jars labeled with: 10;
> on the third shelf, jars labeled with: 100;
> on the fourth shelf, jars labeled with: 1000;
> on the fifth shelf, jars labeled with: 10000.

---

[2]A cheap way to re-create the Button Dealer system is with a couple of boxes of tooth picks to use as buttons and paper cups instead of jars.

Suppose a customer shows up who requires 6038 buttons. To satisfy this order the dealer looks at the notation, 6038, and proceeds as follows. Since Arabic is read right-to-left, the dealer starts with the right-most digit. This digit is an 8, so he goes to the first shelf and gets *eight* jars marked with a 1. The second digit, one to the left of the 8, is a 3, so he goes to the second shelf and gets *three* jars, each one of which is marked with a 10 and puts them with the jar marked 8. The third digit is a 0, so he gets no buttons off the third shelf. The fourth digit is a 6, so he goes to the fourth shelf and gets *six* jars, each marked with 1000 and puts them with the other jars. Since there are no more digits in 6038, he combines the buttons from the seventeen jars into one big jar and gives the buttons to the customer.

Given that the various jars contain the number of buttons specified, it is clear that this simple process will produce a jar containing exactly 6038 buttons. We want to consider why this process will work.

7.1.2 Designing the Button Dealer's System

To understand why the Button Dealer's System system works, we need to analyze the process of designing this system for supplying buttons. We will assume that the designer knows about the numerals 1 to 9 and the cardinal numbers they represent as indicated in the next diagram.

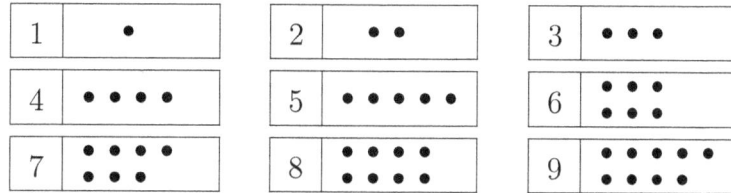

The diagram presents the exact relation between each of the symbols used by the Arabic System and the numerical attribute these symbols represent.

For purposes of discussion, we will assume the designer has to meet the following criteria:[3]

1. the system should use the symbols 1 — 9 and as few others as possible;

---

[3]It would be nice to think our ancestors were clever enough to actually plan things out. History appears to suggest the Arabic System was the result of a long trial and error process that took thousands of years to complete.

2. the system should require a minimum of counting to fill an order;

3. the system should be easy to use and to learn how to use;

4. the system should be perfectly accurate;

5. the system should be able to fill any order for an amount of buttons up to some maximum size.

With these criteria in mind, setting up a system to serve customers where the maximum order size is *nine* is straightforward. The designer simply puts jars containing a single button each on a shelf. To serve a customer, the server has to know and understand the contents of the previous diagram perfectly in order to get the correct number of jars off the shelf. The server never has to count to more than *nine*, since that is the maximum order size. In addition, the server has to know that a collection of the required size can be constructed by taking one jar off the shelf for each dot in the collection associated with the numeral specified by an order, and combining all the buttons from the various jars into one jar which is given to the customer.

Ordering More Than 9 Buttons

Major design questions arise when the system has to accommodate orders of more than *nine* buttons. To be accurate, the system must not skip numbers, which means that the first number that has to be considered carefully after nine is the successor to nine. Obviously an order for *ten* buttons could be accommodated by simply filling an order for *nine* buttons and getting *one* more jar containing a single button off the shelf. Notice that counting to *ten* means our server must come to terms with a diagram that looks like:

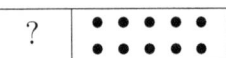

where the question mark indicates an appropriate, but unknown, notation for the successor of *nine*. In other words, as soon as the Button Dealer wants to fill an order for *ten* buttons, he has to have a notation for the number *ten*.

Because our system is comprised of jars of buttons, it is reasonably easy to imagine simply creating a second shelf on which we place jars, **each of which contains ten buttons**. This would satisfy the minimal counting criteria because to fill an order for *ten* buttons we only have to count to *one*. Because *ten* is the successor of *nine*, we know we didn't skip a number. This is the easy part.

The hard part is figuring out a suitable notation for this number, that is, what to write on the jars on the new shelf.

We could, for example, come up with a new, single-digit notation, as in:

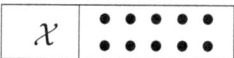

It would be possible to create a system based on this idea that would look something like Roman Numerals. While such a system could provide notations for numbers, it would not support computational procedures in the way the Arabic System does.

The really critical insight that our designer came up with was the realization that when a second shelf with jars containing *ten* buttons was created, the label on these jars must convey **two** pieces of information:

1. get no jars off the first shelf;

2. get one jar off the second shelf.

The question is:

How can we communicate these two pieces of information using the numeration system?

Here is where the system designer got incredibly clever.

First, the designer realized that **communicating two distinct pieces of information would require two symbols**, not one. Further, since one of the instructions was *get no jars from the first shelf*, a new symbol was required that would instruct the user to:

get no jars

off a particular shelf. The result of this realization was the creation of the "do nothing" symbol, namely:

A schematic of the new symbol, 0, that counts the items in an empty jar.

At this point the designer had all the symbols that are needed to convey the two pieces of information identified above. The question left was:

How should these symbols be displayed?

It seems a relatively small step at this point to simply say that the order in which the symbols are read will be the order in which the shelves are accessed. Thus, the first symbol read tells what to get off the first shelf, and the second symbol tells what to get off the second shelf. This being the case, why should

$$10$$

be read as the instruction:

get no jars off the first shelf;
get one jar off the second shelf,

and not the other way around? The answer to this question is that European languages are written and read left-to-right, while Arabian languages are written and read right-to-left. Thus, to the Arabic speaking designer of our button system, the first symbol in 10 is the 0 and the second is 1. Understanding this fact is an essential feature of the Arabic System of notation.

Thus, the notation for the successor of 9 in the Button Dealer system is:

$$10,$$

and it is understood by the user to mean:

get no jars off the first shelf;
get one jar off the second shelf.

Finally, the designer will realize she needs a name for 10, and makes up the word *ten* to correspond to 10. This discussion is summarized in the following digram:

In the diagram, the 9 jars containing one button each are combined with 1 jar containing a single button. The result is equivalent to a single jar containing ten buttons shown on the right-hand side of the equality and identified with the numeral 10.

Finally, we note that the notation 10 is the numeral for the smallest counting number that cannot be expressed with a single-digit numeral.

Given the designer invented 10 and its corresponding instructions, she will know that 11 has to mean:

get one jar off the first shelf,
get one jar off the second shelf,

12 means:

get two jars off the first shelf,
get one jar off the second shelf,

and finally that 19 means:

get nine jars off the first shelf,
get one jar off the second shelf.

So customers requesting 1–19 buttons can now be served. We illustrate serving a customer wanting 17 buttons:

An order for *seventeen* buttons is made up by combining *seven* jars containing one button each from the first shelf and *one* jar containing *ten* buttons from the second shelf.

By making similar diagrams, the reader can verify that orders corresponding to each counting number less than *twenty* can now be filled with perfect accuracy. This truth is entirely due to the fact that each new numeral is a notation for the successor of the previous number for which a numeral had been constructed.

The next question would be what to do about a customer needing the successor of 19 buttons. To deal with this, all the designer has to do is to notice if she combines two jars containing 10 buttons each, that is, two jars from the second shelf, she will have the required number of buttons, as shown below.

A diagram illustrating that *twenty* is realized as two groups of *ten*. To verify that *twenty* is in fact the successor of *nineteen*, construct 19 via the process illustrated above for 17. Then observe that the successor has the same diagram as shown for 20.

This is a critical insight, because it forces the notation for the successor of 19 to be 20, which then translates to the instruction:

get no jars off the first shelf,
get two jars off the second shelf.

This interpretation of 20 exactly extends the meaning of the notation previously created. Moreover, it is clear that we can now represent any number of buttons from 10 — 99, inclusive.

To see why, first observe that numbers of buttons equivalent to

ten, *twenty*, *thirty*, *forty*, *fifty*, *sixty*, *seventy*, *eighty*, *ninty*,

can be obtained by using the notations

10, 20, 30, 40, 50, 60, 70, 80, 90

which mean

get no jars off the first shelf,
get 1 — 9 jar(s) off the second shelf,

respectively. These are all the numbers of buttons that can be obtained using a single digit descriptor of numbers of jars from the second shelf and no jars from the first shelf.

Missing numbers of buttons are obtained by getting the correct number of jars off the first shelf, as in the case of 17, 27, 37, etc., all of which require *seven* jars from the first shelf.

No numbers are missed because, a two-digit number whose numeral ends in 0 is the successor of a number whose right-most digit is 9. For example, 90 is the successor of 89 and as explained in §6.3 there is no counting number strictly between a counting number and its successor. The reader can now see why the successor idea as the **next** counting number is critical to our system of notation.

The naming scheme in English adopted for two-digit numbers greater than twenty is particularly simple, once we have names for twenty, thirty, etc. It simply amounts to reading left-to-right. For example, for 25, we simply say **twenty-five**; in other words, the name is built from the digits in the expression reading left-to-right.

Once the designer has figured out how to deal with numbers from 1 - 99, the design methodology is established. It simply repeats itself. For example, since all possible two-digit combinations of the symbols 0 — 9 are used in making the notations for 0 — 99, the successor of 99 will require using a third shelf and a third digit to communicate this extra piece of information. What goes on the third shelf is jars containing a number of buttons equal to the **successor** of 99 which is the next counting number. The notation for this number is:

100

which reading right-to-left translates into:

get no jars off the first shelf,
get no jars off the second shelf,
get one jar off the third shelf.

Because of its importance, we reiterate the following. Since 10 is the successor of 9, a collection of 10 buttons can be obtained by combining *ten* jars containing a single button each. But this also means that 100, the successor of 99, can be obtained by combining the contents of *ten* jars from the second shelf, each of which contains 10 buttons. While this is something you surely know, you may not have thought about it in quite this way.

If we now pick any three digit number, it is clear what the instruction will be using the Button Dealer's scheme. For example, 836 instructs:

get six jars off the first shelf,
get three jars off the second shelf,
get eight jars off the third shelf,

where the digits are read right-to-left. Combining the buttons from all the jars into one produces a jar containing exactly 836 buttons. We remind our readers that this is a direct consequence of conservation (**CP**).

At this point, the Button Dealer can now fill any order for 1—999 buttons.

Again, there has to be a new name for jars on the third shelf. As you know, it is **one hundred**. This name, for the cardinal number that is the successor of 99, has a different flavor to it. To be specific, it identifies *hundred* as the essential name of the successor of 99 and the *one* tells us that we want exactly *one* unit of this size. So for example, the name associated with 500 is **five hundred** and specifies *five* units of *one hundred.*

Using this naming scheme for three-digit numbers, in speaking of the request for 836 buttons, the Button Dealer would say that

eight hundred thirty six

buttons were supplied. Again notice the right-to-left interpretation of 836 when getting the actual buttons, as opposed to the left-to-right interpretation when speaking English.

Recall, when we reached 9 we needed a new name and numeral for the successor of 9; similarly, when we reached 99, we needed a new name and symbol for the successor of 99. For the same reason, namely, when we get to 999, we will have used all possible three-digit combinations of our symbols, and so we will need a new shelf, a new name and a new symbol for the successor of 999. Each jar on this new shelf will contain the combined contents of *ten* jars of 100 buttons each. As we know, the numeral on the jars is 1000 and its name is **one thousand**. The naming scheme at this point follows that used for hundreds. For example, 7000 is named **seven thousand**, and corresponds to the instruction to the button dealer to:

get no jars from the first shelf;
get no jars from the second shelf;
get no jars from the third shelf;
get seven jars from the fourth shelf.

The introduction of **thousands** on the fourth shelf permits the Button Dealer to accommodate all orders up to 9999 buttons, including our original example:

6038.

Once again, the designer will need a new numeral and name for the successor to 9999. The notation is 10000, and the name is **ten thousand**, which is a combination of the previous names *ten* and *thousand* and reflects their placement in the numeral 10,000, where we have inserted a comma to emphasize the point.

It is now possible to provide names and notations for all counting numbers up to 99999. It is obvious that to a button server, a request for

75756

buttons, gets translated to:

get six jars from the first shelf;
get five jars from the second shelf;
get seven jars from the third shelf;
get five jars from the fourth shelf;
get seven jars from the fifth shelf.

A little thought will convince you that the Button Dealer could extend this system to accommodate arbitrarily large orders simply by adding more shelves as needed to the system. The required notation is built in, although there would have to be some new names created.

Let us recall the design criteria specified at the beginning of this section to see if we satisfy the requirements:

1. the system should use the following symbols, 1 — 9 and as few others as possible;

2. the system should require a minimum of counting to fill an order;

3. the system should be easy to use and to learn how to use;

4. the system should be perfectly accurate;

5. the system should be able to fill any order for an amount of buttons up to some maximum size.

The completed system uses one additional symbol beyond the symbols $1-9$. Since there has to be a symbol associated with *getting no buttons*, and all the other symbols are associated with getting some number of buttons, any system will have to have this additional symbol, 0, whence this addition cannot be viewed as a failure of the first requirement. A user has to count at most *nine* jars on any shelf, which is minimal.

In order to use the system, one needs to know only two things. The first is the relationship between each single digit numeral and its standard collection, in other words, how to count to nine. The second is how the position of a digit in a numeral specifies a shelf to go to.

7.1.3 The Problem of Accuracy

There are two aspects to the problem of accuracy:

1. numeral to collection;

2. collection to numeral.

The first aspect is that given any cardinal number and its numeral, the size of a collection of objects produced described by that numeral must always be the same. The test of this is whether two collections generated from the same numeral always admit an exact pairing between them. That such a pairing should always exist is a consequence of **CP**. Another way of stating this aspect is that the process that takes numerals to collections of the specified size is **reproducible**.

The second aspect of accuracy is more complicated. Consider that we have a pre-existing collection of buttons, A, that has less than $10,000$ buttons in it. For our system to satisfy the accuracy requirement, it must be the case that there is a numeral which when given to the Button Dealer will produce a collection, B, that has the same size as A. Again, the test is whether there is an exact pairing between the members of B and the members of A.

Determining whether the Button Dealer System satisfies these two conditions is a matter of experiment. That we all believe that it does and that the Arabic System does as well is a matter of experience. Ultimately the truth of these assertions comes down to conservation of cardinal number.

In terms of instructing a child, a system like the Button Dealer can easily be modeled using counters. But you might not want to use 10,000 as the maximum!

7.2   Base and Place in the Arabic System

We want to summarize the key components of the Arabic System:

1. a short list of symbols and names for an initial set of counting numbers;

2. a symbol and name for the number associated with the empty set;

3. a recognition that the symbol combination 1 followed by one, or more, zeros has to be the notation for the successor of the largest number that can be written using fewer symbols.

In the actual Arabic System of numeration, the symbol for the largest counting number having a single digit notation is 9. Thus, the initial list contains nine individual symbols denoting numbers associated with the first nine collections, namely, the collections having 1 — 9 members. We also have a special symbol for zero. Thus, in the Arabic System, the short list of symbols has ten members.

The number *ten* is referred to as the **base** of the Arabic System. Alternatively, we speak of the Arabic system as a **base ten** system. You can think of the base as the number of symbols in the short list. Ten is also the successor of nine and the smallest number for which there is no single digit notation.

Consider now a multiple digit number in Arabic notation, say:

$$42027.$$

If we read this as an instruction to the Button Dealer, we know it means:

get seven jars from the first shelf;
get two jars from the second shelf;
get no jars from the third shelf;
get two jars from the fourth shelf;
get four jars from the fifth shelf.

As we know, the position of a digit tells us which shelf to use, starting at the right and proceeding to the left.

The only difference between Button Dealer interpretation and the Arabic System interpretation is that the position or place of a digit now directly conveys a quantity associated with that position as illustrated below:

4	2	0	2	7
ten thousands	thousands	hundreds	tens	ones

Thus reading from right-to-left, the first digit tells us how many ones to use, the second digit tells us how many groups of ten to use, the third tells how many groups of one hundred to use, the fourth how many groups of one thousand to use, and the fifth how many groups of ten thousand to use.

As before, the symbol 0 is essential as a **place-holder** when no groups are used. Thus, 0 has two functions. It is the numeral for the counting number of the empty set, and it prevents other digits from being assigned the wrong value based on their place (position) in multi-digit numerals. The second function is really subsumed by the first once we understand that we must say that we want no hundreds, in the numeric expression for *forty two thousand twenty seven*, 42027.

Lastly, each new group, tens, hundreds, thousands, and so forth, specifies the successor of the largest number that can be written in fewer symbols. Thus,

$$10 \text{ is the successor of } 9;$$
$$100 \text{ is the successor of } 99;$$
$$1000 \text{ is the successor of } 999;$$

and so forth. In each case, the smallest number representable by each new grouping, that is, a 1 followed by some number of zeros, is the successor, of the largest number expressible using fewer symbols. This fact ensures no numbers are missed by the notational scheme and is an essential feature of the system.

There is one additional feature of the base ten system that needs to be emphasized. It is that each place value numeral, *ten*, *one hundred*, *one thousand*, and so forth, is made up of 10 units of the next lowest place value. We express this as statements about collections:

- a collection having 10 members is comprised of *ten* collections having 1 member each;

- a collection having 100 members is comprised of *ten* collections having 10 members each;

- a collection having 1000 members is comprised of *ten* collections having 100 members each;

- a collection having 10000 members is comprised of *ten* collections having 1000 members each;

and so forth. We will recall these facts in Chapter 8 on multiplication.

The discussion to date has not included arithmetic computations. That will happen in successive chapters, and at that point we will see the amazing utility of the Arabic system of numeration.

7.2.1 Equality Between Numerals

In §5.2.1 we gave a specific experimental procedure for determining whether two counting numbers were equal (see Equality Principle). At that time, we did not have numerals for counting numbers available, so the procedure did not depend in any way on the notation for counting numbers. Now we have the Arabic System of notation and it is easy to say when two Arabic numerals denote the same counting number.

To be clear, given an Arabic numeral for any counting number, we know from the discussion in this section how to construct a collection having the cardinal number denoted by the given numeral. Thus, given any two such numerals, we can construct the two collections specified in §5.2.1 and check whether there is an exact pairing. Clearly this would be a tedious process to use on an every-day basis. A simple method for determining equality is the first clever feature of the Arabic System we shall identify.

Two numerals denoting counting numbers will denote the same counting number exactly if:

- the digits in the two numerals, starting at the left and taken in pairs moving to the right, are identical.

Alternatively stated, different numerals denote different numbers.

7.3   The Utility of Names

Wherever possible mathematicians tried to make the names of things convey meaning beyond merely providing an identifier. For example, the name *square root of two* identifies a certain number. But more than this it tells us exactly what property that number has, namely that

$$\sqrt{2} \times \sqrt{2} = 2.$$

So when you come across a new name, try to ask yourself: What additional meaning does this name convey? If there is additional meaning, it can be very helpful and we will see this in much of what follows.

7.4   What Your Child Needs to Know

The focus of this chapter has been on understanding the Arabic System of Numeration. The Arabic system of numeration is the **bedrock** on which all the computational algorithms of arithmetic rest. So complete mastery of this notation for counting numbers is essential.

There are two principal ways we use the names and notations for counting numbers. They are:

1. given a collection, we find the name of the counting number that tells us the cardinality of the collection;

2. given a numeral for a counting number, we construct a collection having that counting number as its cardinality.

Children need to attain complete comfort and facility with both these uses for numbers less than 100. To do this, your child needs to know and be able to use the information in the following diagrams:

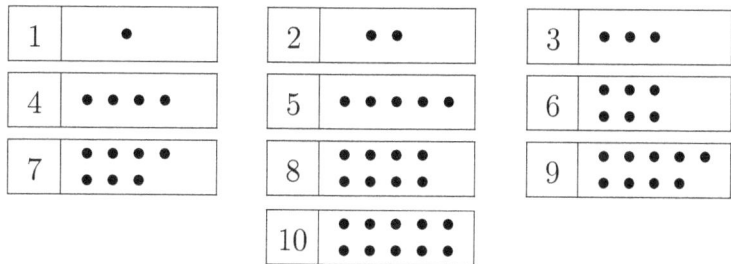

A critical feature of these diagrams is the fact that the numbers are in sequence, that is, 2 comes after 1, 3 comes after 2, and so forth. This means your child understands that each counting number is followed by a **next** counting number, or **successor**. Assimilating these ideas is a critical goal of Pre-K and we direct your attention to activities like the **counting up to** games at the SS website, the number recognition worksheets (click on *Kindergarten* button) at the M-A website and Bunny Count at the EDM website for building these skills.

Here is a list of specific things your child ought to be able to do by age 4–5:

1. given a numeral from $1-10$, form a collection with that number of members;

2. given a collection having ≤ 10 members, identify the correct numeral associated with that collection;

3. given multiple representations of collections having the same size, recognize that they have the same total number of members;

4. understand for numbers less than 10 that the next number in the sequence is associated with a collection containing a single additional element; for example, the collection associated with 6 has a single more dot than the collection associated with 5; (This is how your child comes to terms with the **successor** process.)

5. understand that to count a collection of objects we select the objects **one-at-a-time** and assign a number starting with 1, then 2, then 3, and so forth, until the objects are exhausted, with the largest number so assigned being the number of objects in the collection;

6. count collections containing as many as twenty objects;

7. starting at a given counting number, count-on by a specified further amount.

The third point is illustrated below:

Both collections are assigned the same counting number, namely 5.

The fourth point, finding the next number after 8, is illustrated below:

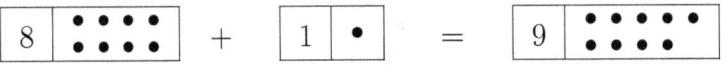

The successor of 8, namely 9, is obtained by adding one element to an existing collection having 8 elements; alternatively, see diagram in §6.3.

71

Board games involving counting are a fun way for children to practice counting skills. The fifth point is illustrated by counting a collection containing *seven* dots as follows:

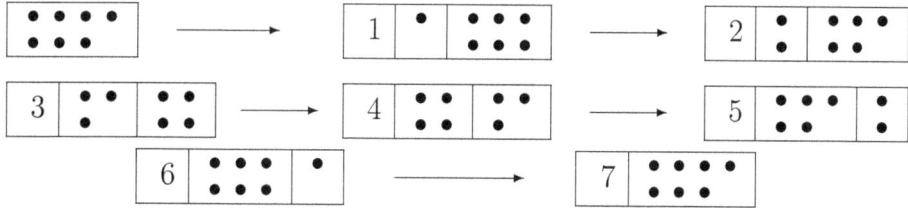

The **one-at-a-time** process for counting a collection containing *seven* objects is illustrated. As each object is counted, the object is moved to the left and the numeral is incremented by *one*. Arrows indicate the order that is followed. When all objects have been counted, the last numeral indicates the number of objects in the collection. See also §6.4.1.

To achieve the sixth goal, you need to encourage your child, as early as possible, and before they start school, to learn to count. The purpose of this is not merely so the child learns their numbers. Equally important is that the child develops an intuitive sense of:

How much is five, or six, or eleven, etc?

Attaining this sense of the quantity attached to a number requires experience.

The seventh goal can be achieved by posing questions like: Find the number that is five more than twelve. This requires the child to start at 12 and count 13, 14, finishing at 17. Tasks like this anticipate addition, in this case finding the sum of 12 and 5.

7.4.1 Goals for Kindergarten

The following list is derived from the CCSS-M informed by the A+ curriculum and repeats many of the Pre-K goals:

1. Count to 100 by ones and skip-count (10, 20, 30, 40, ...) to 100 by tens (see M-A *Kindergarten* for worksheets).

2. Write the numbers from 0 to 20. Write the counting number of a given collection of objects up to 20. Represent a count of no objects by 0.

3. Fully understand the relationship between counting numbers and size of collections (cardinality).

4. When counting, say the numbers in proper order.

5. Understand that successive counting numbers correspond to quantities that are one larger.

6. Understand that the size of a collection is independent of its arrangement — conservation of size attribute.

7. Compare single digit numbers based on their written numerals by using the terms 'less than' and 'the same'.

8. Compose and decompose numbers from 11 to 19 into 10 *ones* and some further number of *ones*. For example, 12 is 10 *ones* and 2 *ones*, where this can be expressed with dots, or physical objects.

Counting to higher numbers is initially learned by **rote**. The websites M-A, SS and EDM all contain activities designed to achieve this and the other learning goals. Skip-counting by tens as in 10, 20, 30, ... etc. is a good way to begin to come to terms with 10 as the base of our number system.

7.4.2 Goals for Grade 1

By the end of Grade 1 your child will be able to do the following in respect to place-value in the Arabic System.

1. Extend rote counting skills to 120 starting from any number less than 120.

2. Read and write numerals less than 120 and represent numbers of objects with a written numeral.

3. Know that a two digit number represents groups of *ten* and groups of *one* and that places in the Arabic System are determined starting at the right and working to the left. Understand the following special cases:

   (a) 10 can be thought of as a bundle of 10 *ones* that we call *ten*;

   (b) Be able to completely describe the place-values assigned to digits in the numbers 11–19, e.g., 18 is 1 *ten* and 8 *ones*;

   (c) know the meaning of 10, 20, ..., 90 as 1, 2, ..., 9 groups of *ten*, respectively, and be able to describe the role of 0 in each numeral.

73

4. Compare two-digit numerals using the terms **less than**, **the same as** and **greater than**.[4]

7.4.3 Goals for Grade 2

By the end of Grade 2 your child will be able to do the following in respect to place-value in the Arabic System.

1. Understand that the three digits of a three-digit number represent amounts of *hundreds*, *tens*, and *ones*; e.g., 706 equals 7 *hundreds*, 0 *tens*, and 6 *ones*. Understand the following as special cases:

   (a) 100 can be thought of as a bundle of 10 *tens* called a *hundred*;

   (b) the numbers 100, 200, 300, 400, 500, 600, 700, 800, 900 refer to 1, 2, 3, 4, 5, 6, 7, 8, or 9 *hundreds* and 0 *tens* and 0 *ones*.

2. Count within 1000; skip-count by 5s, 10s, and 100s.

3. Read and write numbers to 10,000 using base-ten numerals and number names; identify the digits in each place of a four-digit number, and know the value of the digit.

4. Be able to correctly compare two three-digit numbers based on the value of the various digits in their numerals and to correctly record this information using the terms **less than**, **the same as** and **greater than**. (See §??.)

Every child must master the Arabic System of notation. Achieving this takes time and is implicitly recognized in the by grade goals listed above. There are many activities you can find on the websites we have listed that will aid in achieving these goals. After reading this chapter, you should have sufficient knowledge of the Arabic System to enable you to pick and choose activities suited to your child's learning requirements. But you should understand that coming to terms with this system of notation will take time. As can be seen from the goal structure, a four-year process is contemplated.

---

[4]The CCSS-M uses the symbols $\leq$, $=$ and $\geq$. In line with the A+ injunction to avoid early algebra, we stick with words until grade 3.

Chapter 8

Addition of Counting Numbers

Addition builds on all our previous knowledge of counting numbers. For this reason, we take a moment to review the essential facts that will be applied in this chapter.

In Chapter 4, we studied real-world collections and observed we could compare collections in respect to **more** or **less**. Most importantly, we discovered that the number of items in an isolated collection does not change, so that the numerical attribute of **counting number** is a **conserved attribute**. We will refer to this property of collections as (**CP**) and appeal to it as the reason why various things must be true about addition.

In Chapter 5 we used the notion of pairing as a basis for Pre-K development of cardinal (counting) numbers. There we developed a procedure for determining when two collections contained the same number of members, namely when there was an exact pairing of the members of one with the members of the other. In such a case, the two collections must be assigned the same cardinal number and this is the basis for **equality**.

In Chapter 6 the properties of the first nine counting numbers were studied in the context of what children in Pre-K need to learn. The numerals for the first nine counting numbers were given together with a defining standard collection. We found that there were smallest collections characterized by the property that removal of any member destroyed them. Such collections contained a **single** element, and were used to define the cardinal number 1 (see §6.3). The counting process was studied and used to determine the number of elements in collections and to construct collections having a specified number of elements. It was found that collections could be constructed by successively adding single elements to the collection being constructed and we referred to this as the **one-at-a-time** process which is the basis of counting.

In Chapter 7, we developed the Arabic System for naming counting numbers. This system has **zero** as the number denoting the cardinality of the **empty set**. The system also uses place as a means to differentiate the value assigned to a single digit in multi-digit numerals. The discussion surrounding the Arabic System was based on an analogy of a Button Dealer in the real world. In what follows, it will be convenient to refer to that analogy as a means for demonstrating how and why various procedures of arithmetic work.

8.1 What is Arithmetic?

Arithmetic is an abstract model of the behavior of counting numbers as they apply to real-world collections. Models make predictions. As we will see, we can think of $n + m$ as the predicted value of the counting number that will be observed when a collection having n members is combined with a collection having m members. Since we can check this by combining two real-world collections of appropriate size, we have an experimental procedure to verify our abstract model. This leads to a very important conclusion:

> **the computations of arithmetic must agree with our experience of the real world.**

Our purpose in the remainder of this book will be to develop the properties of arithmetic that you will need to help your child succeed. So let's begin by considering what arithmetic is.

> **Arithmetic consists of numbers, operations on those numbers and the procedures for performing the operations.**

We should remember that numbers are abstractions that do not exist in the world. A consequence of this fact is that we will have to use 'set'[1] instead of collections to describe ensembles of numbers. For example, we will use $\mathcal{N}$ to denote the ensemble of all counting numbers. Thus, we could write

$$\mathcal{N} = \{1,\ 2,\ 3,\ 4,\ 5,\ \ldots\}$$

The notation on the right uses curly braces to enclose a list of all the things that are members of the set. Since everything in the set is a number, none of these things actually exist in the world; they are only ideas in our heads. Also, the ... tells us the list goes on forever. So this would be another reason why nothing that is written in this equation can exist in the world.

---

[1] **Sets** are ubiquitous in modern mathematics and will certainly be part of every teacher's vocabulary.

In spite of the fact that everything in arithmetic will be an abstraction, we need to find ways to make things **concrete**, that is, connect them to the world. Here's one way to do that for counting numbers and $\mathcal{N}$. Whenever we speak of $\mathcal{N}$, the equation tells us to think of counting numbers. What we mean by a counting number is the cardinal number of an actual collection in the world, e.g., a jar of buttons. Further, we know that given the numeral for that counting number, we could fill an empty jar with exactly the number of buttons specified by the given counting/cardinal number by using the counting process. Lastly, we know that there would be an exact pairing between the buttons in the jar we filled and the standard collection for the numeral we were given. That's how we make these ideas **connect to the world** and it is how we want our children to think about these things.

By an **operation** we mean any procedure that takes numbers as input and produces a number as an output. Each operation has a physical meaning in that it answers a question. Thus, when thinking about any operation of arithmetic, the thinker should always have in mind the underlying question to which that operation provides an answer.

One may think that any process that takes numbers as inputs and produces a number as output is an operation. This is not so. Operations are special in that an operation **always produces the same output when given the same input**. To make the operation idea concrete, think of addition on a calculator: it takes one or more numbers as inputs and produces one number as an output. Moreover, given the same numbers as inputs, when you press that addition button, you always get the same number for an answer and even if you used a different calculator, you would still get the same output as the answer. This is the key property that defines an operation and simply means:

the result of an operation is universally reproducible.

Although the calculator idea clarifies the criteria for what property something has to have to qualify as an operation, it tells us nothing about what an operation actually does and why it might be interesting. In respect to arithmetic, operations will take numbers as inputs. At this point, this means counting numbers. So if we expect to make an operation on counting numbers concrete, we must expect it to tell us something about counting and collections.

8.2   What is Addition?

Addition is a **binary operation** on the counting numbers. It is **binary** because it takes as input two counting numbers. To have an operation, we need a procedure that identifies a unique counting number as output. We also know that the computation should answer a question about the world.

>**Addition Procedure.** Given two counting numbers, n and m, construct a collection having n elements and a collection having m elements; combine the two collections into a single collection; count the number of members that are in the combined collection; the resulting counting number is $n + m$.

The quantity $n + m$ is called the **sum** of n and m. The inputs n and m are called **summands**, or **addends**. A particular example of the addition procedure is pictured below.

A diagram showing *seventeen* is the sum of *nine* and *eight*.

In Chapter 2, we pointed out that every computation in arithmetic answers a question. It is clear from the addition procedure for counting numbers that addition answers the following question:

>Given two collections having n and m members respectively, if the two collections are combined, how many members will the combined collection contain?

Recall the two key types of understanding identified by Usiskin (see §2.2.2), namely, procedural and use-application. Knowing the procedural definition tells a child not only how to answer computational questions, but also what the use-application of addition is. Thus, this small amount of information deals with the two most important types of understanding identified by Usiskin as applied to arithmetic.

8.2.1 Why Addition is an Operation

The addition procedure defined above clearly takes two numbers as inputs and produces another number as an output. So it could potentially be an operation, but to qualify it has to meet the criteria that

for identical inputs the process always produces the same output.

Let's go through why the addition procedure satisfies this criteria.

Suppose we are given counting numbers n and m. Step 1 of the procedure requires construction of two collections having n and m members, respectively. Being given the counting number n means we are given the numeral for n. So the issue here is whether we can reliably get from a numeral to a collection that is equivalent to the standard collection for that numeral. Here is where the Button Dealer's system is instructive; it shows us exactly how to get from a numeral to a standard collection for that numeral. So as long as we know how to correctly interpret an Arabic numeral, we know we will reliably arrive at the standard collection for n, and similarly for m.

Thus, at this point we may consider that we have a jar with n buttons and a jar with m buttons. Now combine the contents of the two jars into one jar and recall that however we divided up the contents of a jar of buttons, the number of buttons remained the same (see §4.2). So in this case, whether the buttons are in one jar, or in two jars, so long as we are careful, **CP** guarantees the total number of buttons will be constant. Thus, the sum $n + m$ is fixed as illustrated below:

The jar on the left contains 17 buttons. **CP** demands that however the buttons are divided between the two jars on the right, the total number of buttons on the right-hand-side (RHS) of the equality must be the same as the original number on the left-hand-side (LHS). This is why addition is an operation.

The diagrams and ideas presented above should be the bed-rock foundation on which every child's thinking about addition is based.

8.2.2   Use of CP

As we work our way through the theory of arithmetic you will see that conservation of cardinal number (**CP**) is given as the reason why something must be true. Indeed, we just used this principle as the reason why the quantity $n + m$ is fixed and unique. Appealing to **CP** as a primary reason why something must be true does not make the given rationale a **mathematical proof**. Rather it is a statement that for our system of arithmetic to have value, it must agree with this fact about the real world: **cardinal numbers are conserved quantities**; they do not spontaneously change.

8.3 Why Use Collections Instead of Sets in the Addition Procedure?

There is one subtle aspect about making addition concrete that we need to emphasize. It is the point concerning the difference between sets and collections discussed in §4.3.1. Consider two collections, for example, two jars of buttons in the real world. Suppose we pour the buttons into a third jar. There is no possibility that two of the buttons coalesce to become one because we cannot have a single button that occupies both jars simultaneously in the real world. This is a physical fact about the world.

Because mathematics is abstract, this is not so in the mathematical world. For example, we can have two sets, A and B, each having 3 members as shown:

$$A = \{1, 2, 3\} \text{ and } B = \{2, 3, 4\}.$$

When we combine the two sets by forming their **union**, we get[2]

$$A \cup B = \{1, 2, 3, 4\},$$

which has only four members, not the six we would expect from two physical collections of buttons, each of which has three members. This is one reason why dealing with sets is confusing. But the folks who invented our arithmetic did not have these abstract ideas around. They understood the focus was on real-world collections and what such collections had to say about arithmetic. This is why we have spent so much time emphasizing that our ideas about arithmetic should be guided by the real world.

8.4 Equality Properties

A+ curricula avoid algebra before Grade 3. The principal reason that convinces me to subscribe to this approach is that the more focused A+ curricula succeed better than curricula which introduce algebra in Kindergarten as evidenced by test data. However, this book is targeted at adults and for this reason we use the power of algebraic notation which includes the equality symbol. We have already suggested how to substitute for these symbols using language.

The equality relation is an essential part of arithmetic and has an important place in A+ curricula in late elementary. For this reason, we need to discuss its important features.

---

[2]The symbol $\cup$ denotes taking the **union** of two sets. The result is a new set containing all members in either, or both, sets.

In §5.2.1, we provided a procedure using pairing for determining when two collections had to be assigned the same counting number. This procedure is the entire basis for equality between counting numbers. Thus, any properties we might state must be consistent with this procedure.

Further, given a general statement about counting numbers that involves equality, if we replace any variables by particular counting numbers, we can perform an experiment with collections and counting that will verify the resulting instance of the equation. For example, using $n = 12$ and $m = 3$, the equation

$$n + (m + 1) = (n + m) + 1$$

becomes

$$12 + (3 + 1) = (12 + 3) + 1.$$

To witness the LHS combine a jar with 12 buttons with a jar containing $3 + 1 = 4$ buttons and do the count. To witness the RHS combine jars containing 12 and 3 buttons, respectively. Then add a single button and count. Both counts must produce the same cardinal number, in this case 16. Does anyone doubt that the counts will agree? Of course not. But the fact that we know that both processes will give the same count is the basis for our choice of what properties we must take to be true in our formulation of arithmetic.

We state the following four Equality Properties using the word *quantity* to denote any mathematical object, in particular numbers. We let A, B, C and D be any mathematical quantities:

E1: $A = A$;

E2: if $A = B$, then $B = A$;

E3: if $A = B$ and $B = C$, then $A = C$;

E4: if $A = B$ and $C = D$, then $A + C = B + D$.

The following are examples using counting numbers:

E1: $5 = 5$;

E2: if $3 = 2 + 1$, then $2 + 1 = 3$;

E3: if $4 + 3 = 7$ and $7 = 5 + 2$, then $4 + 3 = 5 + 2$;

E4: if $2 = 1 + 1$ and $3 = 2 + 1$, then $2 + 3 = (1 + 1) + (2 + 1)$.

Notice the use of parentheses to indicate groups of symbols that are to be treated as single quantities. Thus, $(1 + 1)$ is treated as a unit or **term**.

For clarity, we consider how the principles are used. In our minds, and indeed children's minds, we should think:

the equals symbol means: *is the same as.*

Indeed, 'is the same as' is exactly what we want children in K-2 to be thinking when they consider the quantities $5 + 4$, $6 + 3$ and 9 is:

$5 + 4$ is the same as $6 + 3$ is the same as 9.

Later, when 'is the same as' is ingrained, the equality symbol is introduced and its meaning is clear.

As well, we can think of a statement like $5 = 5$ as expressing a mathematical fact. Similarly, $5 = 4 + 1$ expresses a mathematical fact about Arabic numerals. What E2 tells us is that given the mathematical fact $5 = 4 + 1$, we immediately know that $4 + 1 = 5$ is also a mathematical fact about Arabic numerals. At the deepest level, what is being asserted by the equality $5 = 4 + 1$ is that

5 and $4 + 1$

are two names for the same counting number in the Arabic System of numeration. That said, you can operate successfully by thinking of equality at the highest level of meaning, namely, *is the same as.*

Again, E1-E4 can be demonstrated by constructing collections. For example, for E2, consider two counting numbers m and n. We know they are equal exactly if when we construct a collection A having m elements, and a collection B having n elements, then there will be an exact pairing of the members of A with the members of B. If such a pairing exists between A and B, a similar pairing will exist between B and A. (Just reverse the arrows.) Thus, if $m = n$, then $n = m$ for any pair of counting numbers.

Similar constructions are possible to support the properties E1, E3 and E4.

We give the Equality Properties their mathematical descriptors and restate them in words:

E1: Equality is **reflexive**, i.e., every quantity is equal to itself;

E2: Equality is **symmetric**, i.e., if one quantity equals a second, the second also equals the first;

E3: Equality is **transitive**, i.e., two quantities equal to the same thing are equal to each other;

E4: Equality has the **addition property**, i.e., equals added to equals are equal.

E4 will have the consequence that:

E5: Equality has the **multiplication property**, i.e., equals multiplied by equals are equal.

Children should become comfortable with the use of $=$ as symbol representing **is the same as** in Grade 3 and equality learning goals are included in the last section.

8.5 Properties of the Addition Operation

There are three properties of addition that will be used repetitively to develop the rules for arithmetic. They are stated as equations A1-A3 and are likely to be very familiar to the reader.

A1-A3, tell you everything you need to know about the theory of addition of counting numbers.

So A1-A3 need to be completely understood.

8.5.1 Addition of 0

Addition was defined in §8.2 as a binary operation on the set counting numbers by appealing to properties of collections and counting. As you know zero is not a counting number, since counting numbers give the size of real-world collections and such collections must have members. Nevertheless, we know zero is a perfectly good number and we must be able to give an answer when we add zero to a counting number. The equation in A1 tells us what the sum of n and 0 will be for any counting number n.

A1: Let n be any counting number, then

$$n + 0 = 0 + n = n.$$

Because 0 satisfies this equation, it is called the **additive identity**. The adjective *additive* refers to the operation, which is addition. The noun *identity* refers to the fact that when this number is combined with any other number, say 5, using the operation of addition, the value returned is 5.

In Chapter 9, we will take the property defined by A1 as the defining property of zero. Namely, zero is the number which when added to any other number gives the original number as the answer. Thus, $0 + n = n + 0 = n$. Proceeding in this manner would be the mathematician's approach. At this point our guide is the real world in the form of collections, so let's look there for why this equation must hold in our system of arithmetic.

Experience with collections together with **CP** should readily convince you that only doing nothing to a collection has the property that it leaves the attributes unchanged. Any process that adds or removes members changes the attribute of size. For this reason, we should expect that zero is the **only** cardinal number which will act as an additive identity. In respect to this property (behavior), zero is **unique**.[3]

8.5.2 The Commutative Law

A2: Let m and n be any two members of $\mathcal{N}$, i.e., counting numbers. Then

$$n + m = m + n.$$

As the reader knows, this equation is called the **Commutative Law** of addition.

To see why it must be true, again we turn to the real world. Consider a jar with n buttons in it, and a second jar with m buttons in it. Ask yourself:

> Will it make a difference to the total number of buttons whether the buttons from the first jar are poured into the second jar, or the buttons from the second jar are poured into the first?

Experience tells us that so long as we are careful not to lose buttons, the total number of buttons will be the same in either case. So the equation above must be true, again as a consequence of conservation **CP**.

What every child should know about the Commutative Law is that the order in which summands are added has no effect on the sum. This applies no matter how many summands there are.

8.5.3 The Associative Law

A3: Let m, n and p be members of $\mathcal{N}$, i.e., counting numbers. Then

$$n + (m + p) = (n + m) + p.$$

---

[3]Mathematicians consider zero to be a cardinal number. But it is not a counting number. There are many cardinal numbers that are not counting numbers, but they do not occur in arithmetic.

This property is called the **Associative Law** of addition.

To see why it must be true, we again turn to the real world. Consider that we have three jars of buttons. The first contains n buttons, the second m buttons, and the third p buttons. The parentheses in the equation are used to tell us in which order the operations are carried out. So the left-hand side instructs us to first combine the jar containing m buttons with the jar containing p buttons to obtain a jar containing $m + p$ buttons. Only when this is done do we complete the task by combining the result with the jar containing n buttons to obtain a jar containing the total, $n + (m + p)$ buttons. The right-hand side forces us to do things in the other order, namely first combine the jars containing n and m buttons, respectively, to obtain a jar containing $n+m$ buttons. Only then add in the p buttons contained in the third jar. **CP** tells us that the number of buttons in the combined collections cannot be affected by the order in which the collection was assembled. So the equation above is true in the real-world, and so we must make sure it is true about our arithmetic.

Notice that if any of m, n and p are 0, then the equation reduces to an identity. For example, if $m = 0$, then the equation becomes $n + p = n + p$ after we apply A1 above.

What a child needs to understand about the Associative Law is that it tells us that the process that is used to combine summands in any computation involving addition cannot affect the sum.

As you can see, the theory underlying addition is small, in the sense that only three equations are required. As we shall see, these three equations have powerful effects, which is why they are so important.

We stress that while the rationale for why these three equations must be true are not proofs in the mathematical sense, they are the essential reasons why the given statements must be facts about arithmetic. They are all derived from our understanding of counting and conservation and that is how they should come to be understood by children. It is almost certainly the case that studying the behavior of counting numbers as applied to collections is how humans discovered there was such a thing as mathematics and that numbers could provide useful information about the world.

8.5.4   Addition From Successor

As we have seen, the one-at-a-time process was critical to the Arabic System of numeration. We want to explore this notion in the context of addition. Given the number n, there is a **next largest counting number**, $n + 1$ which we called the **successor** of n. This fact enables us to define how we add any other counting number to a given counting number n. This process is so fundamental, we explain it in detail. As always, our reasoning is based on collections.

The following diagram illustrates the process of counting a collection containing *seven* dots:

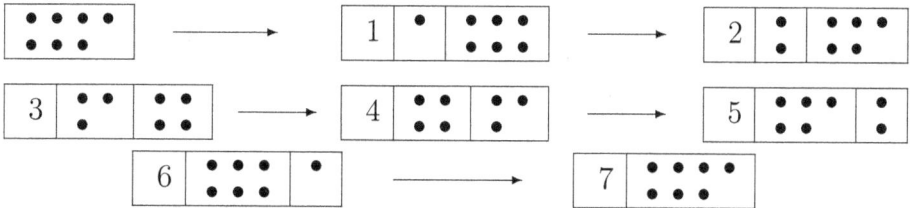

The step-by-step process of counting a collection containing *seven* objects is illustrated. As each object is counted, the object is moved to the left and the numeral is incremented by *one*. Arrows indicate the order that is followed. When all objects have been counted, the last numeral indicates the number of objects in the collection.

We can take this diagram and interpret it in a different way. Suppose we want to find the sum of 4 and 3. According to the concrete procedure we would construct two collections, one with 4 members and one with 3. We would then combine the two collections by moving members one-at-a-time from the collection having 3 to the collection having 4 to obtain a collection having 7. This process is illustrated below:

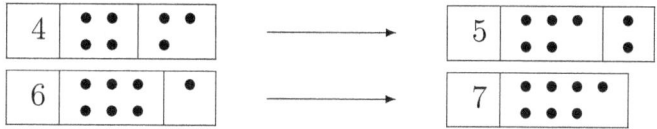

The step-by-step process of adding $4 + 3$ by starting with a collection containing 4 members and a collection containing 3 members and transferring members one-at-a-time from the collection with 3 members to the collection that initially contains 4 members.

This sequence below illustrates computationally the one-at-a-time process:

$$4 + 3 = 4 + (2 + 1) = 4 + (1 + 2) = (4 + 1) + 2 = 5 + 2$$
$$5 + 2 = 5 + (1 + 1) = (5 + 1) + 1 = 6 + 1$$
$$6 + 1 = 7$$

where we have made use of the Commutative and Associative Laws.

The process is directly related to **counting-on**. In the case above, we are counting-on 3 units from 4. To do so, one simply starts with 4 and continues:

$$5, \quad 6, \quad 7$$

stopping when the next 3 numbers have been listed (counted). Counting-on can begin at any number and continue for any number, although it is most useful when counting-on by single digit numbers as in counting-on by 6 beginning at 37:

$$38, \quad 39, \quad 40, \quad 41, \quad 42, \quad 43.$$

Again, all counting-on does is produce the next six counting numbers after 37. It is useful because, as illustrated above, it provides a simple procedure for finding a sum as in: $37 + 6$. We will use counting-on in what follows.

8.6 Making Addition Useful

While the properties of addition discussed above are interesting, particularly to mathematicians, they are not very helpful to folks who simply need numbers to keep track of things. For example, think of a rancher who has 175 cattle on the north forty, another 203 cattle on the high pasture, and who needs to know the total, so he can order winter feed. Even in a situation where the rancher has available an effective naming scheme for numbers, like the Arabic system, the equations A1-A3 above, tell him nothing about how to actually find the required sum. And waiting until he can combine the two herds so he can find the sum by doing a count may take too long. What the rancher needs is a procedure that connects the operation of addition to the naming scheme, namely, the Arabic number system. That there is a connection is one of the amazing facts about the Arabic system of numbering.

To make this connection, we will make use of our Button Dealer analogy. So let us give the rancher's problem to the Button Dealer, namely, find the sum of 175 and 203.

> **How the Button Dealer Does a Sum**. To fill an order for 175 buttons, the Button Dealer goes to the first shelf and gets *five* jars marked with 1, goes to the second shelf and gets *seven* jars marked with 10, and goes to the third shelf and gets *one* jar marked with 100. To fill an order for 203 buttons, the Button Dealer goes to the first shelf and gets *three* jars marked with a 1, gets no jars marked with 10 from the second shelf, and goes to the third shelf and gets *two* jars marked with 100. Since the Button dealer knows that addition corresponds to combining collections

and counting, she proceeds as follows. She counts the jars from the first shelf, and determines there are *eight*. She observes that she still has *seven* jars marked with 10. Lastly, she counts the jars marked with 100, and finds there are *three*. So in the Button Dealer's naming scheme, this corresponds to 378.

Look how simple this process is! To determine the total, the Button Dealer only has to count up, and record, the number of jars coming from each shelf, performing the addition process one shelf at a time. One part of this process is absolutely critical, namely that we only combine jars from the same shelf. In other words, we only count up jars labeled with 1, followed by counting up jars labeled with 10, followed by counting up jars labeled with 100. This is the key that makes the process work.

Let's see if we can adapt this idea to the Arabic system for which the Button Dealer system is an analogy.

Example 1

Suppose you are confronted with finding the sum of: 23 and 54. How are you to find the answer? First we write down 23 and then under it we write down 54, as shown below.

$$
\begin{array}{r}
23 \\
+54 \\
\hline
\end{array}
$$

When we do this, we have to remind ourselves that in the Arabic system, the **position of a digit in the numeral determines its value**. So when we write down the 54 under the 23, we must make sure that the right-most digit in 54 lines up under right-most digit in 23. Thus, we have lined up the *ones* digits in one column and the *tens* digits in another column. We refer to this process as **setting up** the problem.

Once the problem is setup, we are ready to perform the addition. All we have to do is find the sum of the numbers in each column, starting with the *ones* column. These sums involve single digits.

The required sum of the digits in the *ones* column is: $3 + 4$. To find this sum, we create two collections, the first having three members, the second four members, and combine their contents. The size of the resulting collection is the sum. Finding this sum is illustrated in the diagram below.[4]

---

[4]Finding a sum such as $3 + 4$ is where counting-on is really useful.

An equation implementing the addition procedure for $3 + 4$. The M-A website has worksheets based on adding dots.

The result is 7 and this numeral is entered in the *ones* column below the line, as shown below:

$$
\begin{array}{r}
23 \\
+54 \\
\hline
7
\end{array}
$$

The same procedure is used to find the sum of the digits in the *tens* column, namely, $2 + 5$. Again, the procedure is illustrated below:

$$2 \;\boxed{\,\bullet\;\bullet\,}\quad + \quad 5 \;\boxed{\,\bullet\;\bullet\;\bullet\;\bullet\;\bullet\,}\quad = \quad 7 \;\boxed{\begin{smallmatrix}\bullet\;\bullet\;\bullet\;\bullet\\ \bullet\;\bullet\;\bullet\end{smallmatrix}}$$

A graphical illustration of finding the sum $2 + 5$.

The result is again 7, and the result is recorded below the line in the *tens* place as shown:

$$
\begin{array}{r}
23 \\
+54 \\
\hline
77
\end{array}
$$

The procedure described above is one of the great beauties of the Arabic system for naming numbers because it supports procedures for performing the operations of arithmetic that are so simple they can be mastered by elementary school children. Consider that in order to add any two counting numbers, a person only has to know the sums of the digits in pairs, and mechanically follow the rules regarding place. Not only is this procedure so simple it can be mastered by children, it is so simple it can be built into a machine that can be held in the palm of your hand.

8.6.1 The Addition Table

Even though the procedure described above is straight forward to implement, constructing diagrams each time we want to add some numbers would be tiresome and time consuming! Indeed, even the best counting-on procedures are really inefficient because they use working memory. But here is where another one of the clever features of the Arabic system comes into play. There is a short list of single digits; so there is a relatively short list of pairs having sums we need to know as shown in the following table:

+	0	1	2	3	4	5	6	7	8	9
0	0	1	2	3	4	5	6	7	8	9
1	1	2	3	4	5	6	7	8	9	10
2	2	3	4	5	6	7	8	9	10	11
3	3	4	5	6	7	8	9	10	11	12
4	4	5	6	7	8	9	10	11	12	13
5	5	6	7	8	9	10	11	12	13	14
6	6	7	8	9	10	11	12	13	14	15
7	7	8	9	10	11	12	13	14	15	16
8	8	9	10	11	12	13	14	15	16	17
9	9	10	11	12	13	14	15	16	17	18

The **Addition Table**. To find the sum $7 + 6$ go to the row having 7 at the far left. Follow this row across to the column headed by 6. The table entry in this cell is 13. Notice as we move across the row, we get the sums $7 + 0 = 7$, $7 + 1 = 8$, $7 + 2 = 9$, and so forth, in order as cell entries, thereby illustrating the process of counting six more starting at 7.

Requiring children to commit the facts in the table to memory may seem unreasonable. Let's consider this. Since there are ten symbols in our short list, there are 100 possible pairs $m + n$, where both m and n correspond to a single digit. So in principle, to completely master all possible sums requires learning one hundred items. Recall, that

$$n + 0 = 0 + n = n.$$

This fact reduces 19 of the entries to a triviality. Also, we know that $n + m = m + n$. How does this show up in the table? Pick any cell not on the main diagonal, that is, a cell having a different row and column number. Check the value in the cell. Now observe that if you interchange the row header with the column header, the value in the new cell having this row and column header will be the same. This is how $n + m = m + n$ shows up in the table. In terms of facts to be learned, the Commutative Law **reduces the total number to less than fifty**.

Think of it! Knowledge of fifty addition facts, together with an understanding of the role of place in the addition process, enables anyone to master the addition of an infinite number of pairs of counting numbers. In the marketplace of learning, this seems like a pretty good bang for the buck! The various websites provide plenty of different ways to encourage your child to commit the facts in this table to memory.

Reading the Table

Table (cell) entries are the numbers to the right and below the double lines. Each cell entry is labeled with a row header followed by a column header. Row headers are to the left of the double lines. Column headers are above the double lines. The first row is the row having 0 at the far left. The tenth row is the row having 9 at the far left. Similarly, the first column has a 0 at the top, and the tenth column has a 9 at the top.[5]

To find the sum $n + m$, simply find the cell at the intersection of the row having n as its row header and the column having m as its column header. So the cell entry in row headed by 5 and column headed by 3 is the sum of 5 and 3, namely, 8.

Each entry in the body of the table results from a computation of the form:

$$7 \boxed{\begin{smallmatrix} \bullet \ \bullet \ \bullet \ \bullet \\ \bullet \ \bullet \ \bullet \end{smallmatrix}} \quad + \quad 6 \boxed{\begin{smallmatrix} \bullet \ \bullet \ \bullet \\ \bullet \ \bullet \ \bullet \end{smallmatrix}} \quad = \quad 13 \boxed{\begin{smallmatrix} \bullet \ \bullet \ \bullet \ \bullet \\ \bullet \ \bullet \ \bullet \\ \bullet \ \bullet \ \bullet \\ \bullet \ \bullet \ \bullet \end{smallmatrix}}$$

A diagram verifying the entry in the table above for $7 + 6$. The sum can be directly found by counting the total **dots** on both sides.

Learning the Table

Your child needs to know the sums in this table as a matter of recall. Achieving this means transitioning from procedure-based problem solving to memory based problem-solving. To help your child, you need to understand that the learning starts with a procedure. This procedure needs to **always produce a correct answer**. Without a procedure children will simply **guess**. Guessing never works and children need to learn this right away. If your child doesn't know the answer, you should encourage him/her to say so because it is at that point you can go over the procedure that leads to the answer. Having a procedure that works is key because unlearning incorrect information is far harder than learning it in the first place. In addition, the procedure should reinforce the conceptual understanding of the operation. Let's recall:

> In the real world addition corresponds to combining two collections of the required sizes and counting the total to find the sum.

Because we understand this as the process, we know that it doesn't matter which collection is poured into the other in the combining process. The result will be the same. In other words the process is commutative.

---

[5]Entries in arrays of this type are usually referenced by row and column number. In this context, because of the row and column headed by 0, the entry in the intersection of the fourth row and fifth column would not be the sum of $4 + 5$. For this reason we chose to label rows and columns with the headers and not in the usual manner.

We can turn these ideas into an effective procedure for adding two single-digit numbers from the Table. The procedure is based on the process of **counting-on**[6] which your child should become familiar with in Kindergarten. It works as follows:

Step 1: choose the larger of the two numbers;[7]

Step 2: beginning the count with this number, have your child **count-on** raising one finger as each successive number is counted;

Step 3: stop when the number of fingers raised matches the smaller number;

Step 4: the last number counted is the required sum.

As an example, consider finding $4 + 9$. To apply the procedure, the child starts counting at 9 and continues:

$$10, \quad 11, \quad 12, \quad 13.$$

As each new number is counted, the child raises one additional finger. At 13 your child will raise his fourth finger and the count stops with 13 as the answer. This procedure always produces the right answer and is the most efficient because, as given, the count always starts with the larger number.

Once you know your child can correctly use the procedure, you can work on transition to memory-based problem solving. Like everything else in life, it comes down to practice. Here, flash cards can be helpful. You can buy them, or you can make then using index cards. Five or ten minutes of practice each day should make your child an expert in no time.

8.6.2 More Addition Examples

Example 2

For a second example we find the sum of 143 and 25. The setup is:

$$
\begin{array}{r}
143 \\
+25 \\
\hline
\end{array}
$$

Notice that one number has three digits while the other has only two. There is no digit at the left of 25 because we choose not to write down unnecessary zeros. Thus, 025 and 25 name the same number. We could write the required sum as:

---

[6]Exercises involving counting-on can be found at the websites.

[7]This merely makes the process efficient; it does not change the answer.

$$143$$
$$+025$$

but choose not to because mathematicians are lazy. However, we can see that if we are going to be lazy, we have to be careful and make sure the columns are properly aligned so that there is no confusion as to which column a digit is in. It may be helpful for your child to set up the problem as above when starting out.

Again, we simply add the columns starting at the right with the *ones* column. Using the table, we find $3 + 5 = 8$, and this is recorded below the line in the *ones* place:

$$143$$
$$+25$$
$$\overline{8}$$

The next step is to add the digits in the *tens* column. The required sum is $4+2 = 6$, which is found in the table and recorded in the *tens* place below the line:

$$143$$
$$+25$$
$$\overline{68}$$

The final step is to sum the digits in the *hundreds* column, one of which, as discussed above, is 0. In effect, all that has to be done is to write the single *hundreds* digit below the line as:

$$143$$
$$+25$$
$$\overline{168}$$

which completes the process.

It is evident that the efficiency of this process is vastly increased if the sums of the single digits are available in the child's memory.

8.6.3   Carrying

Let us review what we know. In the examples above, the procedure for adding two numbers, n and m, was to write the numerals, one above the other, so that the *ones* digit in the numeral for n was directly above the *ones* digit in the numeral for m, the *tens* digit in the numeral for n was above the *tens* digit in the numeral for m, the *hundreds* digit in the numeral for n was above the *hundreds* digit in the numeral for m, and so forth. The sum of the digits in each column was then found using values obtained from the table (or memory) to produce $n+m$. We know this process worked because addition corresponds to counting combined collections as shown by the Button Dealer analogy.

In all the examples considered so far, the sum in any column could be written as a single digit. This leaves us to wonder if the process still works when the sum is a two digit number, as in $9+7=16$? The answer is yes, but we have to revise our procedure. The revision is referred to as **carrying**.

Once again we use the Button Dealer analogy to find the sum of 28 and 54. As before the Button Dealer proceeds as follows:

> **How the Button Dealer Carries.** From the first shelf take *eight* jars marked with 1 and from the second shelf take *two* jars marked 10, to obtain 28 buttons. Then from the first shelf take *four* jars marked with 1 and from the second shelf *five* jars marked 10 to obtain 54 buttons. To find the total number of buttons, the Button Dealer starts by counting the jars marked with 1 and finds *twelve*. What the Button Dealer knows is that *ten* jars marked with 1 contain the same number of buttons as *one* jar marked with 10, so the Button Dealer simply replaces *ten* of the jars marked with 1, with *one* jar marked with 10. At this point, the button dealer has *two* jars marked with 1 and a total of *eight* jars marked with 10. Hence, there are 82 buttons in the combined collection.

Two jars of 10 come from 28, and *five* come from 54, for a total of $2+5=7$ jars marked with 10. The extra jar, comes from the fact that $8+4=12$ which in Button Dealer terms is *two* jars marked 1, and *one* jar marked 10. The process, by which we turn *ten* jars marked with 1 into *one* jar marked 10, is called **carrying**.

The above example in the context of the Button Dealer shows us how, even when we have to carry, the addition procedure comes down to counting the contents of two collections. We now consider several examples of the procedure that illustrate what children are taught and how the process is supported by the Arabic System of numeration. We start with the sum we just found.

Carrying: Example 1

As before, the setup consists of writing one numeral above the other, making sure the columns are properly aligned, *ones* above *ones* and *tens* above *tens*:

$$\begin{array}{r} 28 \\ +54 \\ \hline \end{array}$$

The sum of the numbers in the *ones* column is 12, and this is recorded in the following intermediate way:

$$\boxed{1}$$
$$28$$
$$+54$$
$$\overline{2}$$

In the above, the *ones* digit from the sum of the first column, namely 2, is recorded in the *ones* place in the answer. The *tens* digit from this sum, which can only be a 1 when summing any pair of single digit numbers, is recorded in a new row in the *tens* column as shown. (This 1 has been placed in a box for clarity and corresponds to the 10 *ones* being combined in a single extra jar marked with 10 in the Button Dealer example.) We then sum all the numbers in the *tens* column, namely, $(1+2)+5=8$ and record the result in the *tens* column below the line:

$$1$$
$$28$$
$$+54$$
$$\overline{82}$$

This example illustrates why the columns are summed right-to-left. It is because if the sum of the digits in a column produces a two-digit numeral, then the digit on the left gets added to the digits one column to the left. Thus, in a two-digit sum of *ones* digits, the left-hand digit is a *tens* digit. In a two-digit sum of *tens* digits, the left-hand digit will be a *hundreds* digit. These facts are best understood in the context of the Button Dealer example where 10 jars of 10 buttons each combine to produce 1 jar of 100 buttons.

Carrying: Example 2

We find the sum of 999 and 1, which we know is the successor of 999. Finding this sum will further illustrate how the Arabic System supports computations. The setup is shown below where there is more space between columns for exposition purposes:

$$9\quad 9\quad 9$$
$$+\qquad\quad 1$$

Since the sum of the digits in the *ones* column is 10, we write 0 in the *ones* place in the answer and carry a 1 into a new row at the top of the *tens* column as shown below. Note this 1 has been placed in a box and it is important to understand that it arises as the sum of 10 *ones* in the computation $9+1$.

$$\boxed{1}$$
$$9\quad 9\quad 9$$
$$+\qquad\quad 1$$
$$\overline{\qquad\quad 0}$$

95

The sum in the *tens* column, which now has a carried 1 at the top, is again 10. So we write 0 in the *tens* place in the answer, and carry a boxed 1 (representing 10 *tens*) to the top of the *hundreds* column, as shown below.

$$
\begin{array}{cccc}
\boxed{1} & 1 & & \\
9 & 9 & 9 & \\
+ & & 1 & \\
\hline
& 0 & 0 &
\end{array}
$$

The sum in the *hundreds* column is now 10, so we write 0 in the *hundreds* place in the answer. This time there is no *thousands* column, so we create one by carrying a boxed 1 (representing 10 *hundreds*) to the top of the empty *thousands* column in the same row as the other carried 1 s, as shown.

$$
\begin{array}{cccc}
\boxed{1} & 1 & 1 & \\
& 9 & 9 & 9 \\
+ & & & 1 \\
\hline
& 0 & 0 & 0
\end{array}
$$

Summing the newly created column produces:

$$
\begin{array}{cccc}
1 & 1 & 1 & \\
& 9 & 9 & 9 \\
+ & & & 1 \\
\hline
1 & 0 & 0 & 0
\end{array}
$$

which we recognize as the correct notation for the successor of 999.

The procedures described provide an ability to perform the operation of addition on all counting numbers, and indeed, on all decimal numbers after that extension has been made. This is truly remarkable for several reasons. First, it is not at all clear why there should exist a good system of notation for counting numbers that corresponds so well to the one-at-a-time process for constructing collections. Second, it is even more remarkable that this system of notation should mesh so completely with the arithmetic operation of addition. Not only does it mesh, it is so inherently simple it can, and is, successfully taught to children around the world.

8.7   What Your Child Needs to Know

In respect to this question, there is good news and bad news. The good news is there isn't all that much your child needs to know. The bad news is, what your child needs to know, your child needs to know perfectly at the level of **instant recall**.

Let's deal with the good news first. There are at most fifty separate items in the addition table, provided one is thoroughly aware of the Commutative Law:

$$m + n = n + m.$$

In comparison to the number of words a First Grader would be expected to be able to read, this is a very small number. The role of place in the numbering scheme and how it is used as part of the addition algorithm is more complex.

As noted, the bad news is these things need to be known perfectly. What do we mean by this? In respect to things that are in the realm of mathematical facts, for example, the sum of $5+4$, the correct value should be available as instant recall. With respect to things that can be thought of as computational processes, for example, the procedure used to find the sum of $973 + 858$, it is expected that the child would be able to correctly execute all aspects of the procedure with no hesitation, or in the jargon used in curriculum guides, fluently.

8.7.1   Why We Demand Instant Recall

Most of the parents who read this book will be familiar with computers and/or calculators. Very likely, they worked with a calculator that could produce values for trigonometric functions like sin. The point I want to make is that such a calculator doesn't compute values for this function, it remembers them because that is much more efficient than calculating the value each time it is used.

The brain is a biological computer. It also has working memory which it uses when it wants to solve problems. Suppose we are asked to solve the following problem:

> Johnny is on a hockey team that has 20 players. If each player has three hockey sticks, how many sticks total do team members have?

To solve this problem, eventually you get to finding: 20×3. Remembering the product is far more efficient than any actual calculation process. Moreover, every procedure utilizes nerve cells in the brain's problem-solving space and this depends on short-term memory. The part of your brain available to accept and process this information is limited and varies from individual to individual. What is clear is that the more nerve cells an individual can apply to a given problem, the more likely that individual will be able to find a solution. Thus, any part of the problem solving process that can be dealt with by retrieving information quickly from long-term permanent memory releases working memory to address the given problem more effectively.

We liken this to muscle memory, which is what skilled athletes use on the field of play. For example, great skaters are not thinking about how they skate. In **Nike** terms, they *just do it!* Similarly, musicians do not think about what note to play when they sight-read music; if they had to, they couldn't possibly do it. Their responses are automatic, and that's what we want to achieve here in respect to the facts and procedures of arithmetic.

8.7.2 How to Achieve the Learning Goal

First we must recognize that acquiring these skills takes time and practice. This is why a famous mathematician once remarked: *Mathematics is not a spectator sport.* That said, for these skills to be well-learned by children, the process should be fun. And nothing is quite as much fun as **getting the right answer**. This is why we have suggested board games that require counting. In the beginning, playing the games is probably more engrossing for the child as opposed to generating the equivalent worksheet with 30 problems on it. The key websites we have listed, EDM, M-A, SS and SKids all offer worksheets and activities at all levels.

We have stressed counting activities for children as a learning device. Board games using a pair of dice effectively teach counting two collections up to 6, which, as we know, is addition. So the addition table up to 6 can be learned using a pair of dice. As well, use of dice in board games makes children very familiar with the visual verification of equations like:

$$ \boxed{3 \;\; \boxed{\bullet\,\bullet\,\bullet}} \;\; + \;\; \boxed{5 \;\; \boxed{\bullet\,\bullet\,\bullet\,\bullet\,\bullet}} \;\; = \;\; \boxed{8 \;\; \boxed{\begin{smallmatrix}\bullet\,\bullet\,\bullet\,\bullet\\ \bullet\,\bullet\,\bullet\,\bullet\end{smallmatrix}}} $$

An equation verifying the addition table entry for $3 + 5$. See Learning Addition Game under Pre-K at the SS website.

Successfully learning of all the items in the addition table may require your help, e.g., by using flash cards. Involving yourself in this way provides you with knowledge about the current state of your child's learning that is independent of school assessments. As well, it offers a chance to praise your child for achieving learning goals.

8.7.3 Addition Goals for Kindergarten

All use of relation symbols like $=$, $<$ and $>$ is to be avoided; use words like combined, total, is the same as, etc. Use language to make thinking as concrete as possible.

1. Be able to count-on within 20.

2. Apply counting-on to addition of single digit numbers as follows: given n, $m \leq 10$, starting with the larger member of the pair, count-on until the second member is exhausted thereby finding $n + m$. For example, To find $5 + 6$, count-on from 6 by 5 to obtain 11.

3. Be able to represent the addition of numbers within 10 by combining appropriately sized collections of objects such as fingers, drawn dots, other physical objects as in diagram above. See Learning Addition Game under Pre-K at the SS website.

4. Be able to solve word problems finding sums within 10 by using objects.

5. Use counting-on as a tool to solve word problems finding sums within 10.

6. Decompose numbers within 10 in multiple ways; for example, a collection having 8 members has the same total number as two collections containing 4 each, and so forth.

7. Given any single digit, know what is required to count-on to 10 as in: A bag contains 4 marbles. How many more marbles would be needed to make 10?

8. Fluently (by recall) add within 5.

8.7.4   Addition Goals for Grade 1

Again, children in Grade 1 should be using words, not symbols, to express relationships.

1. Use counting-on as a primary tool for finding sums and differences within 20.

2. Use addition and subtraction within 20 to solve word problems involving situations of adding to, taking from, putting together, taking apart, and comparing by using objects and drawings to represent the problem.

3. Solve word problems that call for addition of three whole numbers whose sum is less than or equal to 20, e.g., by using objects, drawings to represent the problem.

4. Know that addition is associative and commutative and be able to express these properties in words. Be able to use these properties in situations involving numbers. For example,

$$5 + 7 + 3 \quad \text{is the same as} \quad 5 + (7 + 3) \quad \text{is the same as}$$
$$5 + 10 \quad \text{is the same as} \quad 15.$$

and

$$5 + 10 \text{ is the same as } 10 + 5.$$

5. Know that when 0 is added to any counting number that counting number is unchanged and be able to express this in words, as in:

$$0 + 9 \text{ is the same as } 9 \quad \text{and} \quad 9 + 0 \text{ is the same as } 9.$$

6. Relate counting to addition and subtraction using the counting-on process.

7. Understand subtraction as an unknown-addend problem. For example, subtract $10 - 8$ by finding the number that makes 10 when added to 8, again using counting-on.

8. Determine the unknown in an equation like: $8 + \boxed{?} = 17$ by using counting-on; For children in Grades 1 and 2 this would be expressed: What number when added to 8 makes 17?

9. Demonstrate fluent knowledge of the addition table based on recall.

10. Add two digit numbers having a sum within 100 using the standard procedure.

11. Given a two digit number, be able to say what number is 10 more, or 10 less than the given number without counting.

8.7.5 Addition Goals for Grade 2

1. By the middle of Grade 2, know subtraction table by recall.

2. Add groups of equals as foundation for multiplication, e.g., $4 + 4 + 4 + 4 = 16$.

3. Use addition to find total cells in rectangular arrays having up to 5 rows and 5 columns. For example, in a 4 by 3 array, each column has 4 cells, so the total number of cells is found by summing $4 + 4 + 4$.

4. Add up to four two-digit numbers using standard procedure.

5. Add within 1000 using the standard procedure and explain in terms of a concrete model (Button Dealer).

6. Understand the role of place in adding or subtracting three-digit numbers; one adds or subtracts hundreds and hundreds, tens and tens, ones and ones; and sometimes it is necessary to carry tens or hundreds.

7. Mentally add 10 or 100 to any three digit number.

8. Be able to explain why place value procedure works — understanding Button Dealer model addresses this.

9. By the end of Grade 2 children should have mastered addition of two-digit numbers using the standard algorithm.

8.7.6 Addition Goals for Grade 3

1. Fluently add within 1000 using standard procedure.

8.7.7 Equality Goals for Grade 3

These goals should be achieved early in the year.

1. Children should understand the meaning of $=$ in numerical expressions as **is the same number as** and know;

    (a) $=$ is reflexive and be able to state the meaning in words;

    (b) $=$ is symmetric and be able to state the meaning in words;

    (c) $=$ is transitive and be able to state the meaning in words;

    (d) $=$ is additive (E4) and be able to state the meaning in words;

    (e) $=$ is multiplicative (E5) and be able to state the meaning in words;

    (f) be able to correctly use and interpret equations like $2 + 5 = 5 + 2$, $9 = 11 - 2$;

    (g) understand subtraction as an unknown-addend problem, e.g., $\boxed{?} + 3 = 8$ has solution $8 - 3 = 5$;

    (h) understand that any number subtracted from itself yields 0, for example, $8 - 8 = 0$; $17 - 17 = 0$, etc;

    (i) be able to solve simple equations like $2 + 5 = \boxed{?}$, $7 = 10 - \boxed{?}$ and $\boxed{?} + 4 = 9$.

2. understands that $+$ is Commutative and Associative and can explain this through counting;

3. can correctly apply the Commutative Law to quantities like $5 + 4$;

101

4. can correctly apply the Associative Law to quantities like $5 + (7 + 4)$ and $(3 + 8) + 6$; notice that this means the child functionally has to understand the use of parentheses;

5. can explain how the Commutative and Associative Laws are applied in particular situations, e.g., since $12 + 28 = 40$, $12 + 28 + 5 = 45$;

6. understands that 0 acts as the additive identity in the sense that $0 + 12 = 12$.

7. understand the use of formulae like the formula for area of a rectangle: $\mathcal{A} = \mathcal{L} \times \mathcal{W}$.

8.7.8 Strategies

If you look at modern curriculum documents (including CCSS-M), you will find the prominent mention of **strategy**.[8] For example, you may find that a child is requested to display two strategies for performing some computation. At this point you may be wondering what I'm talking about, so let me give an example in the current context.

Suppose your child is asked to add $19 + 41$. One **strategy** for doing this computation is to make the numbers simpler without changing the sum. Thus,

> 19 is almost 20 and we could make it 20 by taking 1 from 41 which would mean we get the same sum in $20 + 40$. Then all we have to do is add $2 + 4$ and put it in the *tens* place.

A strategy is simply a procedure for performing a specific calculation. Given the creative properties of the human mind, it is clear that there will be nearly as many strategies as there are children. This book narrowly focuses on a single procedure for performing each calculation. Each chosen procedure has the property that it will have application in the future hierarchy and that it **always works** for the calculation in question. For addition, that procedure is **counting-on**. This always works and it can be used in the context of the standard algorithm to check single-digit sums. Moreover each time it is used it reminds the user of the connection between addition and counting. Finally, limiting the number of procedures considered implements the essence of what is means for a curriculum to be **coherent and focused** in the sense of ACC.

---

[8]Links to the CCSS-M documents and to the Newfoundland curriculum documents can be found at my website — http://www.math.mun.ca/ hsgaskill/ — as well as many other documents relevant to K-6 math curricula.

Chapter 9

Subtraction of Counting Numbers

In the last chapter, we discussed the binary operation of addition applied to counting numbers. We found that the addition of the counting numbers n and m corresponded to finding the total number of elements when two real-world collections having n and m members, respectively, are combined. Thus, the abstract mathematical operation of addition answered the real-world question: How many total members result when two collections are combined?

In a like manner, the operation of **subtraction**, or **take away**, is also founded in the real world.

9.1   Subtraction as a Real-World Operation

Consider a collection A that has n members. Now suppose we **remove**, or **take away** some of the members. If we put the members we removed into a new collection, and call it B, then we can count the members in B. Call this number m. What can we say about this situation with complete certainty?

The certain fact is that the collection B can not have more members than were originally in A. Thus, the counting number m must be smaller or the same as the counting number n. Further, if m is less than n it must be because there are members still left in A, and if m is the same as n it must be because all the members in A were taken and A no-longer exists.

Again consider a collection having n members. Suppose we are given another counting number m that is not larger than n. Then we know we could physically remove m members from the collection A. The question that subtraction answers is:

> Given a collection having n members, if we **take away** m members, how many will be left?

Again, we stress, this question only makes sense in the world if m is less than or equal to n.

It is clear that we can remove items from a collection in the world and count what is left. But if n and m are fixed, does this process always result in the same number of items being left over? The following diagram illustrates why subtraction is an operation.

A diagram showing 7 is the result of subtracting 10 from 17. Notice that the collection having 10 items and the collection having 7 items sum to 17.

In the diagram we see that the total number of items in the collection having 17 have been divided between two collections, a fact that can be established by counting. **CP** guarantees that in this situation, the total number of items is fixed. That being the case, since 17 and 10 are fixed, the third number is also fixed and must be 7. Thus, take away does qualify as an operation.

Given n and m with m less than or equal to n, a procedure for finding n take away m is as follows:

Starting at m, count-on to n; the number of steps taken is the number of items left when m items are removed from a collection containing n items.

For example consider a collection having 12 items. If 7 are removed, how many are left? To solve this problem, we count on from 7 as follows:

$$8, \quad 9, \quad 10, \quad 11, \quad 12.$$

The counting-on process stops at 12 because that is how many were in the original collection. The first number written is 8, because 7 items were removed. There are 5 numbers written down so 5 is how many will be left.

Using the procedure above, we say the number p is the result of performing the **subtraction**, n **minus** m, and write:

$$n - m = p.$$

The operation of subtraction (take away) is denoted by a **centered dash** as shown.

9.2   A Mathematical Definition of Subtraction

As shown, subtraction is founded in the real world using the concept of take away. As such, it gives the appearance that subtraction is an entirely separate operation from addition. However, as shown by use of counting-on as part of the take-away procedure, subtraction is a closely related to addition. We explore that relationship in this section.

Let n and m be two counting numbers such that $m \leq n$. (We use the standard symbol for *less than or equal to* for convenience.) Because of the requirement for subtraction that $m \leq n$, we know we can find a number that makes the equation

$$\boxed{?} + m = n$$

true. The number that solves this equation will either be a counting number or 0 and in the future we will refer to the set containing the counting numbers and 0 as the non-negative whole numbers. The reason it must exist follows from the discussion in the previous section. For example, $6 \leq 9$, so the equation

$$\boxed{?} + 6 = 9$$

has a solution, 3, found by counting-on from 6 to 9 and results in:

$$\boxed{3} + 6 = 9.$$

Thus, given $m \leq n$, if p is a number with the property that:

$$p + m = n,$$

we will write

$$p = n - m$$

and say, p is the result of **subtracting** n from m. Alternatively, we say p is the **difference** between n and m.

Approaching subtraction in this manner has the effect of giving primacy to the operation of addition. To obtain a deeper understanding of these ideas let's focus for a minute on a particular counting number, say 11. Then let's ask the question: In how many ways can we decompose 11 as a sum? Notice that each such decomposition provides two counting numbers for the following equation:

$$\boxed{?} + \boxed{?} = 11.$$

For example, we know one pair of numbers that gives us a true equation is 7 and 4, as in

$$7 + 4 = 11 = 4 + 7.$$

These equations give us two subtraction equations, namely,

$$7 = 11 - 4 \quad \text{and} \quad 4 = 11 - 7.$$

The equation is easily made concrete by thinking of the various ways a collection of 11 items can be split up into two collections, in the case above, one having 4 members and the other 7.

Thinking about decomposing counting numbers as sums is one of the ways subtraction is introduced to children. A benefit of this approach is that the requirement to verify that subtraction is an operation disappears because the work in that regard has already been performed for addition. However, in taking this approach we lose the concrete aspects of subtraction as *take away*. But this loss will be more than made up when we discuss the integers in Chapter ??.

So that you are aware of the nomenclature, in $n - m$, n is referred to as the **minuend** and m is referred to as the **subtrahend**. One way to remember this is minuends get smaller and subtrahends subtract.

Thinking of subtraction in terms of addition as in:

$$n - m = p \quad \Longleftrightarrow \quad p + m = n$$

gives us a procedure for checking any subtraction result, p. To check we are told to perform the addition, $p + m$, to see whether the sum is n. We may all remember that when we learned arithmetic in school, the method for checking whether a subtraction had been done correctly, was to perform exactly this addition.

If we think of the trivial decomposition of the counting number n into 0 and itself, we have

$$0 + n = n = n + 0$$

from which we conclude that

$$n - n = 0.$$

If we reflect on why $n - n = 0$ should be true, we know it simply corresponds to the fact that when we remove all the elements from a real-world collection, we are left with nothing. This is obvious. However, we will return to this observation later and see that it leads to a major step forward in the development of numbers (see Chapter ??).

When we think of subtraction in terms of addition, as discussed above, we are implicitly finding the solution to an equation which looks like:

$$x + m = n$$

where x is the unknown quantity and n and m are given. The use of variables like x, m and n in equations does not occur until after Grade 3. Younger children are most likely to be confronted with equations like:

$$\boxed{?} + 7 = 12$$

in early grades which means Grade 3 in an A+ curriculum. Prior to Grade 3 the question represented by the last equation should be stated in words as:

> Find $12 - 7$ by finding the counting number that when added to 7 makes a total of 12.

Using language instead of symbols leaves no room for ambiguity and reinforces the connection between subtraction and addition.

9.3 Making Subtraction Useful

The discussion above tells us what subtraction is, and how to check whether a given number is the correct answer to a subtraction problem. However, it tells us almost nothing about how to perform computations using the Arabic number system. Since, understanding the role of place is essential, we develop the subtraction procedure in detail, partly as a means to review our ideas.

As discussed above for $12 - 7$, solving the subtraction problem $n - m$ requires finding an unknown number p having the property:

$$p + m = n.$$

So that these ideas are really concrete, we develop the standard procedure using the Button Dealer analogy.

> Suppose the Button Dealer has just filled an order from a customer for 897 buttons. The customer then says that 534 buttons are for his wife and the remainder for his mother-in-law. Since the customer is not good at arithmetic, he asks the Button Dealer to divide the buttons. How should the Button Dealer proceed?
>
> The original order consists of *seven* jars marked with a 1, *nine* jars marked with 10, and *eight* jars marked with 100. To solve the customer's problem, the Button Dealer needs to split up the buttons into two collections, one of which will contain *four* jars marked with a 1, *three* jars marked with 10, and *five* jars marked 100.

To accomplish the split, the Button Dealer proceeds as follows. To obtain buttons for the wife, he takes *four* jars from the *seven* marked with 1; he takes *three* from the *nine* jars marked with 10; and from the *eight* jars marked with 100, he takes *five*. This gets 534 buttons for the wife. The remainder, or **difference** are for the mother-in-law, and there are *three* jars marked with 1, *six* jars marked with 10, and *three* jars marked 100, or 363 buttons left for the mother-in-law.

How easy was that?! What's more, we are guaranteed that the result is correct because of **CP**. In mathematical terms we would express this as:

$$897 - 534 = 363.$$

The representation of this problem in Arabic notation is shown below.

$$\begin{array}{r} 897 \\ -534 \\ \hline 363 \end{array}$$

We can check that the result 363 is correct by computing $534 + 363$ using the standard addition procedure. Further examples will be discussed below to illustrate the subtraction process in detail.

What the button analogy shows us with clarity is, that like addition, subtraction is again a real-world process. That this must be so follows from the fact that the answer to a subtraction problem is also the answer to an addition problem.

To be specific, recall that the answer, p, to the subtraction problem, $n - m$, is also the solution to the addition equation, where for given n and m, we must find p such that

$$p + m = n.$$

Since subtraction is really a form of addition, having the addition table available for reference would be useful. (Of course, if the table has already been committed to memory, this is unnecessary. But, even in this case, its presence is useful for discussion.)

+	0	1	2	3	4	5	6	7	8	9
0	0	1	2	3	4	5	6	7	8	9
1	1	2	3	4	5	6	7	8	9	10
2	2	3	4	5	6	7	8	9	10	11
3	3	4	5	6	7	8	9	10	11	12
4	4	5	6	7	8	9	10	11	12	13
5	5	6	7	8	9	10	11	12	13	14
6	6	7	8	9	10	11	12	13	14	15
7	7	8	9	10	11	12	13	14	15	16
8	8	9	10	11	12	13	14	15	16	17
9	9	10	11	12	13	14	15	16	17	18

To explain how the table is used in subtraction, suppose we are given m and n subject to:

$$n \leq 18 \quad \text{and} \quad m \leq 9$$

and we want to find $n - m$. Observe that the conditions merely mean that n can be found as an entry in the body of the table, and m can be found in the top row of the table as a column heading. Further, we require that n can be found in a cell in the table having m as its column header. For example, $n = 17$ and $m = 9$ satisfy these conditions, whereas $n = 16$ and $m = 4$ do not.

Now to use the table to find $n - m$, simply go to the column headed by m, proceed down the column to the cell containing n, and thence to the far left to find the number heading that row. Call this number p. Because the table is the **addition table**, we know

$$p + m = n,$$

which means $n - m = p$. So the addition table provides answers to all subtraction problems that meet the conditions listed above.

Let's do an example using Arabic notation.

Example 1.

Find $854 - 623$. As with addition the first step (setup) is to write the minuend 854 above the subtrahend 623 so that *units* are above *units*, *tens* are above *tens* and *hundreds* are above *hundreds*. The result is displayed below with a subtraction sign and a line to indicate where the answer goes.

$$
\begin{array}{r}
854 \\
-623 \\
\hline
\end{array}
$$

The only requirement here is that $m \leq n$, which we see is true. As in the case of addition, we perform the computation column-by-column starting at the right. The first subtraction we have to perform is $4 - 3$. Since 4 and 3 satisfy the requirements for using the table, using the procedure outlined, we find $1 + 3 = 4$, so $4 - 3 = 1$. This result is recorded below the line in the *units* place as shown:

$$
\begin{array}{r}
854 \\
-623 \\
\hline
1
\end{array}
$$

Similarly, we find $5 - 2 = 3$ and $8 - 6 = 2$, which are recorded below the line in the *tens* and *hundreds* places, respectively. The resulting solution is:

$$
\begin{array}{r}
854 \\
-623 \\
\hline
231
\end{array}
$$

This example is like the example solved by the Button Dealer, namely, the subtraction required in each column satisfies the condition, $m \leq n$.

9.3.1   Borrowing

Example 2.

Consider the subtraction problem setup below:

$$
\begin{array}{r}
62 \\
-45 \\
\hline
\end{array}
$$

First observe $45 \leq 62$, so we can indeed do the subtraction. However, the first subtraction to be performed, namely $2 - 5$, does not meet the criterion: $m \leq n$. Recall, addition problems required us to carry into the next column to the left when the sum became a two digit number. In subtraction, when $n < m$ **for a given column**, (that is to say, the subtrahend m is bigger than the minuend n in a given column) we borrow from the column to the left as shown below.

$$
\begin{array}{cc}
5 & \\
\not{6} & 12 \\
-4 & 5 \\
\hline
\end{array}
$$

When we borrow from the *tens* column, we reduce the *tens* digit in the number we are subtracting from by 1. This is indicated by crossing out the 6 and writing the replacement digit at the top of the *tens* column.

We put the borrowed 1 to the left of the *ones* digit in the number being subtracted from, as shown. This has the effect of putting an extra 10 *ones* in the *ones* column.

The purpose of borrowing is to ensure that the required conditions on n and m are satisfied in each column, that is, $m \leq n$ and n occurs as a cell value in the column headed by m in the addition table. The reader can check that after borrowing this will be the case since 12 occurs in a cell in the column headed by 5. At this point, we can find the difference using the table. The first step is to perform the subtraction in the *units* column by finding $12 - 5$. The result, 7, is recorded as shown:

$$
\begin{array}{rr}
5 & \\
\not{6} & 12 \\
-4 & 5 \\
\hline
& 7 \\
\end{array}
$$

The second step is to perform the revised subtraction in the *tens* column, $5 - 4$, with the result recorded as shown:

$$
\begin{array}{rr}
5 & \\
\not{6} & 12 \\
-4 & 5 \\
\hline
1 & 7 \\
\end{array}
$$

The subtraction process is completed by first finding $12 - 5$, then $5 - 4$. Note, the 5 replaces the crossed out 6 in the calculation.

Example 3.

We consider one more example of borrowing, namely, $804 - 578$. We begin by noting that $578 \leq 804$, so we know we can perform this computation in the form of removing 578 members from a collection having 804 members. We set up the problem below:

$$
\begin{array}{rrr}
8 & 0 & 4 \\
-5 & 7 & 8 \\
\hline
\end{array}
$$

Notice, neither the *ones* column, nor the *tens* column satisfies the condition that $n \leq m$, which enables one to find of $n - m$ using the addition table. Moreover, the *tens* place in 804 is a 0 which further complicates the problem.

The problem has been properly expressed in Arabic notation. As before, start with the *ones* column at the far right. We observe that since $4 < 8$, we will have to borrow. However, when we try to borrow from the *tens* place in 804, we are confronted with a 0. While this seems problematic, we know we must be able to perform the computation, because $578 \leq 804$. The solution is to borrow from the *hundreds* column which we illustrate step-by-step below.

$$
\begin{array}{rrr}
7 & & \\
\not{8} & 10 & 4 \\
-5 & 7 & 8 \\
\hline
\end{array}
$$

In the first step, 100 is borrowed from the *hundreds* column and placed in the *tens* column as 10 *tens*.

The borrowed one hundred has now made the *tens* place non-zero, so we can borrow 1 *ten* from the *tens* place. This borrowing is the second step and produces the revised problem pictured below where we now have an extra 10 *ones* in the *ones* column.

$$
\begin{array}{rrr}
7 & 9 & \\
\not{8} & \not{10} & 14 \\
-5 & 7 & 8 \\
\hline
\end{array}
$$

In the second step, a unit of 10 is borrowed from the *tens* column and placed in the *ones* column. Note that all the columns now satisfy the requirements on m and n to perform subtraction using the table.

We can now perform the computation, starting with the column at the right. This gives:

$$
\begin{array}{rrr}
7 & 9 & \\
\not{8} & \not{10} & 14 \\
-5 & 7 & 8 \\
\hline
2 & 2 & 6 \\
\end{array}
$$

The subtractions are performed column-wise starting at the right: $14-8$, $9-7$, and $7-5$.

All subtraction follows this pattern, including subtraction of decimal numbers which are essential for keeping track of your bank balance.

9.4   What Your Child Needs to Know

Your child needs to develop an understanding of the what subtraction is in respect to Usiskin's criteria. This means first, using the standard algorithm to fluidly perform the computations. Second, your child needs to know the definition of subtraction in terms of **take away**. And third, your child should be able to visually represent subtraction in a diagram like that in §9.1. Given the connection between addition and subtraction, the goals for addition identified in §8.7 are also relevant.

As illustrated in the examples, the standard procedure for subtraction has two parts:

1. setting up the subtraction using Arabic numerals with minuend above subtrahend, and figuring out what borrowing is required;

2. performing the subtraction $n - m$ for n and m, where n is in a cell of the addition table in a column headed by m.

The first step in the process is a matter of practice and an understanding of the Arabic System of notation.

The second step, performing the column-by-column subtractions, depends on thorough knowledge of the addition table.[1] Note that in learning the table for purposes of addition, the child must solve problems like:

$$4 + 9 = \boxed{?}.$$

In accessing the table for subtraction, the child is solving problems like:

$$\boxed{?} + 9 = 13$$

as explained in §9.2. This second type requires a deeper level of learning, hence a more thorough knowledge of the table. Again, flash cards may be useful and working with your child provides you with an independent means of assessing your child's progress.

Fluency with the subtraction algorithm is should be achieved by the end of Grade 2. In working with subtraction, as with addition, algebra is to be avoided.

9.4.1   Subtraction Goals for Kindergarten

The primary goal in Kindergarten in respect to subtraction is for the child to develop an understanding that subtraction amounts to **taking apart** or **taking away**. As such the focus is on the concrete. Thus, by the end of Kindergarten your child should be able to:

---

[1]See also the expectations for addition.

1. Represent addition and subtraction within 10 with objects, fingers, mental images and drawings.

2. Solve addition and subtraction word problems within 10. For example, 8 children went to the store. If 6 were girls, how many were boys?

3. Decompose numbers ≤ 10 in multiple ways as sums.

4. For any number < 10 know what to add to make 10.

5. Fluently add and subtract within 5.

9.4.2 Subtraction Goals for Grade 1

By the end of Grade 1[2] it is expected that your child will be able to:

1. represent and solve problems involving subtraction;

2. understand the relation between addition and subtraction;

3. solve subtraction problems involving numbers ≤ 20;

4. recognize that each counting number less than 20 has multiple representations, for example,

$$2 + 5 \text{ is the same as } 7 \text{ is the same as } 4 + 3 \text{ is the same as } 9 - 2;$$

5. use place value and the standard algorithms to add and subtract;

6. subtract multiples of 10 that are < 100 from other multiples of 10 < 100 using concrete models, for example, sticks of length 10.

9.4.3 Subtraction Goals for Grade 2

By the end of Grade 2 it is expected that students are able to

1. use addition and subtraction within 1000 to solve one- and two-step word problems involving situations of adding to, taking from, putting together, taking apart, and comparing;[3]

---

[2]See also Grade 1 Addition Goals.

[3]A two-step word problem would be: Jane has 20 marbles. She gave four to Bill and eight to Mary. How many did Jane have left?

2. determine whether a group of ≤ 20 objects has an odd or even number of members by pairing or counting by $2\,$s;[4]

3. fluently mentally add and subtract within $20\,$;[5]

4. understand and be able to fluently subtract numbers ≤ 1000 using the standard place value algorithm;

5. mentally add or subtract 10 or 100 from numbers in the range 100 to $1000\,$;

6. solve word problems involving money using dollars, quarters, dimes, nickels and pennies and the symbols for dollars and cents;

7. tell time from digital and analog clocks.

---

[4]The reader will recall that **even** numbers are divisible by 2. A list of even counting numbers is: 2, 4, 6, 8, 10, 12, ... Numbers that are not even are called **odd**.

[5]Within 20 means the result is $\leq 20\,$.

Chapter 10

Length as a Numerical Attribute

Length is an attribute of physical objects. Exactly when this attribute was identified by human being is unknown. What is known is that human beings in the Indus Valley Civilization (c. 7600 BCE) had well-developed processes for manufacturing bricks in standard sizes while Mesopotamians were constructing pyramid-like structures during the fourth millennium BC.[1] Constructing pyramids of any kind requires a well-developed geometry and engineering skill. In particular, such constructions require an ability to measure and use the attribute of length.

In an A+ curriculum only one topic is introduced in Grades K-2 other than the meaning of whole numbers meaning and the operations of addition and subtraction, namely, **measurement units**. Measurements and their units are important as the basis of modern science. **Length** and **units of length** are important because lengths provide another real-world interpretation of arithmetic.

10.1   The Length Attribute

For collections and counting numbers, the key concepts were more, less and the same. The method of determination was pairing. For length the key concepts are **longer**, **shorter** and the same. Pairing is replaced by simply laying one length down next to another one and seeing which one has greater extent.

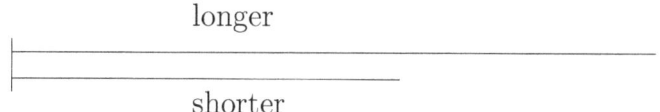

longer

shorter

Two line segments illustrating the shorter/longer relations.

---

[1]See Wikipedia entry for Indian Mathematics and *pyramid*.

As an attribute, length is useful only if it is conserved. For an inanimate object this is true; its length does not change without cause. Thus, we can think of the length of most things as being a stable attribute. And even when length does change it is still useful, for example, as in the height of a child.

If you think about it, length in the context of longer and shorter comparisons suffers from exactly the same limitations as collections compared by more or less. So in order to make this attribute useful a length has to have a name that identifies it and can be used for purposes of comparison. It is a remarkable fact that lengths can be named using the same tool that we used for collections as a starting point.

10.2 Units of Length and Measurement

While the history of how geometry developed is unknown, what seems likely is that counting predated the development of geometry and that human beings noticed that if one took a fixed length as a standard, one could identify longer units of length by using the process of counting. This idea leads directly to the concept of **unit of length** and **measurement**. Two common units of length are shown next.

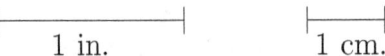

1 in.   1 cm.

The two units of length to which children in North America are typically exposed. Each line segment specifies a unit of length with the units shown below that segment. For the segment on the left, the unit of length is inches; for the segment on the right, the unit of length is centimeters.

10.2.1 Measuring a Length

The measurement process begins by starting at one end of the item whose length is to be measured and marking off units of length as shown below:

A line segment to be measured. The unit of length used is 1 inch. Each vertical arrow marks one unit of length. No space is left between consecutive measurements, i.e., the next begins where the previous one stopped.

Once the marks have been determined, it is simply a matter of counting. There are 5 arrows, each one marking a one-inch line segment, so the total length is 5 inches.

Of course the line segment can be measured using the units of centimeters as shown below:

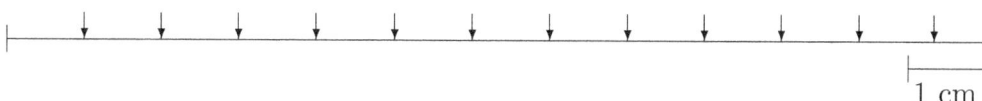

A line segment to be measured. The unit of length used is 1 centimeter. Each vertical arrow marks one unit of length. This time there are 12 and a piece is left over.

The length of the line segment in centimeters is 12 and a bit. The bit at the end is shorter than 1 centimeter in length as indicated in comparison to the standard unit. Since the measurement process must begin at one end or the other and continue with no spaces, it is clear that an exact measure for this line segment in centimeters is not possible, unlike the previous measurement in inches.

These two diagrams illustrate the fundamental difference between counting numbers used to measure how many objects are in a collection and counting numbers used to measure lengths. As a measure of collection size, the count is fixed and all who make the count will get the same answer. But correct measurements of length of the same object can and will be different when different units of length are used. It is essential that children come to terms with this fact and how the intrinsic length of the unit will affect the outcome. In this case, we see that fewer inch units are required that centimeter units because a centimeter is shorter than an inch.

10.3 Properties of Length

There are there are three important properties of length that have nothing to do with the units on a particular measuring tape. These properties are so fundamental that we don't even think about them, but children have to come to terms with them.

Consider measuring the line segment pictured in the last section. There are three things we can say irrespective of the measuring instrument we choose and the units:

1. The result of measuring this segment will be a positive number, that is, it will require an actual length of tape.

2. If we start the measurement at the left end of the segment, we get the same result as if we start the tape at the right end of the segment.

3. If we were to start the measurement at the left end and measure to any point not on the segment, and then measure from that point to the right end of the segment, the sum of the two lengths will be greater than the length of the segment, i.e., the shortest distance between two points is a straight line.

These are three facts that children should learn by the process of making measurements of length.

10.4 Representing Addition

In our opening remarks we indicated that it was possible to use length and measurement as a means to represent arithmetic.

First we assume we have a large collection of rods each having the same identical length. What this length actually is is of no consequence for representing addition.

Now suppose we want to perform an addition problem, say find $7+9$ for example. To replicate addition, using the standard pieces we construct two line segments, one having total length 7 units and the other having total length 9 units as shown:

The required two line segments of 7 and 9 units in length, respectively.

The essential feature of the construction process is that the line segments are placed end-to-end so that there is no space separating the end of one from the beginning of the next. In this way the total length of each constructed segment is obtained by counting the number of segments. So for example, the first segment has 7 units of length, where for the reasons discussed above, acknowledging the units is essential.

Addition is simply the process of combining, which in this case means producing a segment having a total length of $7+9$ units. Thus to add simply bring the two segments together so that they form a continuous segment with no breaks and no overlaps as shown below:

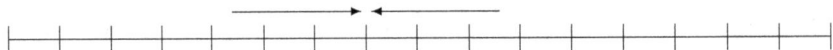

The two line segments brought together to form a continuous segment that is 16 units long.

The total length of the resulting line segment is 16 units which can be found by counting. In this way, addition can be represented within geometry. That said, it still comes down to counting even as we try to obtain more accuracy by using ever shorter units of length.

10.5 What Your Child Needs to Know

Children should come to terms with as many numerical attributes of physical objects as there objects that they regularly experience in the world. These would include length, area, volume, weight, height, time, speed, etc. Where it is possible, a child's experience with these attributes should include making measurements. For example, learning to use a measuring tape and the fact that there are different units, e.g., centimeters and meters.

Making measurements can begin in pre-K and K as a play activity. More focused activities around measurement and units begin in Grade 1/2 in an A+ setting.

10.6 What Your Child Needs to Know

10.6.1 Goals for Grade 2

It is expected that your child will be able to:

1. measure the length of an object by selecting and using appropriate tools such as rulers, yardsticks, meter sticks, and measuring tapes;

2. measure the length of an object twice, using length units of different lengths for the two measurements;

3. describe how two measurements using different units relate to the size of the unit chosen;

4. measure to determine how much longer one object is than another, expressing the length difference in terms of a standard length unit.

Chapter 11

Multiplication of Counting Numbers

The operation of multiplication was known to the ancient Egyptians and Babylonians.[1] Since keeping track of land areas, which requires multiplication, was certainly of importance in ancient Egypt, it is clear why knowledge of multiplication would be useful.

11.1   What Is Multiplication?

Like addition, multiplication is a binary operation on counting numbers. Given two counting numbers, n and m, referred to as **factors**, we form a new counting number $n \times m$ called the **product** of n and m.

Multiplication applied to counting numbers is simply **repetitive addition**. What is meant by this is that 4×5 signifies adding 4 to itself 5 times, as in:

$$4 \times 5 = 4 + 4 + 4 + 4 + 4.$$

To give another example:

$$7 \times 3 = 7 + 7 + 7$$

whereas:

$$3 \times 7 = 3 + 3 + 3 + 3 + 3 + 3 + 3.$$

We may think of the **first counting number as specifying what is to be added**, and the **second counting number as specifying how many of the first number are to be added** together. If we think about the operations of arithmetic as answering questions, then multiplying n by m answers the following question:

---

[1] See Wikipedia entry for Multiplication.

What is the total sum when n is repetitively added to itself m times?

This is the question that children should have in their minds when they multiply two counting numbers if only because it provides a process for finding the answer in the case of single digits.

Because multiplication is defined in terms of addition, the properties of addition are transferred to multiplication. For example, that multiplication is a binary operation follows from the fact that addition is a binary operation. In fact, as we proceed through arithmetic, we will see that there is really only one operation from which all else is derived and that is addition. Addition can be founded on successor, and that in turn is concretely based on counting collections. Thus, in the final analysis, almost all of arithmetic comes down to counting.[2]

The ideas underlying multiplication are most easily made concrete using diagrams like the following for 4×5:

Each of the five rectangular columns contains four dots corresponding to the counting number 4. It is clear from the diagram why the process would be referred to as **repetitive addition**. Since the process is addition, in the end it all comes down to counting. A+ curricula expect that children in Grade 2 can perform the addition required to find the total dots in ≤ 5 by ≤ 5 arrays and relate these ideas to multiplication.

11.2 Area as an Attribute

In an agricultural society the size of a field would be important so that having a way to account for differing field sizes would be important. Again, exactly when geometric measures were discovered is unknown, but we do know that knowledge of geometry sufficient to recognize length, area and volume as numerical measures of things in the world existed around 7600 BCE during the Indus Valley Civilization.[3] What seems most likely is that as humans settled into communities based on agriculture, the need for geometrical measures to support agriculture, building construction and the manufacture of building materials would become ever greater. Length was discussed in Chapter 10, here the focus is on **area**.

---

[2]We say *almost all* because when we discuss the multiplication of unit fractions we will have to find another interpretation of multiplication.

[3]See Wikipedia entry for *Indian Mathematics*.

11.2.1 Area in Relation to Length

Consider the two boxes in the following figure:

The box on the left is bigger, but what does this mean?

We simply do not know how our ancestors went about finding the answer to the question posed, but we can think about the problem experimentally. Consider the following three figures:

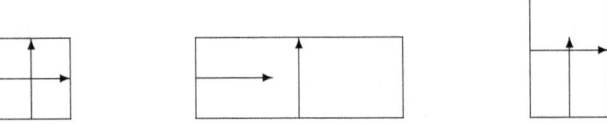

All the arrows in this figure have the same fixed length.

In the discussion of these figures, we will use the standard names, even though we know these names and concepts would not have been available to our ancestors.

The smallest figure in terms of area is the one on the left. The middle figure has the same vertical extent as the figure on the left, but greater horizontal extent. The greater horizontal extent is therefore the source of larger area of the middle figure in comparison to the figure on the left.

The figure on the right is also larger than the figure on the left. However in this case, the source of the additional area is the greater extent of the figure on the right in the vertical direction; the horizontal extent of both figures is the same as shown by the arrows.

From comparisons like these our ancestors could arrive at the following facts about the world:

- to have area, a figure must have extent in two mutually perpendicular directions simultaneously, e.g., horizontal and vertical;

- extent in one direction only results in a line having no area, only length;

- each figure having extent in two directions has a stable numerical attribute called its **area**;

- area increases as the extent of a figure increases in either direction and the measure of this extent is length.

11.2.2 Standard Measure of Area

To make the area attribute useful would require a system of numeration. The main problem our ancestors would have had to sort out was how to relate length to area. The notion of **unit of length** was discussed in the last chapter and it seems reasonable to believe our ancestors would have arrived at the notion of a unit of area as having 1 unit of length in each of the two mutually perpendicular directions which leads directly to a **square** having four sides of length 1 unit in some unit of measure. Recalling the properties of area, defining property for squares is that it is a four-sided figure in which the all sides have the same fixed length and adjacent sides extend in mutually perpendicular directions. (Alternatively, the four angles between adjacent sides have the same measure.) For example, the figures pictured below and the figure on the left above are squares. Each of the squares that follow can be thought of as defining one unit of area, although these units are obviously different.

1 in.   1 cm.

Each box satisfies the criteria of being a square and each has an area of 1 **square unit**. For the box on the left, the unit of length is inches; for the box on the right, the unit of length is centimeters.

11.3 Area and Multiplication

To see the relationship between area and multiplication, consider the following grid (array):

A standard square unit is pictured at the left. The large square has been divided into squares of this size so that each column contains four squares. There are and five columns. Counting the total gives the area of the large square in units of the small square.

Consider the combined figures:

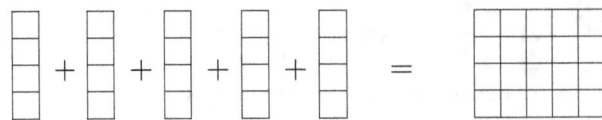

Starting with the five columns of four on the left-hand-side (LHS) of the equality, imagine removing the plus signs and moving the columns together to form the array on the right-hand-side (RHS). **CP** once again, see §4.2, guarantees the total squares on either side of the = are the same. So area has to be **conserved**.

Rather than thinking of the last diagram as finding the sum of five 4 s, we can concentrate on the array on the RHS and think of finding the total number of cells (unit squares) in a rectangle of width (vertical dimension) 4 cells and length (horizontal dimension) 5 cells. Considerations such as these lead to the well known formula for the **area of a rectangle** of length $\mathcal{L}$ and width $\mathcal{W}$:

$$area = \mathcal{L} \times \mathcal{W}.$$

Every product of two numbers can be represented as an area. Because of **CP**, we know that area must be conserved and as shown above, we see that finding areas essentially comes down to counting. Understanding that the product of two numbers may be thought of as **area** is an essential part of learning about multiplication. While the formula for finding area given above does not display it explicitly, area is always expressed in **square units**, for example, square inches. This results from the fact that the two factors in the product, length and width, both have units of length attached to them. Since the unit of area is based on squares, the length and width must be expressed in the same units and this fact must be emphasized to children.

Area is a numerical attribute that arises from geometric aspects of objects in the real world. Beginning in Grade 3 methods for finding values for for such attributes are given in terms of formulae such as the one above. Another such formula is the formula for perimeter. Dealing with formulae for simple geometric shapes is a good way to introduce the equality symbol and simple equations.

11.4   Properties of Multiplication

Since multiplication is repetitive addition, it must be the case that all properties of addition carry over to multiplication. Two of these are:
the **Commutative Law**

$$n \times m = m \times n$$

and the **Associative Law**

$$(n \times m) \times p = m \times (n \times p).$$

where n, m and p are any three counting numbers, or 0. These laws are critical to our understanding of arithmetic. Indeed, these and a very few others are the basis for all our computations. For this reason it is essential that readers achieve comfort with their content and their use! They need to become *old friends* and we intend to make them so.

We know that arithmetic is a model for counting processes taking place in the world. Since multiplication is merely repetitive addition, and hence a counting process, we could take this to be justification for the truth of these laws. But this would not achieve our goal of deep understanding. For this reason, we examine these laws in detail.

11.4.1 The Commutative Law

Here, we show why the Commutative Law must be true for multiplication using considerations about area. The reader should remind themselves that any time we speak of area, the numerical measure will have units attached.

We consider a specific equation

$$5 \times 4 = 4 \times 5.$$

The next diagram shows the LHS of this calculation as an instance of finding the area of a figure with $\mathcal{L} = 5$ and $\mathcal{W} = 4$.

Counting the cells in the grid on the RHS tells us that $5 \times 4 = 20$. So the area of the rectangle is 20 square units.

Compare the rectangle on the RHS of the following diagram with the rectangle above. The rectangle diagramed below has has $\mathcal{L} = 4$ and $\mathcal{W} = 5$ and so has an area given by 4×5, which is the RHS of our Commutative Law equation.

Counting the cells in the grid on the RHS tells us that $4 \times 5 = 20$ as well. So the area is again 20 square units.

As the reader can see, both rectangles have the same number of cells, namely 20. While counting the cells gives us this result, we don't need to count because the RHS's are actually the same diagram! To see this, note the grid in the RHS of the current figure is the identical grid to that in the previous diagram, but rotated $90°$ clockwise. **CP** ensures that rotating a figure cannot change the number of grid squares! It is expected that children will understand that rotating a figure, as in the last two diagrams, cannot change the number of cells comprising the figure. Hence area is **fixed** under rotations.

In the end, every child should know that the **Commutative Law** for multiplication ensures that for every pair of counting numbers, n and m,

> **the product of n times m has the same value as the product of m times n.**

11.4.2   The Distributive Law

Before proceeding to the Associative Law for multiplication, we consider the **Distributive Law**. It states that for any counting numbers n, m and p:

$$p \times (n + m) \ = \ p \times n \ + \ p \times m.$$

Recall, the LHS of this equation uses parentheses to indicate that the operation of addition is to be performed **before** the operation of multiplication. The parentheses are on the LHS to remove the ambiguity that exists in the expression

$$p \times n + m.$$

To clarify the nature of this ambiguity, consider the expression $5 \times 4 + 2$. Do we mean

$$20 + 2 \quad \text{or} \quad 5 \times 6?$$

The parentheses in $p \times (n + m)$ inform us that the sum $n + m$ must be computed before any multiplication occurs. Thus, p is multiplied by the sum $n + m$.

There are no parentheses on the RHS of the equality:

$$p \times m \ + \ p \times n.$$

127

A Precedence Rule

Parentheses are not required on the RHS because mathematicians employ what are called **rules of precedence**. Rules of precedence specify in which order operations are to be performed when there are different operations occurring in an expression, as in the Distributive Law. According to one of these rules, in the absence of parentheses, multiplication has higher precedence than addition, which means multiplication must be performed before addition. So the effect of this rule is that

$$p \times m \ + \ p \times n \ = \ (p \times m) + (p \times n)$$

and no parentheses are required on the RHS of the $=$ in the Distributive Law.

The Distributive Law Continued

To understand why the Distributive Law must be true, consider the product

$$4 \times 8 = 4 + 4 + 4 + 4 + 4 + 4 + 4 + 4.$$

We focus on the RHS and recall what the Associative Law tells us about addition: however we introduce parentheses into the RHS of the last equation, we must get the same answer. We understand this as an instance of **CP**. Thus, for example,

$$
\begin{aligned}
4 \times 8 \ &= \ 4 + 4 + 4 + 4 + 4 + 4 + 4 + 4 \\
&= \ (4 + 4 + 4) + (4 + 4 + 4 + 4 + 4) \\
&= \ 4 \times 3 + 4 \times 5,
\end{aligned}
$$

because multiplication is repetitive addition. Since $8 = 3 + 5$, we we can rewrite the LHS as shown to obtain

$$4 \times (3 + 5) = 4 \times 3 + 4 \times 5,$$

which is just the Distributive Law with $n = 3$, $m = 5$ and $p = 4$. The parentheses on the LHS in $(3 + 5)$ force us to compute $3 + 5 = 8$ before we multiply, so the LHS is still 4×8.

Another instance of the Distributive Law, this time with $n = 6$, $m = 2$ and $p = 4$ is

$$
\begin{aligned}
4 \times 8 \ &= \ (4 + 4 + 4 + 4 + 4 + 4) + (4 + 4) \\
&= \ 4 \times 6 + 4 \times 2.
\end{aligned}
$$

Since $8 = 6 + 2$, we have

$$4 \times (6 + 2) = 4 \times 6 + 4 \times 2,$$

where the parentheses still force the LHS to be 4×8.

Recall again the general statement of the Distributive Law:

$$p \times (n + m) = p \times n + p \times m.$$

For any counting numbers, we can formulate both sides as repetitive addition and recognize that the RHS is obtained from the LHS by the insertion of parentheses exactly as in the examples above. Thus, although the Distributive Law looks new, it is really only an application of the Associative Law for addition.

We can also apply the Commutative Law for multiplication to the Distributive Law. When we do this, we obtain:

$$(n + m) \times p = n \times p + m \times p.$$

Thus, on the LHS we replaced $p \times (n + m)$ by $(n + m) \times p$ On the RHS we replaced $p \times n$ by $n \times p$ and $p \times m$ by $m \times p$. Children should understand that this form of the Distributive Law is a **consequence** of the Commutative Law for multiplication (see §11.6).

Children should also understand why this form of the Distributive Law has to be true in the real world. To do this we consider

$$(3 + 4) \times 8 = 3 \times 8 + 4 \times 8$$

as an area computation. The last equation is represented pictorially by:

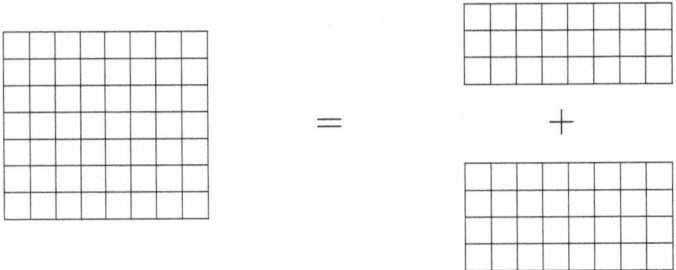

The grid on the LHS of the diagram consists of *eight* columns, each of which has *seven* cells. As such, it represents the product

$$7 \times 8 = (3 + 4) \times 8.$$

The RHS consists of two grids which are to be summed. In the top grid, there are *eight* columns of *three* cells each, corresponding to the product

129

3×8. In the bottom grid there are *eight* columns of *four* cells each, corresponding to the product 4×8. The total number of cells in any single vertical column (top and bottom) is still *seven*. Thus, we still have *eight* columns, each of which has *seven* cells as shown on LHS.

The grid on the LHS may be thought of as a collection of $7 \times 8 = 56$ tiles. If we split this collection into two collections, one containing 24 tiles, and the other containing 32 tiles, as is shown on the RHS, then conservation of real world objects tells us that the sum must preserve the original total. This is why the Distributive Law must be true. It is simply a version of the Conservation Principle (**CP**). So any system of arithmetic we create, as we have argued, is merely an abstract model for the process of counting real-world collections and must satisfy the Distributive Law.

Using the Distributive Law

Because multiplication is commutative, the Distributive Law is **two-sided**:

$$p \times (n + m) = p \times n + p \times m \quad \text{and}$$
$$(n + m) \times p = n \times p + m \times p.$$

Ordinarily, the Distributive Law is applied in the following way:

$$4 \times (3 + 7) = 4 \times 3 + 4 \times 7 = 12 + 28.$$

In this application, multiplication by the 4 on the outside of the parentheses on the LHS is *distributed* across the addition on the inside to produce a sum of products. As a descriptive term, *distributive* seems appropriate for this process.

There is a second equally important use of the Distributive Law illustrated by the following equation:

$$4 \times a + 3 \times a = (4 + 3) \times a.$$

In this use, we start with a sum of products that have a **common factor**, in this case a as shown on the LHS. We then use the Distributive Law, **backwards** if you will, to **pull the common factor** a outside of the sum thereby producing the RHS. This second use is at least as important as the first use and needs to be recognized and understood by children early on because it is particularly important tool for manipulating symbolic quantities.

11.4.3 The Associative Law of Multiplication

Recall the general statement of the Associative Law:

$$(n \times m) \times p = m \times (n \times p).$$

We want to consider why we should take this equation as a law. We could simply appeal to the fact that multiplication is repetitive addition and addition is associative. But we want to dig deeper and look for a physical explanation that we could present to children.

Products of Three Numbers as Volume

We have already pointed out that the product of two numbers representing lengths can be taken to represent **area** (see §11.2). This leads directly to the question: How should we think of the product of three numbers each of which is a length?

In Figure 11.1 we picture a cube measuring 1 unit on a side. We refer to such a cube as a **unit cube** and say it has 1 **cubic unit of volume**. The property that makes this figure a cube is that each of its faces is a square. Since each face of

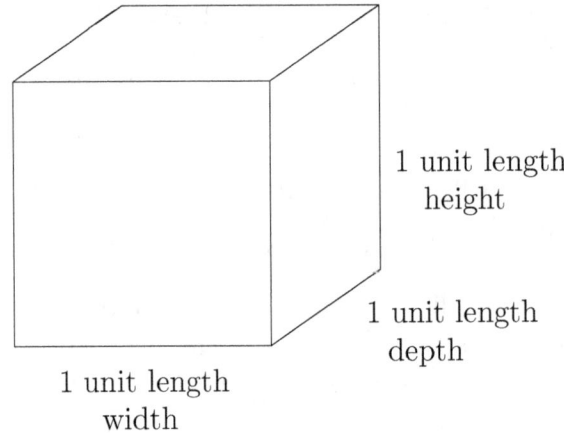

Figure 11.1: A single cube of 1 unit length on each side. Its volume is 1 cubic unit.

the block is a square and the length along every edge is 1 unit of length, we know that the area of each face is $1 \times 1 = 1$. But the block exhibits more than mere areal extent of its faces because the faces extend in three mutually perpendicular directions. Thus, the block has **depth** too. Since for the unit cube, all the edges have the same length, namely, 1, the length, the width and the height of the cube are all 1, and if we multiply these three numbers together, the volume of the unit cube is $1 \times 1 \times 1 = 1$ cubic units, which is why we refer to this cube as a **unit** cube. In summary, when objects have physical length in each of three mutually perpendicular directions thereby enclosing space, we say that such objects have **volume** and like area, this is a physical attribute of objects in the world.

131

We can think of larger volumes as being comprised of such unit cubes in the same way we can think of areas as being comprised of unit squares. Thinking about measuring volumes by counting unit cubes comprising a whole illustrates in concrete terms why volumes involve the product of three numbers. To make this discussion in respect to the Associative Law as concrete as possible, we consider two stacks of blocks as pictured in Figure 11.2.

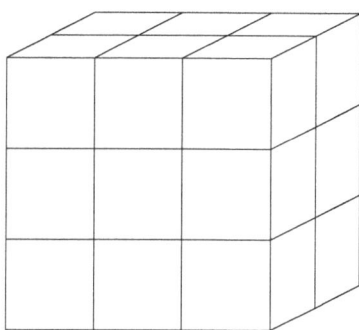 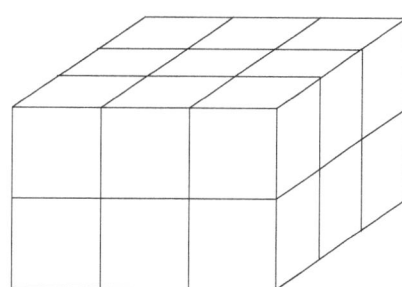

Figure 11.2: Two stacks of blocks made from 18 blocks each of which has unit volume. In the text the stack on the LHS is referred to as Stack 1, the stack on the RHS as Stack 2.

Stack 1 on the LHS is made by stacking 3 blocks to form a column. Three of these columns placed side-by-side form the front face. As the reader can see, there are a total of $3 \times 3 = 9$ blocks forming the front face. As indicated in the figure, there is a second identical row of 3 columns behind the front face which must also contain 9 blocks, so the total blocks used is $(3 \times 3) \times 2 = 18$. Now, if we simply ask how many columns of blocks are in Stack 1, we see number is $3 \times 2 = 6$, which is the area of the top face. Since each column contains 3 blocks, the total blocks used counting this way is $3 \times (3 \times 2)$. Since both computations yield 18 blocks, we record this as

$$(3 \times 3) \times 2 = 3 \times (3 \times 2)$$

which is an exact instance of the Associative Law of multiplication.

Imagine that the blocks on the LHS are stuck together, and someone rolls Stack 1 so that it is now lying on its side with its front face on the top. That is the arrangement of Stack 2 on the RHS of the figure. As the reader can see, each column now contains 2 blocks. Since there are 3 columns making up the front face, we know there are a total of 2×3 blocks in the front face. But now the stack on the RHS is 3 columns deep, so there is a total of $(2 \times 3) \times 3 = 18$ blocks in the entire stack. Alternatively, the area of the top face has an area of 3×3, and each block in the top face represents a column of 2, so the total number of blocks is $2 \times (3 \times 3)$. Since the total number of blocks is fixed, these computations again are an exact instance of the Associative Law:

$$(2 \times 3) \times 3 = 2 \times (3 \times 3).$$

The reader knows that however we find the number of blocks making up these stacks, the number of blocks is fixed at 18. **CP** is the guarantor of this fact and once again, we conclude that a fundamental law of our arithmetic is derived from a truth about counting in the real world.

In respect to the Associative Law, every child should understand that in any computation involving **only** multiplication, all ways of carrying out the multiplication must give the same result.

The discussion above makes clear why we consider the product of three lengths to represent volume. In respect to the stacks of blocks, we would say the total volume of the entire stack is:

$$V = (3 \times 3) \times 2 \text{ cubic units} = 18 \text{ cubic units},$$

where 1 cubic unit is the volume of one block. Every child should come to identify **volume** as one outcome of taking the product of three numbers, just as **area** is one outcome when we take the product of two numbers.

11.4.4 The Multiplicative Identity

The reader will recall the following fact about addition:

$$0 + n = n + 0 = n$$

for any counting number n. Because of this equation, we refer to zero as the **additive identity**.

The reader may wonder whether there is an analogous equation for multiplication in which we have a fixed number whose product with any other number leaves that number unchanged. In fact, there is, namely:

$$1 \times n = n \times 1 = n$$

where again, n is any counting number or 0. To be consistent with our nomenclature we refer to 1 as the **multiplicative identity**.

To see why this equation is true, we simply remind ourselves what $1 \times n$ means. The first number, in this case 1, tells us that 1s are what must be added together. The second number, in this case n, tells us how many to add, in this case, n. We could think of this in concrete terms by having each 1 correspond to a single button. We would have a total of n buttons.

Alternatively, $n \times 1$ specifies n is to be added exactly once, again giving n.

The following is an numerical example of the multiplicative identity equation:

$$1 \times 638 = 638 \times 1 = 638.$$

11.4.5 Multiplication by 0

Recall the equation that makes 0 the additive identity: $0 + n = n + 0 = n$. As we will see, this equation applies to all numbers. For our discussion here, we want to apply this fact when $n = 0$, in which case we would have

$$0 = 0 + 0 = 0 \times 2.$$

Changing the sum to a product merely uses the fact that multiplication is repetitive addition. Since the sum of any number of zeros is still 0, we can write

$$0 \times n = 0 = n \times 0$$

for any counting number n or 0. Simply stated,

when we multiply any number by 0, we must get 0.

This is a fact that every child should know.

11.4.6 Products of Counting Numbers are Not 0

Consider that we have two counting numbers n and m. As such, these numbers are cardinal numbers of real-world collections and therefore cannot be 0. Now any sum of counting numbers is again a counting number. We know this from the definition of addition as a real-world process. Since multiplication is repetitive addition, the product of these two counting numbers n and m cannot be 0. We state this as

$$n \times m \neq 0.$$

This is another seemingly simple fact. However, it has an important consequence, namely

if a product is 0, one of the factors comprising the product must be 0.

Again, this is a fact that every child should come to know as they learn arithmetic.

11.5 Making Multiplication Useful

In order to realize the utility of multiplication, we have to develop procedures for finding products of numbers written in some system of numerals. As before, we turn to the Arabic System of notation.

Let us recall the general situation for addition. The procedure really consisted of two parts, first, an ability to find the sum of two single digits, and second, a procedure for using this information to find sums of multi-digit numbers expressed in the Arabic system of notation. Developing a procedure for multiplication will involve similar steps.

11.5.1 Reinterpreting the Arabic System of Notation

Our original interpretation of the Arabic System of notation was based on the Button Dealer analogy. To be specific, given a multi-digit numeral, say:

$$8952,$$

the Button Dealer would get the following jars:

two jars marked 1 from the first shelf;
five jars marked 10 from the second shelf;
nine jars marked 100 from the third shelf;
eight jars marked 1000 from the fourth shelf.

Recall that each jar marked 1 contains *one* button, each jar marked 10 contains *ten* buttons, each jar marked 100 contains one hundred buttons, and each jar marked 1000 contains one thousand buttons. The rest of the Button Dealer process involves **combining all the buttons into one jar**, in other words addition. So let's rewrite things in terms of addition. If we do this, we find the Button Dealer gets:

$$1 + 1 = 1 \times 2$$

buttons from the first shelf;

$$10 + 10 + 10 + 10 + 10 = 10 \times 5$$

buttons from the second shelf;

$$100 + 100 + 100 + 100 + 100 + 100 + 100 + 100 + 100 = 100 \times 9$$

buttons from the third shelf;

$$1000 + 1000 + 1000 + 1000 + 1000 + 1000 + 1000 + 1000 = 1000 \times 8$$

buttons from the fourth shelf.

Each of the sums has been reinterpreted as a product. We express this more succinctly as the Button Dealer gets:

1×2 buttons from the first shelf;
10×5 buttons from the second shelf;
100×9 buttons from the third shelf;
1000×8 buttons from the fourth shelf.

Now we apply the Commutative Law of multiplication to rewrite things as follows:

2×1 buttons from the first shelf;
5×10 buttons from the second shelf;
9×100 buttons from the third shelf;
8×1000 buttons from the fourth shelf.

Now to complete the process, we know that in effect, the Button Dealer combines the buttons into one jar, in other words, adds the counting numbers. Thus, we have

$$8952 = 8 \times 1000 \ + \ 9 \times 100 \ + \ 5 \times 10 \ + \ 2 \times 1.$$

We remind the reader that we do not need to use parentheses on the RHS because multiplication takes **precedence** over addition and must be performed first.

This example tells us how to reinterpret Arabic notation using multiplication and the notion of **place**. In Arabic notation, we have been referring to the digit at the far right as the *ones* digit and the place as the *ones* place. This is because the right-most digit is multiplied by 1 as in our expanded form for 8952. The digit immediately to the left of the *ones* digit is referred to as the *tens* digit, and the place as the *tens* place, because in the above expression it gets multiplied by 10. The next digit to the left is called the *hundreds* digit and the place called the *hundreds* place because it gets multiplied by 100. The last digit in our four digit number, at the far left, is called the *thousands* digit and its place called the *thousands* place because it is multiplied by 1000. In a five digit number, the left-most digit will be the *ten thousands* digit and the place the *ten thousands* place, and so forth.

Applying these remarks to an arbitrary four-digit number in Arabic notation, we would have

$$n_{1000} n_{100} n_{10} n_1$$

where each of the symbols n_{1000}, n_{100}, n_{10}, n_1 is a single digit from the list 0, ..., 9 and the subscript on n specifies the **place** determined by starting at the right and moving to the left. Thus, n_1 is in the *ones* place, n_{10} is in the *tens* place, n_{100} is in the *hundreds* place and n_{1000} is in the *thousands* place, in 8952:

$$n_{1000} = 8, \quad n_{100} = 9, \quad n_{10} = 5, \quad \text{and} \quad n_1 = 2.$$

The interpretation of such an arbitrary numeral is that it stands for the result of the following computation:

$$n_{1000} \times 1000 \; + \; n_{100} \times 100 \; + \; n_{10} \times 10 \; + \; n_1 \times 1$$

or, in equation form:

$$n_{1000}n_{100}n_{10}n_1 = n_{1000} \times 1000 \; + \; n_{100} \times 100 \; + \; n_{10} \times 10 \; + \; n_1 \times 1.$$

In §?? we pointed out an important relation that holds between adjacent place values, for example, between 100 and 10. The relation is that the

> the value of any place is obtained by summing *ten* units having the value of the adjacent place to the right.

So, for example, the *hundreds* place and the *tens* place satisfy,

$$100 = 10 + 10 + 10 + 10 + 10 + 10 + 10 + 10 + 10 + 10.$$

Obviously, this relation begs for the use of multiplication. Stating the various place relations using the power of multiplication, we have:

$$
\begin{aligned}
10 &= 1 \times 10, \\
100 &= 10 \times 10, \\
1000 &= 100 \times 10, \\
10000 &= 1000 \times 10,
\end{aligned}
$$

and so forth. So a digit in any place has 10 times the value of the **same digit** positioned one place to the right. A+ curricula expect all children to come to a complete understanding of these relationships which is why the subject of whole-number (counting number) meaning extends into Grade 5.

As the reader can see from the discussion above, this interpretation using multiplication still implements the Button Dealer analogy, but in a more compact and sophisticated way. The original interpretation was directly tied to **counting**. It becomes more abstract as we reinterpret counting as addition. The current reinterpretation, in which addition is replaced by multiplication introduces an even higher level of abstraction. But, nothing essential has changed and the notation still comes back to **counting**. That's where a child has to start, with understanding place as being groups of a fixed size coming off the shelf as in the Button Dealer analogy.

The multiplication as an operation was known to the ancient Egyptians. So in principle, it should have been possible to invent the Arabic System of notation as soon as the operation was known. But the Arabic System did not come into use until the Middle Ages. Why so long? The only possible reason is that these ideas were difficult to generate, particularly the concept of zero as a number. However, once known, the ideas are so simple, they can be universally taught to children.

11.5.2 The Multiplication Table

To make multiplication useful, children need to know the products of single digits. In a like manner to addition, these products can be placed in a **multiplication table** which is easily constructed using the definition of multiplication in terms of repetitive addition and the addition table. It can also be constructed directly by counting areas using grids of the appropriate size.

The table is read in exactly the same way as the addition table. For example, suppose we want to know the result of 7×5. Start on the row having 7 at the far left, proceed along this row to the column headed by 5. The entry in that cell is 35, which is the required value.

There are two rows and two columns in the table that are trivial, namely those containing products where one factor is either 0 or 1. There are only a total of 36 remaining products that have to be learned! This number is reduced from 64 due to the Commutative Law.

×	0	1	2	3	4	5	6	7	8	9
0	0	0	0	0	0	0	0	0	0	0
1	0	1	2	3	4	5	6	7	8	9
2	0	2	4	6	8	10	12	14	16	18
3	0	3	6	9	12	15	18	21	24	27
4	0	4	8	12	16	20	24	28	32	36
5	0	5	10	15	20	25	30	35	40	45
6	0	6	12	18	24	30	36	42	48	54
7	0	7	14	21	28	35	42	49	56	63
8	0	8	16	24	32	40	48	56	64	72
9	0	9	18	27	36	45	54	63	72	81

The **Multiplication Table**. To find the value of 4×3, for example, find the cell in the row having a 4 at the far left and in the column having 3 at the top. Notice that each row/column can be obtained by skip counting by the number at the left/top starting at 0 and counting across/down.

A+ curricula intend that the study of whole number operations will be completed in Grade 4. Twenty percent of A+ countries complete this part of the curriculum by the end of Grade 3. This means that every parent should make sure that their child knows the Multiplication Table as a matter of recall by the middle of Grade 3.

Learning the Multiplication Table

To master the table, children need to start with a procedure for finding the cell entries that always works. Without a procedure, children will guess and each wrong guess creates a wrong memory, however fleetingly, that has to be unlearned.

The procedure we suggest for finding cell values is skip-counting because each row and each column in the table can be obtained by skip-counting. Since skip-counting is just addition, using skip-counting to find table values reinforces thinking about multiplication as repetitive addition.

Consider the row headed by 4 taken from the table:

$\times$	0	1	2	3	4	5	6	7	8	9
4	0	4	8	12	16	20	24	28	32	36

The header row and the interior row with row-header 4 from the Multiplication Table.

Observe that the left-most interior entry is 0, corresponding to computation:

$$4 \times 0 = 0.$$

Moving one cell to the right, the value is 4 corresponding to the computation:

$$4 \times 1 = 4 = 0 + 4.$$

Moving another cell to the right, the value is 8 corresponding to the computation:

$$4 \times 2 = 8 = 4 + 4.$$

Moving still another cell to the right, the value is 12 corresponding to the computation:

$$4 \times 3 = 12 = 8 + 4.$$

This process is simply skip-counting by 4s starting at 0. To see why, we apply the Distributive Law to obtain the value in the cell next to $4 \times n$ on the right:

$$4 \times (n + 1) = 4 \times n + 4 \times 1 = 4 \times n + 4.$$

In other words, to move one cell to the right we simply add 4. Now let's apply these ideas to other rows in the Table.

The first thing your child needs to know is that:

multiplication of counting numbers is repetitive addition.

This definition is why skip-counting works and is the bedrock on which the child goes forward. Applying the definition in the form of skip-counting is why your child needs to have mastered addition within 100 by the end of Grade 2.

Consider the row headed by 2, so that every product has the form:

$$2 \times n = 2 + \ldots + 2 \qquad n \text{ times.}$$

Once again the value in the first cell will be 0 and we skip-count by 2s starting there. The result is the following sequence

$$0 \to 2 \to 4 \to 6 \to 8 \to 10 \to 12 \to 14 \to 16 \to 18.$$

This row will be easy for children who have mastered skip-counting by 2s. Thus, to find 2×5, the child simply repeats:

$$2, \quad 4, \quad 6, \quad 8, \quad 10,$$

counting on their fingers, if necessary, until five 2s have been added.

Now let's skip to the row having 7 at the left with the assumption that your child has learned the products in the rows above. The critical equation is:

$$7 \times n = 7 + \ldots + 7 \qquad n \text{ times}$$

and skip-counting by 7s starting at 0 produces:

$$0 \to 7 \to 14 \to 21 \to 28 \to 35 \to 42 \to 49 \to 56 \to 63.$$

Consider how to find 8×7. Given one knows the skip-count sequence, one simply recalls:

$$7, \quad 14, \quad 21, \quad 28, \quad 35, \quad 42, \quad 49, \quad 56,$$

stopping at the eighth step. Alternatively a child can pick up the count by recalling the row for 6. The child recalls $42 = 6 \times 7 = 7 \times 6$ and count-on by 7s from 42, as in

$$42, \quad 49, \quad 56.$$

Here again is where previous mastery of addition within 100 makes finding the last two sums straightforward.

To use skip-counting effectively, your child must know the definition of multiplication. The process of using skip-counting reinforces learning of the definition because each time it is used the child has to correctly think out exactly how many 7s am I adding up here (eight in the example above).

Once you are sure that your child can correctly find each product in the table, it is a matter of **practice makes perfect** to accomplish the transition from procedure to recall. Practice problems are available at websites. Flash cards will be helpful to make recall instantaneous. They can be used to learn the skip-count sequences by simply pulling out the cards having multiples of any specific number and asking them in order once a child knows the relation between multiplication and addition.

11.5.3 The Distributive Law, Multiplication and the Arabic System

The target in an A+ curriculum for using the multiplication algorithm for children is that they will be able to multiply a four-digit number by a two-digit number with facility without using a calculator. So that our explanations are not too unwieldy, we will begin our explanations using three-digit numbers.

As discussed above, a three-digit counting number in the Arabic System satisfies

$$n_{100}n_{10}n_1 = n_{100} \times 100 \ + \ n_{10} \times 10 \ + \ n_1 \times 1.$$

To take a specific counting number as an example:

$$357 = 3 \times 100 \ + \ 5 \times 10 \ + \ 7 \times 1.$$

Now suppose we want to multiply 357 by another number, say 8. Applying the Distributive and Associative Laws to obtain the second and third lines, respectively, we would have

$$
\begin{aligned}
8 \times 357 &= 8 \times (3 \times 100 \ + \ 5 \times 10 \ + \ 7 \times 1) \\
&= 8 \times (3 \times 100) \ + \ 8 \times (5 \times 10) \ + \ 8 \times (7 \times 1) \\
&= (8 \times 3) \times 100 \ + \ (8 \times 5) \times 10 \ + \ (8 \times 7) \times 1.
\end{aligned}
$$

The key thing to notice in this sequence is that as a consequence of the Distributive Law, each digit in the Arabic numeral gets multiplied by 8. Now there is nothing special about 8. Indeed, if we were multiplying by an arbitrary counting number, say m, we would just have:

$$
\begin{aligned}
m \times 357 &= m \times (3 \times 100 \ + \ 5 \times 10 \ + \ 7 \times 1) \\
&= m \times (3 \times 100) \ + \ m \times (5 \times 10) \ + \ m \times (7 \times 1) \\
&= (m \times 3) \times 100 \ + \ (m \times 5) \times 10 \ + \ (m \times 7) \times 1.
\end{aligned}
$$

As you can see, each digit is now multiplied by m. And there in nothing special about 357 either. If we take an arbitrary three-digit number in its usual form, multiplication

141

by 8 would yield:

$$8 \times n_{100}n_{10}n_1 \ = \ 8 \times (n_{100} \times 100 \ + \ n_{10} \times 10 \ + \ n_1 \times 1)$$
$$= \ 8 \times (n_{100} \times 100) \ + \ 8 \times (n_{10} \times 10) \ + \ 8 \times (n_1 \times 1)$$
$$= \ (8 \times n_{100}) \times 100 \ + \ (8 \times n_{10}) \times 10 \ + \ (8 \times n_1) \times 1.$$

And of course, if we multiply $m \times n_{100}n_{10}n_1$, we would have:

$$m \times n_{100}n_{10}n_1 \ = \ m \times (n_{100} \times 100 \ + \ n_{10} \times 10 \ + \ n_1 \times 1)$$
$$= \ m \times (n_{100} \times 100) \ + \ m \times (n_{10} \times 10) \ + \ m \times (n_1 \times 1)$$
$$= \ (m \times n_{100}) \times 100 \ + \ (m \times n_{10}) \times 10 \ + \ (m \times n_1) \times 1.$$

These calculations are typical. We will see many instances of applications of the Distributive and Associative Law in what follows, so it is important you are comfortable with these manipulations. Most importantly, once you recognize that a particular manipulation is an application of one of these laws, you don't have to worry about why it is true, because, as you now know, it all comes back to counting and **CP**.

11.5.4   Multiplication by 10, 100, 1000, etc.

The rest of the multiplication procedure depends on the relationship of the Arabic System of numeration to the operation of multiplication. The basis of this relationship is the effect of multiplying by 10, 100, etc. Thus to go forward, you must completely understand how and why this works. The important theoretical facts we will apply are discussed above.

Multiplying a Two-digit Number by 10

To examine what happens when we multiply by 10, we start with a two-digit example, say, 42. Following the scheme developed in §11.3.3 above with an extra step applying the Commutative Law to obtain line 4, when we multiply 42 by 10 we get the following:

$$10 \times 42 \ = \ 10 \times (4 \times 10 \ + \ 2 \times 1)$$
$$= \ 10 \times (4 \times 10) \ + \ 10 \times (2 \times 1)$$
$$= \ (10 \times 4) \times 10 \ + \ (10 \times 2) \times 1$$
$$= \ (4 \times 10) \times 10 \ + \ (2 \times 10) \times 1$$
$$= \ 4 \times (10 \times 10) \ + \ 2 \times (10 \times 1)$$

$$= 4 \times 100 + 2 \times 10$$
$$= 4 \times 100 + 2 \times 10 + 0 \times 1$$
$$= 420.$$

Observe, when we start, the 4 in 42 is a *tens* digit. When we finish, the 4 in 420 is a *hundreds* digit. **The 4 has been moved one place to the left.** Similarly, the 2 in 42 is a *ones* digit. In 420, the 2 is now a *tens* digit, so it also has been moved one place to the left.

Thus, after multiplying 42 by 10, we no longer have a *ones* digit, so we have to create one. We did that by putting in 0×1 on the next to last line. We can insert 0×1 exactly because the product is 0 and since 0 is the additive identity, it has no effect on the sum but it does fill the *ones* place. This place has to be filled if we are going to correctly interpret the digits 4 and 2 as part of a three-digit number.

The essence of multiplying by 10 is:

$$10 \times 42 = 42 \times 10 = 420.$$

Because we write language left-to-right, when we think about multiplying 10 times 42, we picture in our minds

$$10 \times 42.$$

However the Arabic number system reads right-to-left, and operations are performed right-to-left. Thus, although we think of 10 as the multiplier in 10×42, we write the product as 42×10 here because we want to emphasize the relation of multiplying by 10 on the right which merely introduces an extra zero on the right in the Arabic numeral as the equation

$$42 \times 10 = 420$$

shows.

We can emphasize the right-to-left aspect of the Arabic System if we apply the Commutative Law in the first step as in:

$$
\begin{aligned}
10 \times n_{10}n_1 &= n_{10}n_1 \times 10 \\
&= (n_{10} \times 10 + n_1 \times 1) \times 10 \\
&= (n_{10} \times 10) \times 10 + (n_1 \times 1) \times 10 \\
&= n_{10} \times (10 \times 10) + n_1 \times (1 \times 10) \\
&= n_{10} \times 100 + n_1 \times 10 \\
&= n_{10} \times 100 + n_1 \times 10 + 0 \times 1 \\
&= n_{10}n_1 0.
\end{aligned}
$$

143

The calculation is one step shorter, and the 10 is on the right, which is where we want it to be in the end. As in the numerical example, in the next to last line, the reader will notice the addition of the 0×1 so that the final result has a *ones* digit. The result of the computation is the numeral

$$n_{10}n_{1}0,$$

which has three digits, the right-most of which is 0. The digit n_{10}, which originated as a *tens* digit, is now a *hundreds* digit, i.e., it has been moved one place to the left. Similarly, n_1, which originated as a *ones* digit, is moved one place to the left and is now a *tens* digit.

Before continuing, let's summarize:

When any two digit number is multiplied by 10, the result is the same two digits, in the same order, followed by a 0 on the right.

The following are numerical examples:

$$25 \times 10 = 250, \quad 63 \times 10 = 630, \quad \text{and} \quad 71 \times 10 = 710.$$

A very important special case of this result is:

$$10 \times 10 = 100,$$

although we know this from the basic properties of the Arabic System.

Multiplying a Three-digit Number by 10

Consider the three-digit number 357. Using the Laws and the fact that $1000 = 100 \times 10$ on line 3 of what follows, multiplying 357 by 10 gives:

$$
\begin{aligned}
10 \times 357 = 357 \times 10 \;&=\; (3 \times 100 \;+\; 5 \times 10 \;+\; 7 \times 1) \times 10 \\
&=\; (3 \times 100) \times 10 \;+\; (5 \times 10) \times 10 \;+\; (7 \times 1) \times 10 \\
&=\; 3 \times (100 \times 10) \;+\; 5 \times (10 \times 10) \;+\; 7 \times (1 \times 10) \\
&=\; 3 \times 1000 \;+\; 5 \times 100 \;+\; 7 \times 10 \\
&=\; 3 \times 1000 \;+\; 5 \times 100 \;+\; 7 \times 10 \;+\; 0 \times 1 \\
&=\; 3570.
\end{aligned}
$$

The essential result is

$$357 \times 10 = 3570.$$

In other words, once again the result is the original digits in the same order with a 0 tacked on at the right.

You can start with any three-digit number whatsoever and put it into the calculation above for 357 and repeat the sequence of steps. The end result will be the three-digit number you started with, with a 0 added on as the right-most digit. Some additional examples are:

$$275 \times 10 = 2750, \quad 382 \times 10 = 3820, \quad \text{and} \quad 701 \times 10 = 7010.$$

We might now guess that to multiply any counting number in Arabic notation by 10, we can express the result by writing the digits in the same order, left-to-right followed by one additional 0 on the right. This is indeed correct, and the exact reasons why are contained in the computational scheme laid out above.

Multiplying a Two-digit Number by 100

Consider now the problem of multiplying a two digit number by 100. For example, 56×100. Since we know $100 = 10 \times 10$ we can use the Associative Law to obtain,

$$
\begin{aligned}
56 \times 100 \quad &= \quad 56 \times (10 \times 10) = (56 \times 10) \times 10 \\
&= \quad 560 \times 10 = 5600.
\end{aligned}
$$

The net effect most simply stated is

$$56 \times 100 = 5600.$$

Observe that the 5, which starts as a *tens* digit in 56, is moved to the *thousands* place in the answer. Similarly, the 6, which starts in the *ones* place in 56, moves to the *hundreds* place in the answer. In other words, each original digit has been moved two places to the left, one place for each of the zeros in 100.

Once again the result above is true in general. using the Associative Law we have

$$
\begin{aligned}
n_{10}n_1 \times 100 \quad &= \quad n_{10}n_1 \times (10 \times 10) \\
&= \quad (n_{10}n_1 \times 10) \times 10 \\
&= \quad n_{10}n_10 \times 10 \\
&= \quad n_{10}n_100.
\end{aligned}
$$

Thus, for clarity, we have:

$$n_{10}n_1 \times 100 = n_{10}n_100.$$

As the reader can see, each digit from the original two digit number, $n_{10}n_1$, has been moved two places to the left in the answer so that n_{10} is now in the *thousands* place and n_1 is now the *hundreds* place.

We can summarize the above as:

When any two digit number is multiplied by 100**, the result is the same two digits, in the same order, followed by two** 0**'s on the right**.

The following are numerical examples:

$$38 \times 100 = 3800, \quad 65 \times 100 = 6500, \quad \text{and} \quad 92 \times 100 = 9200.$$

Multiplying a Two-digit Number by 1000

Consider now the problem of multiplying a two digit number by 1000. For example, 84×1000. Since we know $1000 = 100 \times 10$ using the Associative Law and the previous rules for multiplying by 100 and 10 gives,

$$
\begin{aligned}
84 \times 1000 &= 84 \times (100 \times 10) \\
&= (84 \times 100) \times 10 \\
&= 8400 \times 10 \\
&= 84000.
\end{aligned}
$$

To restate this result in its simplest form gives

$$84 \times 1000 = 84000.$$

Observe that the 8, which started as a *tens* digit in 84, is a *ten thousands* digit in the answer; in other words, it has been moved three places to the left, one place for each of the zeros in 1000. Similarly, the 4, which starts as a *ones* digit in 84, becomes a *thousands* digit in the answer. Again, it has shifted three places to the left, one place for each of the zeros in 1000.

Once again the computation does not depend on the particular number. Thus,

$$n_{10}n_1 \times 1000 = n_{10}n_1 000$$

by repeated application of the Associative Law and the rules for multiplying by 100 and 10. As the reader can see, each digit from the original two digit number, $n_{10}n_1$, has been moved three places to the left in the answer, so that n_{10} is now in the *ten thousands* place and n_1 is now the *thousands* place.

We can summarize the above as:

When any two digit number is multiplied by 1000**, the result is the same two digits, in the same order, followed by three** 0**'s on the right**.

The following are numerical examples:

$$45 \times 1000 = 45000, \quad 18 \times 1000 = 18000, \quad \text{and} \quad 30 \times 1000 = 30000.$$

The general rule for multiplying by a multiple of 10 is:

to multiply a two digit number by a 1 followed by some number of zeros, write the two digits of your original number, then add on the right, the same number of zeros as follow the 1.

This is the rule. It's simple to use, and the above discussion explains why it works.

11.5.5   Multiplying an Arbitrary Number by a Single Digit

Recall, the purpose stated at the outset of this section, §11.3, was to develop methods for performing multiplication based on the Arabic System of numeration. At this point the reader may be wondering what all the details in the preceding subsections have to do with that problem. So let's consider exactly where we are. To keep things reasonably simple, we work with a three-digit number, 952.

Based on the reinterpretation of the Arabic System (see §11.3.1), we know that

$$952 = 9 \times 100 \; + \; 5 \times 10 \; + \; 2 \times 1.$$

Suppose we want to compute the product of 7 and 952. Using the Distributive and Associative Laws as in §11.3.3-4, we have

$$
\begin{aligned}
7 \times 952 \; &= \; 7 \times (9 \times 100 \; + \; 5 \times 10 \; + \; 2 \times 1) \\
&= \; 7 \times (9 \times 100) \; + \; 7 \times (5 \times 10) \; + \; 7 \times (2 \times 1) \\
&= \; (7 \times 9) \times 100 \; + \; (7 \times 5) \times 10 \; + \; (7 \times 2) \times 1.
\end{aligned}
$$

Notice that what is now required is to find three products, each of which consists of two single-digits, namely,

$$7 \times 9, \quad 7 \times 5 \quad \text{and} \quad 7 \times 2.$$

Each of these products can be found using the Multiplication Table in §11.3.2, and the results are

$$7 \times 9 = 63, \quad 7 \times 5 = 35 \quad \text{and} \quad 7 \times 2 = 14,$$

all of which are two-digit numbers. Incorporating these facts into the original computation and using the rules for multiplying by 10, etc., gives

$$
\begin{aligned}
7 \times 952 \; &= \; 63 \times 100 \; + \; 35 \times 10 \; + \; 14 \times 1 \\
&= \; 6300 + 350 + 14.
\end{aligned}
$$

To find the answer, we have to perform the indicated sums. Let's do this using the standard procedure. We write the numerals in columns as follows:

147

$$14$$
$$350$$
$$+\ 6300$$

As we know, each of the three numbers contributing to this sum arises as the product of two single-digits and a multiple of 10. One of the single-digits is 7 which is the single-digit multiplier. The remaining single-digits are the digits of 952 and the products are,

$$14 \times 1 \ = \ 7 \times (2 \times 1)$$
$$35 \times 10 \ = \ 7 \times (5 \times 10)$$
$$63 \times 100 \ = \ 7 \times (9 \times 100).$$

The multiple of 10 occurring in the product is the place multiplier of each individual digit in 952. To be clear, the place multiplier on 2 is 1 which tells us that 2 is the *ones* digit in 952; the place multiplier on 5 is 10 which tells us that 5 is the *tens* digit in 952; and the place multiplier on 9 is 100 which tells us that 9 is the *hundreds* digit in 952. In each case, when we form the product of the single digit with 7 as shown above on the LHS, we get a two-digit number times the same place multiplier that is on the RHS. Thus, 4 is a *ones* digit in 14 and arises from the 2 in 952 which is also a *ones* digit; 5 is a *tens* digit in 350 and arises from the 5 in 952 which is also a *tens* digit; and 3 is a *hundreds* digit in 6300 and arises from the 9 in 952 which is also a *hundreds* digit.

Each of the single-digit products results in a two-digit answer, 14, 35, and 63. The left-most digit, considered as part of a two-digit numeral, will always be a *tens* digit, hence it will always contribute to a column one place to the left of the *ones* digit. This positioning is shown in the sum of $14 + 350 + 6300$ written in columns above.

We turn these ideas into a concise procedure as follows.

Example 1.

The setup amounts to writing 952 above 7 as shown:

$$
\begin{array}{ccc}
9 & 5 & 2 \\
\times & & 7 \\
\hline
\end{array}
$$

The 7 is a *ones* digit and its place in the setup is directly under the *ones* digit in 952. We have three single digit products that have to be computed,

$$7 \times 2, \quad 7 \times 5, \quad 7 \times 9$$

in the order shown (right-to-left). The remaining issue is how to record the results of these products. The first product is 14, so we record the 4 in the *ones* place below the line. The 1 in 14 is a *tens* digit and as such, has to be put in the *tens* column one place to the left of the 4. We do this by carrying the 1 to the *tens* place in a new top row (the **carry row**) above the line, as shown below.

$$
\begin{array}{cccc}
 & 1 & & \\
 & 9 & 5 & 2 \\
\times & & & 7 \\
\hline
 & & & 4 \\
\end{array}
$$

Now, the second product is $35 = 5 \times 7$ and since the 5 is in the *tens* place, we record the 5 in 35 in the *tens* place below the line and carry the 3 to the *hundreds* place in the carry row as shown:

$$
\begin{array}{cccc}
 & 3 & 1 & \\
 & 9 & 5 & 2 \\
\times & & & 7 \\
\hline
 & & 5 & 4 \\
\end{array}
$$

The last product is $63 = 9 \times 7$. Since the 9 is in the *hundreds* place, we record the 3 in 63 in the *hundreds* place below the line and carry the 6 to the *thousands* place in the carry row as shown:

$$
\begin{array}{ccccc}
 & 6 & 3 & 1 & \\
 & & 9 & 5 & 2 \\
\times & & & & 7 \\
\hline
 & & 3 & 5 & 4 \\
\end{array}
$$

What remains is to sum the carried digits and the digits below the line, as shown below:

$$
\begin{array}{ccccc}
 & 6 & 3 & 1 & \\
 & & 9 & 5 & 2 \\
\times & & & & 7 \\
\hline
+ & & 3 & 5 & 4 \\
\hline
 & 6 & 6 & 6 & 4 \\
\end{array}
$$

The process above consists of two steps, multiplication/recording followed by summing the carry row and the row below the line. As we will see in the examples with multi-digit multipliers, we will need to combine these two steps. As well, our examples reflect the fact that the A+ curriculum emphasis is on multiplying two- and three-digit numbers by one- and two-digit numbers.

Before continuing to our next example, we note that multiplication of an arbitrary number by a single digit number requires us to use the various theoretical laws, the data in the multiplication table, together with our knowledge about multiplication by 10, our understanding of place, and what we learned about carrying when we studied addition. In short, multiplication applies all of our previous knowledge.

You can see from this sentence why learning arithmetic is like climbing a ladder. To get to the next rung, you have to stand on a lower rung. To learn the next item, you have to use what you learned previously. If previous knowledge is weak, or incomplete, it will make proceeding difficult. Eventually, progress becomes impossible. This is the central and inescapable fact. What has to be learned is a small amount, but it has to be learned perfectly. If you understand this and help your child achieve a solid foundation, your child will go far. Without a solid foundation, eventually your child starts to struggle, becomes needlessly frustrated, and ultimately is blocked from success.

Example 2.

For simplicity, our next example does not require carrying: 21×4. The setup places the *ones* digits in a single column.

$$\begin{array}{cc} 2 & 1 \\ \times & 4 \\ \hline \end{array}$$

Two single digit products are required: $1 \times 4 = 4$ and $2 \times 4 = 8$. The first of these, namely 4 is recorded below the line in the *ones* column, the same column as the 1 in 21 as shown below.

$$\begin{array}{cc} 2 & 1 \\ \times & 4 \\ \hline & 4 \end{array}$$

Because this product generated a single digit numeral, there is nothing to carry. The second product, namely 8, which is also a single digit numeral, is recorded in the *tens* column, which is the same column as the 2. Again, we show below:

$$\begin{array}{cc} 2 & 1 \\ \times & 4 \\ \hline 8 & 4 \end{array}$$

The multiplication is now complete because there is nothing carried.

Example 3.

Let's find 84×9. The setup aligns the *ones* digits

$$
\begin{array}{rr}
8 & 4 \\
\times & 9 \\
\hline
\end{array}
$$

Two single digit products are required: $4 \times 9 = 36$ followed by $8 \times 9 = 72$. Following the outline above, the 6 from the 36 is recorded below the line in the *ones* column as shown below, and the 3 is recorded above the *tens* place in a new carry row, again as shown below:

$$
\begin{array}{rr}
3 & \\
8 & 4 \\
\times & 9 \\
\hline
& 6 \\
\end{array}
$$

The result of the second single digit product is 72. Since the 8 is in the *tens* column, the 2 should be recorded in the *tens* column below the line. However, since there is a carried 3 in the *tens* column which we know will have to be added to the 72, we can do this now, as part of this step. When we perform this addition, $3 + 72 = 75$, we cross the 3 out as shown and record the 5 below the line. Finally since all single digit products are completed, we write the 7 below the line one place to the left of the 5 in the *hundreds* place as shown:

$$
\begin{array}{rrr}
& \cancel{3} & \\
& 8 & 4 \\
& \times & 9 \\
\hline
7 & 5 & 6 \\
\end{array}
$$

Example 4.

The next example has three digits, 852×4 with the setup shown.

$$
\begin{array}{rrr}
8 & 5 & 2 \\
& \times & 4 \\
\hline
\end{array}
$$

There are three single digit products which, in order right to left, are

$$2 \times 4 = 8, \quad 5 \times 4 = 20 \quad \text{and} \quad 8 \times 4 = 32.$$

The first product, $8 = 2 \times 4$ is recorded below the line in the *ones* column since the 2 is in the *ones* column. There is nothing to carry.

$$\begin{array}{r} 8 \quad 5 \quad 2 \\ \times \quad 4 \\ \hline 8 \end{array}$$

The next single digit product is $5 \times 4 = 20$ where the 5 is in the *tens* column. So the 0 from 20 gets recorded in the *tens* column below the line, and the 2 is carried and put in a new carry row above the *hundreds* place as shown:

$$\begin{array}{r} 2 \quad \quad \\ 8 \quad 5 \quad 2 \\ \times \quad 4 \\ \hline 0 \quad 8 \end{array}$$

The last single digit product is $8 \times 4 = 32$, where the 8 is in the hundreds place in 852. Thus, the 2 in 32 should be recorded in the *hundreds* column below the line. But there is a already a 2 in the *hundreds* place in the carry row which must be added to the 32 at some point. We can do that now giving $2 + 32 = 34$ and record the 4 in the *hundreds* place below the line. The carried 2 is crossed out to show that it has been used. Since 8×4 is the last single digit product, the 3 from 34 is recorded one place to the left in the *thousands* place as shown.

$$\begin{array}{r} \not{2} \quad \quad \\ 8 \quad 5 \quad 2 \\ \times \quad 4 \\ \hline 3 \quad 4 \quad 0 \quad 8 \end{array}$$

Example 5.

We do a last example using a four digit number: 7568×9. The setup is:

$$\begin{array}{r} 7 \quad 5 \quad 6 \quad 8 \\ \times \quad \quad 9 \\ \hline \end{array}$$

Four single digit products are required which are, right to left:

$$8 \times 9 = 72, \quad 6 \times 9 = 54, \quad 5 \times 9 = 45 \text{ and } 7 \times 9 = 63.$$

Since the 8 in 8×9 is in the *ones* column, the 2 from 72 is recorded below the line in the *ones* column and the 7 is recorded in a new carry row above the *tens* place.

$$\begin{array}{r} 7 \quad \\ 7 \quad 5 \quad 6 \quad 8 \\ \times \quad \quad 9 \\ \hline 2 \end{array}$$

The next single digit product is 6×9 where the 6 is in the *tens* place. So the 4 from 54 would be recorded in the *tens* place below the line. However, we have a 7 in the *tens* place in the carry row. When this 7 is added to 54, the result is $7 + 54 = 61$, so that 1 is recorded below the line in the *tens* place and 6 is carried to the *hundreds* place in the carry row. Again the 7 is crossed out to indicate the sum has been recorded:

$$
\begin{array}{ccccc}
 & 6 & \cancel{7} & & \\
7 & 5 & 6 & 8 & \\
 & \times & & & 9 \\
\hline
 & & & 1 & 2 \\
\end{array}
$$

The next single digit product is 5×9 where the 5 is in the *hundreds* place; so the 5 from from the product, 45, has to be recorded in the *hundreds* place below the line. However, we have a carried 6 also in the *hundreds* place. When this 6 is added to 45, it produces $6 + 45 = 51$ and that 1 is recorded below the line in the *hundreds* place and the 6 is crossed out. The 5 is carried to the *thousands* place in the carry row, giving:

$$
\begin{array}{ccccc}
5 & \cancel{6} & \cancel{7} & & \\
7 & 5 & 6 & 8 & \\
 & \times & & & 9 \\
\hline
 & & 1 & 1 & 2 \\
\end{array}
$$

The last single digit product is 7×9 where the 7 is in the *thousands* place along with the 5 in the carry row. So we have to add the product 63 and the carried 5 to get $5 + 63 = 68$; the 8 is recorded below the line in the *thousands* place. Since there are no more single digit multiplications to be performed, the 6 is recorded below the line in the *ten thousands* place.

$$
\begin{array}{cccccc}
 & \cancel{5} & \cancel{6} & \cancel{7} & & \\
 & 7 & 5 & 6 & 8 & \\
 & & \times & & & 9 \\
\hline
6 & 8 & 1 & 1 & 2 & \\
\end{array}
$$

As detailed at the end of this chapter, multiplication of whole numbers is introduced in Grade 2 and concludes in Grade 4 with the expectation that children can skillfully complete calculations like those presented above.

11.5.6 Multiplying by Multi-digit Numbers

Once your child masters multiplying by a single digit number, multiplying multi-digit numbers by other multi-digit numbers is relatively straight forward. To see why, consider that we have a multi-digit number, say $k = 89747$ and we want to multiply it by 27. Using the theory, we know that this product is obtained as:

$$k \times 27 = k \times (20 + 7) = k \times 20 + k \times 7 = (k \times 2) \times 10 + (k \times 7) \times 1.$$

Using the same k and completing the calculation requires we sum the two numbers on the LHS of the following:

$$
\begin{aligned}
(89747 \times 7) \times 1 &= 89747 \times (7 \times 1) \\
(89747 \times 2) \times 10 &= 89747 \times (2 \times 10).
\end{aligned}
$$

In this case, the place multiplier is associated with the digits in 27. In all of this, the only new wrinkle is the existence of the place multiplier 10 on the second line. But we know what the effect of this multiplier is, it simply moves all digits one place to the left. Thus

$$(89747 \times 2) \times 10 = 179494 \times 10 = 1794940.$$

Suppose instead we want to multiply k by a three-digit number, say 975, we would have:

$$k \times 975 = k \times (900 + 70 + 5) = (k \times 9) \times 100 + (k \times 7) \times 10 + (k \times 5) \times 1.$$

Using the same value for k and completing this calculation would require we sum the three numbers on the LHS of the following:

$$
\begin{aligned}
(89747 \times 5) \times 1 &= 89747 \times (5 \times 1) \\
(89747 \times 7) \times 10 &= 89747 \times (7 \times 10) \\
(89747 \times 9) \times 100 &= 89747 \times (9 \times 100).
\end{aligned}
$$

Again, the new feature is the existence of the place multipliers, 10 and 100, in the number we are multiplying by. As discussed, the 10 moves the entire product one place to the left and the 100 moves the entire product two places to the left as the following shows:

$$(89747 \times 9) \times 100 = 807723 \times 100 = 80772300$$

The key to correctly performing this procedure is how we record and add the various products. Even this comes down to a couple of reasonably straightforward rules as our next examples will show.

Example 6.

We want to find 57×46. The setup puts the 57 above the 46 so that the *ones* are above the *ones* and the *tens* above the *tens*:

$$
\begin{array}{ccc}
 & 5 & 7 \\
\times & 4 & 6 \\
\hline
\end{array}
$$

The multiplication process starts by computing 57 times 6. The procedure exactly replicates what is done in Examples 3 and 4 and produces:

$$
\begin{array}{ccc}
 & \not{4} & \\
 & 5 & 7 \\
\times & \boxed{4} & 6 \\
\hline
3 & 4 & 2 \\
\end{array}
$$

Reviewing the material previous to this example leads to the next step which is to find 57×40. Computing this product starts by finding 57×4 where we have written $\boxed{4}$ to distinguish it from the carried 4. To compute the product, we have to find the two single-digit products (written in the manner of Examples 3 and 4 above)

$$
7 \times \boxed{4} = 28 \quad \text{and} \quad 5 \times \boxed{4} = 20,
$$

in that order. The multiplier $\boxed{4}$ is actually $4 \times 10 = 40$ because it is in the *tens* column. This means that the 8 in 28 has to be recorded in the *tens* place, that is, **in the same column as its multiplier**, in this case $\boxed{4}$ and the only way to do this is to start a new row below the row holding $342 = 57 \times 6$. Further, if we have to carry, which in this case we do, what is carried goes in the carry row in the *hundreds* place. These computations are shown as an intermediate step below:

$$
\begin{array}{ccc}
2 & \not{4} & \\
 & 5 & 7 \\
\times & \boxed{4} & 6 \\
\hline
3 & 4 & 2 \\
 & 8 & \\
\end{array}
$$

Notice that by putting the 8 in the same column as the multiplier $\boxed{4}$, we have automatically accounted for the fact that we are multiplying by 40 as opposed to 4. This also ensures that the results of all other products with $\boxed{4}$ are correctly placed.

To complete multiplication by $\boxed{4}$, we need to record the second single digit product, $5 \times \boxed{4} = 20$. In this case, since the 5 is in the *tens* place and the $\boxed{4}$ is in the *tens* place, the 0 from the 20 has to be in the *hundreds* place ($10 \times 10 = 100$), which puts it one column to the left of the previously placed 8. Since we have a 2 in the *hundreds* place on the carry line, we add it to the 0 and cross out the carried 2 as shown:

$$
\begin{array}{ccc}
\cancel{2} & \cancel{4} & \\
& 5 & 7 \\
\times & \boxed{4} & 6 \\
\hline
3 & 4 & 2 \\
2 & 2 & 8 \\
\end{array}
$$

As the reader can see, the 2 from 20 automatically ends up in the *thousands* place. What remains is to sum these two products using the standard procedure to obtain:

$$
\begin{array}{cccc}
& \cancel{2} & \cancel{4} & \\
& & 5 & 7 \\
& \times & 4 & 6 \\
\hline
& {}^{1}3 & 4 & 2 \\
+ & 2 & 2 & 8 \\
\hline
2 & 6 & 2 & 2 \\
\end{array}
$$

Example 7.

Find the product of 683 and 37. Again we know that what is required is finding and summing the two products shown in the equation

$$683 \times 37 = 683 \times (30 + 7) = 683 \times 30 + 683 \times 7.$$

Aligning the *ones* the setup is:

$$
\begin{array}{ccc}
6 & 8 & 3 \\
\times & 3 & 7 \\
\hline
\end{array}
$$

The first required product is 683×7 which produces the following

$$
\begin{array}{cccc}
& \cancel{5} & \cancel{2} & \\
& 6 & 8 & 3 \\
& \times & 3 & 7 \\
\hline
4 & 7 & 8 & 1 \\
\end{array}
$$

using the methods of §11.3.4.

The second required product is 683×30. For clarity, we repeat the last formulation with the 30 multiplier identified by a box.

$$
\begin{array}{cccc}
\require{cancel} & \cancel{5} & \cancel{2} & \\
6 & 8 & 3 \\
\times & \boxed{3} & 7 \\
\hline
4 & 7 & 8 & 1
\end{array}
$$

Finding this product will require computing and recording the single digit products

$$3 \times \boxed{3} = 9, \quad 8 \times \boxed{3} = 24, \quad \text{and} \quad 6 \times \boxed{3} = 18,$$

where the box is being used to remind the reader that the multiplier is in the *tens* place. As observed in the last example, the issue is how to record the results. Since the multiplier is in the *tens* place and $3 \times \boxed{3}$ yields a single digit, we record the 9 below the line in the *tens* column in a new row below the existing row

$$
\begin{array}{cccc}
& \cancel{5} & \cancel{2} & \\
6 & 8 & 3 \\
\times & \boxed{3} & 7 \\
\hline
4 & 7 & 8 & 1 \\
& & 9
\end{array}
$$

Notice that recording the 9 directly below the $\boxed{3}$ in the *tens* column automatically accounts for the fact that the actual product being performed is

$$3 \times \boxed{3}0 = 90$$

and by only recording the 9, we are suppressing the 0. Moreover, the remaining products will now be in their correct places.

The next product is $24 = 8 \times \boxed{3}$. The 4 is recorded in the new row below the line in the *hundreds* place to the left of the 9. As noted, the 4 is in the correct place, since the actual computation being performed is $80 \times 30 = 2400$. The 2 from 24 is a *thousands* digit and is recorded in the carry row in the *thousands* column. We can do that since the *thousands* place in the carry row is empty as shown

$$
\begin{array}{ccccc}
\boxed{2} & \cancel{5} & \cancel{2} & \\
& 6 & 8 & 3 \\
& \times & \boxed{3} & 7 \\
\hline
4 & 7 & 8 & 1 \\
& 4 & 9
\end{array}
$$

The last single digit product is $18 = 6 \times \boxed{3}$ where the 6 is a *hundreds* digit and the $\boxed{3}$ is a *tens* digit. So the 8 in the 18 has to be a *thousands* digit and the 1 has to be in the *ten thousands* place. Since we have a carried $\boxed{2}$ in the *thousands*

157

column already, we add to obtain $\boxed{2} + 18 = 20$. The 0 is recorded below the line as $\boxed{0}$ in the *thousands* place as shown below. The 2, which must be carried, is in the *ten thousands* place. However, since there are no more products to compute, we can record this 2 directly below the line in the *ten thousands* column to the left as shown:

$$
\begin{array}{ccccc}
 & \overset{2}{2} & \overset{5}{5} & \overset{2}{2} & \\
 & & 6 & 8 & 3 \\
 & & \times & \boxed{3} & 7 \\
\hline
 & 4 & 7 & 8 & 1 \\
2 & \boxed{0} & 4 & 9 &
\end{array}
$$

What remains is to sum the two product lines. Note the utility of crossing out items on the carry row as they are used so that we know they have already been included in the numbers below the line. Summing gives:

$$
\begin{array}{ccccc}
 & \overset{2}{2} & \overset{5}{5} & \overset{2}{2} & \\
 & & 6 & 8 & 3 \\
 & & \times & 3 & 7 \\
\hline
 & 4 & 7 & 8 & 1 \\
+ & 2 & 0 & 4 & 9 \\
\hline
2 & 5 & 2 & 7 & 1
\end{array}
$$

Example 8.

A last example finds the product of two three digit numbers. The setup is:

$$
\begin{array}{cccc}
 & 3 & 0 & 5 \\
\times & 6 & 4 & 7 \\
\hline
\end{array}
$$

As the reader knows,

$$305 \times 647 = 305 \times (600 + 40 + 7)$$

so that we have to find the following three products

$$305 \times 7, \quad 305 \times 40 \quad \text{and} \quad 305 \times 600$$

which then have to be summed.

The first computation required is 305×7 where the 7 is in the *ones* place and results in:

$$
\begin{array}{r}
\overset{\not{3}}{} \\
3 \quad 0 \quad 5 \\
\times \quad 6 \quad 4 \quad \boxed{7} \\
\hline
2 \quad 1 \quad 3 \quad 5 \\
\end{array}
$$

The second multiplier is 4 in the *tens* place and results in the following computation.

$$
\begin{array}{r}
\overset{\not{2}}{} \\
3 \quad 0 \quad 5 \\
\times \quad 6 \quad \boxed{4} \quad 7 \\
\hline
2 \quad 1 \quad 3 \quad 5 \\
1 \quad 2 \quad 2 \quad \boxed{0} \\
\end{array}
$$

Notice that by placing the 0 from $5 \times 4 = 20$ directly under the 4 in the *tens* column, the remaining digits in

$$305 \times 4 = 1220$$

are correctly positioned and reflect the actual calculation which is

$$305 \times 40 = 12200.$$

The third multiplier is a 6 in the *hundreds* place and results in the following computation:

$$
\begin{array}{r}
\overset{\not{3}}{} \\
3 \quad 0 \quad 5 \\
\times \quad \boxed{6} \quad 4 \quad 7 \\
\hline
2 \quad 1 \quad 3 \quad 5 \\
1 \quad 2 \quad 2 \quad 0 \\
1 \quad 8 \quad 3 \quad \boxed{0} \\
\end{array}
$$

Notice that by placing the 0 from $5 \times 6 = 30$ directly under the multiplier 6, the product

$$305 \times 6 = 1830$$

is automatically correctly positioned and reflects the actual computation

$$305 \times 600 = 183000.$$

All that remains is to sum the three products

$$
\begin{array}{ccccccc}
 & & & \not{3} & \not{2} & \not{3} & \\
 & & & 3 & 0 & 5 \\
 & & \times & 6 & 4 & 7 \\
 \hline
 & & 2 & 1 & 3 & 5 \\
 & 1 & 2 & 2 & 0 & \\
+ & 1 & 8 & 3 & 0 & \\
 \hline
1 & 9 & 7 & 3 & 3 & 5
\end{array}
$$

Using the Commutative Law to reverse the order of multiplication would produce a simpler, two-step, process as we can see below:

$$
\begin{array}{ccccccc}
 & & \not{1} & \not{2} & & & \\
 & & \not{3} & \not{2} & \not{3} & \\
 & & 6 & 4 & 7 & \\
 & \times & \boxed{3} & 0 & 5 & \\
 \hline
 & & 3 & 2 & 3 & 5 \\
 & & 0 & 0 & 0 & \\
+ & 1 & 9 & 4 & 1 & \\
 \hline
1 & 9 & 7 & 3 & 3 & 5
\end{array}
$$

Observe, we had to open a second carry row above the first because the position that we wanted to use for the 2 from $21 = 7 \times 3$ had already been used. While opening additional rows adds a bit of complexity, it is simply a matter of positioning the carried digit in the correct column and making sure it is crossed out as it is used. The overall computation is simplified by the second product being 0.

The procedure described in these examples is what was taught in schools when I was a child. Given its simplicity, the reader may well wonder why we spent so much time on place value and products with 10? The answer is that if you are going to be able to help your child, you need to know how things work at a fundamental level. If you understand the previous material, you know why this procedure works, and you know why it is really nothing more than a high-powered form of counting. Finally, this level of understanding will make helping your child with decimals easy.

11.6 What Your Child Needs to Know

11.6.1 Multiplication Goals for Grade 2

Multiplication is formally introduced in Grade 2. The focus is on studying addition of equal groups of objects leading to the definition of multiplication. In this respect, a child should be able to:

1. determine whether a group of objects having ≤ 20 members has an odd or an even number of members by skip counting by $2\,$s;

2. skip count by $2\,$s, $5\,$s and $10\,$s to $100\,$;

3. use addition to find the total number of objects in a rectangular array (grid) having ≤ 5 rows and/or columns;

4. be able to represent the total number of grid-squares in a rectangular grid as a product;

5. know from the definition of multiplication that

$$10 = 10 \times 1 = 1+1+1+1+1+1+1+1+1+1;$$

6. be able to represent a two-digit numeral as a sum using products of 1 and $10\,$, e.g.,

$$38 = 3 \times 10 + 8 \times 1$$

and relate this representation to previous representations of two-digit numbers.

11.6.2 Goals for Grade 3

Multiplication is a key focus of Grade 3 in an A+ curriculum. At the highest level, it is expected children will understand: that multiplication of counting numbers is repetitive addition; when multiplication is required to solve problems; and be able to use multiplication in problem solving. We give a simple example:

Bill is the team manager for a hockey team having 20 members. Each team member has to have 3 hockey sticks. How many sticks, total, are required to fulfill this requirement?

The relationship of multiplication and division should be known — division will be discussed in much greater detail in Chapter 12. We think of division only in respect to whole numbers, and only as a process for dividing one counting number into equal parts as specified by another counting number. Thus, 24 is divisible by 4 because 24 can be obtained by adding 6 to itself four times as can be seen by skip-counting by $6\,$s starting at $0\,$.

Multiplication.

On completion of Grade 3 your child should be able to:

1. explain that the meaning of a product, e.g., 5×3, is the sum of 3 added to itself 5 times, or the cardinal number of 5 groups of 3 objects combined in a single collection, or the total number of squares in a 5 by 3 grid;

2. recognize that each row of the Multiplication Table is obtained by skip-counting by the digit heading the row and starting from 0 as in

$$0, \quad 7, \quad 14, \quad 21, \quad 28, \ldots$$

which reproduces the row having 7 as its header;

3. recognize that a product can be partitioned by its factors, e.g., $24 = 3 \times 8$, so a group of 24 objects can be subdivided into 3 groups of 8 and also into 8 groups of 3 and that these groupings can be represented in the form of grids;

4. use multiplication within 100 to solve problems involving equal groups and arrays;

5. demonstrate fluent knowledge of the multiplication table sufficient to solve equations like those shown above;

6. be able to solve problems that require different operations, e.g., If Cindy's class has four girls and five boys and she has 27 pencils, how many will be left if she gives three pencils to each child?;

7. multiply single digit numbers by any single digit multiple of 10, e.g., $8 \times 30 = 240$;

8. be able to correctly execute the standard algorithm for multiplication for two three-digit numbers;

9. fluidly execute the standard algorithm to multiply two two-digit numbers.

Algebra.

An A+ curriculum introduces children to algebra and its symbols in Grade 3. The key symbol is the equality symbol $=$ and its meaning: **is the same as**. By the end of Grade 3 children should be able to correctly use $=$ in formulae and formulate simple equations. As well, your child will:

162

1. be able to express the Commutative, Associative and Distributive properties in both words and symbols using counting numbers;

2. understand the Associative and Commutative laws for multiplication and correctly apply them in the context of equations, e.g., $8 \times 13 = 13 \times 8$;

3. know that 1 acts as the multiplicative identity, that is, that $1 \times n = n \times 1 = n$ for any n.

4. understand the relationship between multiplication and addition, in particular the Distributive Law;

    (a) use parentheses and/or order of precedence where necessary to specify the order of computations;

    (b) understand how the Distributive Law makes multiplication into repetitive addition via equations like $2 \times 7 = (1 + 1) \times 7$;

    (c) understand that by knowing $4 \times 3 = 12$ and $4 \times 5 = 20$, one can obtain $4 \times 8 = 32$ from the Distributive Law.

    (d) understand how the Distributive Laws turns any multiple of an even number into the sum of two equal addends, e.g., $17 \times 6 = 17 \times 3 + 17 \times 3$;

5. know the properties of equality identified in §8.4 and be able to state these in both words and symbols for numerical quantities, as in: $4 + 3$ is the same as 7, so 7 is the same as $4 + 3$, or $4 + 3 = 7$ so $7 = 4 + 3$, and so forth;

6. solve equations like: $3 + 7 = \boxed{?}$, $\boxed{?} + 6 = 14$, $5 = \boxed{?} - 6$, and $25 - \boxed{?} = 17$ by inspection and/or recall;

7. solve equations like: $3 \times 7 = \boxed{?}$, $\boxed{?} \times 6 = 24$, $5 = \boxed{?} \div 3$, and $5 \times \boxed{?} = 80$ by inspection and/or recall;

8. solve two-step word problems involving the three operations, $+$, $\times$ and $-$;

9. be able to formulate an equation that represents problems like the Cindy problem (see above): the number of pencils left over is given by

$$27 - 3 \times (4 + 5) = \boxed{?};$$

10. correctly represent two-step word problems using an equation with the unknown represented by a letter as in: $3n + 7 = 28$ to represent "the product of an unknown and 3 added to 7 is 28";

11. correctly use formulae like $A = \mathcal{L} \times \mathcal{W}$ to solve problems (see below);

12. understand the role of multiplication in the Arabic System of numeration:

    (a) know the multiples of $10 \leq 1,000,000$, e.g.,

    $$10,000 = 10 \times 1,000 = 10 \times (10 \times 100) = 10 \times ((10 \times (10 \times 10))$$

    (b) be able to represent the value of any digit in a six-digit numeral as the product of a single digit and a multiple of 10 as in: the value of 7 in 4795 is 7×100;

    (c) know that the value of any digit in a numeral is ten times the value of that same digit when it occurs one place to the right, e.g., $800 = 80 \times 10 = 10 \times 80$;

    (d) recognize that 4795 is the following sum:

    $$4795 = 4 \times 1000 + 7 \times 100 + 9 \times 10 + 5;$$

Geometry.

Understand and use the following geometric ideas:

1. length as an attribute of a line;

2. the requirement that length must be expressed in units of length;

3. be able to make a simple measurement using a ruler and correctly record the result;

4. recognize area as an attribute of plane figures and understand concepts of area measurement;

    (a) a square with side length 1 unit, called a unit square, is said to have one square unit of area, and can be used to measure area;

    (b) a plane figure which can be covered without gaps or overlaps by n unit squares is said to have an area of n square units;

    (c) measure areas by counting unit squares (square cm, square m, square in, square ft, and improvised units);

5. relate area to the operations of multiplication, addition and counting;

(a) find the area of a rectangle with whole-number side lengths by tiling it, and show that the area is the same as would be found by multiplying the side lengths;

(b) multiply side lengths to find areas of rectangles with whole-number side lengths in the context of solving real world and mathematical problems, and represent whole-number products as rectangular areas in mathematical reasoning.

11.6.3 Goals for Grade 4

Multiplication.

By the end of Grade 4 your child should:

1. be able to interpret a multiplication equation as a **comparison**, e.g., $3 \times 7 = 21$, means 21 is three times the value of seven;

2. be able to solve multi-step word problems involving multiplication and division of whole numbers;

3. be able to find all factor pairs for whole numbers in the range 1 to 100;

4. understand the meaning of terms like: **multiple**, **prime**, **composite**, and **factor** as applied to whole numbers (discussed in a later chapter);

5. list multiples of one digit numbers as in: $4, 8, 12, 16, 20, 24, \ldots$;

6. understand that in the Arabic System the value of any digit in a numeral is ten times the value of that same digit when it occurs one place to the right, e.g., $800 = 80 \times 10$;

7. fluidly multiply four digit numbers by any of the ten single digit numbers;

8. use the standard place value method (see Example 5 in §11.5) to fluidly multiply pairs of two-digit numbers.

9. use the standard place value method to multiply a four-digit number by a two-digit number.

Algebra.

By the end of Grade 4 a child should:

1. be able to express (in equation form) and use the algebraic laws listed below that govern numbers and operations:

    (a) the Additive Identity Law: $0 + n = n + 0 = n$;

    (b) the Commutative and Associative Laws for multiplication;

    (c) the Distributive Law relating addition and multiplication;

    (d) the Multiplicative Identity Law: $1 \times n = n \times 1 = n$;

2. solve multi-step word problems posed with whole numbers and having whole-number answers using the four operations, including problems in which remainders must be interpreted;

3. represent word problems using equations with a letter standing for the unknown quantity.

11.6.4 Multiplication Goals for Grade 5

In the A+ curriculum that we recommend it is expected that all children will be able to fluidly use the four operations of arithmetic on counting numbers and common fractions by the end of Grade 4. Indeed, the study of counting numbers and operations thereon is clearly winding down in Grade 4 and some A+ curricula have completed this material by the end of Grade 3.

Algebra.

By the end of Grade 5 it is expected your child will:

1. show a complete understanding of the role of parentheses, brackets and braces in numerical expressions and correctly evaluate expressions that use these symbols;

2. demonstrate a complete understanding of the place value system and its role in supporting arithmetic procedures;

3. be able to write and interpret arithmetic expressions in language, e.g., $9+(4\times3)$ means *first multiply* 4 *by* 3, *then add* 9; or *add* 8 *and* 5 *and then multiply by* 7 is written arithmetically as $(8+5) \times 7$;

166

4. write simple expressions that record calculations with numbers, and interpret numerical expressions without performing the indicated operations;

5. recognize that $3 \times (18932 + 921)$ is three times as large as $18932 + 921$, without having to calculate the indicated sum or product.

Geometry.

By the end of Grade 5 your child should:

1. recognize **volume** as an attribute of solid figures and understand concepts of volume measurement;

    (a) a cube with side length 1 unit, called a unit cube, is said to have one cubic unit of volume, and can be used to measure volume;

    (b) a solid figure which can be packed without gaps or overlaps using n unit cubes is said to have a volume of n cubic units;

2. measure volumes by counting unit cubes, using cubic cm, cubic in, cubic ft, and improvised units;

3. relate volume to the operations of multiplication and addition and solve real world and mathematical problems involving volume;

    (a) find the volume of a right rectangular prism with whole-number side lengths by packing it with unit cubes, and show that the volume is the same as would be found by multiplying the edge lengths, equivalently by multiplying the height by the area of the base;

    (b) represent threefold whole-number products as volumes, e.g., to represent the associative property of multiplication;

    (c) apply the formulas $V = l \times w \times h$ and $V = b \times h$ for rectangular prisms to find volumes of right rectangular prisms with whole-number edge lengths in the context of solving real world and mathematical problems;

    (d) recognize volume as additive, e.g., find volumes of solid figures composed of two non-overlapping right rectangular prisms by adding the volumes of the non-overlapping parts, applying this technique to solve real world problems.

A+ curricula expend significant time on data. In the primary grades the focus is on making measurements of length. It is critical that when a child learns about length, the child understands that every such measurement involves a specific **unit**

of measure like **inches** or **centimeters** and a standard measuring device would be used. Coming to terms with measurements of length and the units in which they are measured is a prerequisite to understanding the concept of **area**. It is a hands-on task that is accomplished by experiment. One outcome from such experiments is that children recognize that different units of length produce different numerical measures for the same object in the world; for example, the length of the same table expressed in centimeters results in a larger number than its length expressed in inches.

Chapter 12

Division

The last operation on counting numbers introduced in primary and elementary grades is division. Although we have not done so, division is introduced simultaneously with multiplication. The reason for this simultaneous introduction is because in the same way there is a relationship between addition and subtraction, there is also a fundamental relationship between multiplication and division.

12.1   Division as a Concrete Process

The basic idea that children should hold about division is that it is a process for partitioning a fixed number of objects into groups, each of which contains the **same number** of objects. For example:

> There are 18 candy bars in a box. If 6 children are at a birthday party, how many bars go in each loot bag so every child gets the same number?

In this example, the fixed number is 18 and there are 6 children (groups) each of which has to receive the same number of candy bars. The result, as we know, is that each loot bag will contain 3 candy bars.

Now consider the same problem from a different perspective:

> Mom put 3 candy bars in each of 6 loot bags. How many total bars did she use?

This time we know the solution process involves multiplication (repetitive addition of 3) and the total will be 18.

The two problems illustrate the relationship between division and multiplication. The first question requests a solution to:

$$6 \times \boxed{?} = 18;$$

169

the second, a solution to:
$$6 \times 3 = \boxed{?}.$$
Children who know their multiplication table will know that
$$6 \times 3 = 18$$
and therefore that 3 is the required solution to the first question as a matter of recall. This is why we introduce multiplication first.

Ultimately, children should understand that because 18 is a multiple (see §12.3 below) of 6, $\boxed{?} = 18 \div 6$ produces 6 groups of **equal size** and that 3 is the number of items that is placed in any group.

In addition, they are expected to understand the use of remainders in concrete settings:

> There are 30 candy bars in a box. If 8 children are at a birthday party, what is the maximum number of bars that can be put in each loot bag so every child gets the same number? How many bars, if any, are left for pop to have for desert?

Again, this question relates directly to multiplication and specifically the process of skip-counting we used in the last chapter.

12.2 Division < 100

Consider the following equation:
$$m \times \boxed{?} = n$$
subject to the following restrictions: m is a single-digit counting number and $n < 100$. This equation has a solution exactly if m can be found in the row headed by n in the Multiplication Table (reproduced below).

$\times$	0	1	2	3	4	5	6	7	8	9
0	0	0	0	0	0	0	0	0	0	0
1	0	1	2	3	4	5	6	7	8	9
2	0	2	4	6	8	10	12	14	16	18
3	0	3	6	9	12	15	18	21	24	27
4	0	4	8	12	16	20	24	28	32	36
5	0	5	10	15	20	25	30	35	40	45
6	0	6	12	18	24	30	36	42	48	54
7	0	7	14	21	28	35	42	49	56	63
8	0	8	16	24	32	40	48	56	64	72
9	0	9	18	27	36	45	54	63	72	81

For example, suppose we want to find the solution to

$$7 \times \boxed{?} = 56.$$

Go to the row headed by 7 in the table and proceed across that row until you get to 56. The column header is 8 which tells you that

$$7 \times 8 = 56$$

so the solution to the equation is 8. In this case we would say that

when 56 is divided by 7, the result is 8,

and write

$$56 \div 7 = 8.$$

Using the usual set-up for division, we write:

$$
\begin{array}{r}
8 \\
\hline
7 \mid 5\ 6
\end{array}
$$

Setting up a division problem in this manner incorporates the structure of the the Multiplication Table. Notice that the 7 is the row-header at the left, the 56 is a cell entry in the row headed by 7, and if you go up the column in which you found 56, the column-header is 8.

Recall what we know about rows in the Multiplication Table. Each row results from skip-counting by the row-header. What we also know is that unless the row-header is 1, skip-counting does exactly that: it skips. So, for example, skip-counting by 7s to 56 does not list all the counting numbers less than or equal to 56. Thus, for example, $7 + 7 = 14$ so that

$$7 \to 14$$

and the numbers $8, 9, 10, 11, 12, 13$ are all skipped in the process of skip-counting by 7s to 56.

Children need to understand that when we say that n **is divided by** m using counting numbers n and m, and write $n \div m$, we mean that the equation

$$m \times \boxed{?} = n$$

has a whole number solution, which means in turn that a collection having n members could be divided into m groups each of which has the same number of members.

The simultaneous requirements that the m groups all contain the same number of members and that the total number of members in all the groups be n means that the equation

$$m \times \boxed{?} = n$$

does not, in general, have a counting number solution. For example,

$$4 \times \boxed{?} = 15$$

does not have a solution in the counting numbers because there is no way to distribute 15 items into 4 groups and have each group contain the same number of items. A child can verify this by experiment.

12.3   Multiples and Factors

In trying to find a counting number solution to the equation

$$m \times \boxed{?} = n$$

we are asking two questions:

Is n **a multiple of** m?

Is m **a factor of** n?

For any pair of counting numbers n and m, the answer to both questions is always the same, either both yes, or both no. But the focus of each question is quite different.

12.3.1   Multiples

The notion of **multiple of** is defined for any counting number, m. Multiples of m are the numbers:

$$m \times 1, \quad m \times 2, \quad m \times 3, \quad m \times 4, \quad m \times 5, \quad m \times 6, \quad m \times 7, \quad m \times 8, \quad \ldots$$

Thus, for example, the multiples of 6 are:

$$6, \quad 12, \quad 18 \quad 24, \quad 30 \quad 36, \quad 42 \quad 48, \quad \ldots$$

and so forth. In this context the phrase *and so forth* indicates the list of multiples is unlimited; it goes on forever. Moreover, each successive number in the list can be obtained by adding the same number to its predecessor. In the above list we simply add 6 to get the next number in the list, in other words, **skip-count** by 6.

Given a number on the list of multiples of 6, say 96, we know the equation

$$6 \times \boxed{?} = 96$$

has a counting number as a solution. Picking a number not on the list, for example, 103, means the equation

$$6 \times \boxed{?} = 103$$

has no solution in the counting numbers.

Thus, in our original numerical example $\boxed{?} \times 4 = 20$, we were able to find a solution precisely because 20 is a multiple of 4. On the other hand, 19 is not in the list of multiples of 4 and so the equation $\boxed{?} \times 4 = 19$ does not have a solution in the counting numbers.

12.3.2 Factors and Divisors

The notion of **divisor of** is also defined for any counting number n. For example, consider $n = 24$. We know that

$$1, \quad 2, \quad 3, \quad 4, \quad 6, \quad 8, \quad 12, \quad \text{and} \quad 24$$

are all divisors, or **factors**, of 24. Notice that this list is short and includes **all** the factors of 24. Thus, when we are asking about factors, we are asking for a very short list. When we are asking about multiples, we are asking for an impossibly long list.

Consider again the equation

$$m \times \boxed{?} = n$$

where n and m are given.

When we focus on the product on the LHS of this equation, we are asking: Is n a **multiple of** m?

When we focus on the RHS of this equation, we are asking: Is m a **factor** or **divisor of** n?

We consider one last numerical example. Let's think of 5 as a divisor, and ask for which values of n, n a counting number, will the equation

$$5 \times \boxed{?} = n$$

have a solution? We know the list of values of n admitting solutions is:

$$5, \ 10, \ 15, \ 20, \ 25, \ 30, \ 35, \ 40, \ 45, \ 50, \ 55, \ 60, \ 65, \ 70, \ 75, \ \ldots$$

This is exactly the list of multiples of 5 and it is exactly the list produced by **skip-counting** by $5\,$s.

Instead of focussing on the exact multiples of 5, consider numbers not in the list. Specifically, between each successive pair of multiples, there are four numbers. For example, between 55 and 60 are the four numbers, 56, 57, 58, and 59. Let's write these four numbers in terms of 55, which is the largest multiple of 5 less than each of the four. Thus,

$$56 = 55 + 1, \quad 57 = 55 + 2, \quad 58 = 55 + 3, \quad 59 = 55 + 4.$$

Or, consider the gap between 100 and 105. The numbers in this gap can be written as:

$$101 = 100 + 1, \quad 102 = 100 + 2, \quad 103 = 100 + 3, \quad 104 = 100 + 4,$$

where $100 = 5 \times 20$ and $105 = 5 \times 21$. Again, notice that 100 is the largest multiple of 5 that is less than each of the four numbers in the gap. From these two examples we see the gaps between the successive multiples of 5 are all the same. Each such gap contains four counting numbers. Each of these counting numbers is the sum of the largest multiple of 5 less than itself and a **residual** that is either 1, 2, 3 or 4. What's important to realize here is that every counting number that is not a multiple of 5 can be found in such a gap. For example, neither 382 nor 868 is a multiple of 5, so each must sit in a gap between multiples, and in consequence, each can be written as the sum of a multiple of 5 and a residual that is 1, 2, 3 or 4 as shown:

$$382 = 380 + 2, \quad \text{and} \quad 868 = 865 + 3.$$

Recalling the idea of skip-counting, note that if we start at 6 and skip-count by five, every number on the list generated by this process will have the form

$$m \times 5 + 1.$$

Skip-counting in this way produces the following list:

$$6, \ 11, \ 16, \ 21, \ 26, \ 31, \ 36, \ 41, \ 46, \ 51, \ 56, \ 61, \ 66, \ 71, \ 76, \ \ldots$$

You can try this yourself to see that any counting number that is not a multiple of 5 can be written as a multiple of 5 plus a residual counting number $r < 5$. Moreover, the number you pick will be found on a skip-counting list that starts at $5 + r$ and proceeds by adding fives.

To summarize what we have learned about 5 as a divisor, we now know that given any counting number n, there are whole numbers q and $r < 5$ such that

$$n = 5 \times q + r.$$

Notice that if $n < 5$ $q = 0$ and if n is an exact multiple of 5, $r = 0$.

174

12.4 Division of Counting Numbers

Clearly, there should be nothing special about 5. So we reconsider these ideas using an arbitrary counting number, d, as the divisor. Fix in your mind any counting number you want to be the divisor. Now ask: Can we skip-count by d? Of course we can. Doing so produces the following list:

$$0 \times d, \ 1 \times d, \ 2 \times d, \ 3 \times d, \ 4 \times d, \ 5 \times d, \ 6 \times d \ \ldots.$$

What we know is that every counting number that is a multiple of d appears in this list.

Now consider numbers in the gaps, that is the counting numbers, n, that occur between consecutive multiples of d:

$$q \times d < n < (q+1) \times d = d \times q + d.$$

If we consider the first gap, namely when $q = 0$, then we know the list of counting numbers in the gap is:

$$1, \ 2, \ 3, \ 4, \ 5, \ldots, \ d - 1.$$

All of these counting numbers satisfy the inequality:

$$0 < 1, \ 2, \ 3, \ 4, \ 5, \ldots, \ d - 1 < d.,$$

that is, they are all less than d. Further, if we add $q \times d$ to each of the numbers in the gap we would have:

$$q \times d < q \times d + 1, \ q \times d + 2, \ \ldots, \ q \times d + (d-1) < (q+1) \times d$$

where this is a complete list of the counting numbers in the gap between $q \times d$ and $(q+1) \times d$. Thus, n must be on this list, so n can be written as the sum of a multiple of d and a residual counting number that is less than d.

We can summarize this as follows:

> Given counting numbers n and d, we can find whole numbers and r such that $0 \leq r < d$ and $q \times d + r = n$.

The number d is called the **divisor**; it is what we are dividing by. The number n is called the **dividend**; it is the number being divided into. The whole number q is called the **quotient** and r is called the **remainder**. The quotient q turns out to be the **largest whole number** with the property that $q \times d \leq n$ so that $q \times d$ is the largest multiple of d that is less than or equal to n.

For example, suppose $n = 58$ and $d = 5$. Then q will be 11 and r will be 3, so that

$$5 \times 11 + 3 = 58$$

which is exactly what you would find by applying the algorithm you learned in school. We reiterate, 55 is the largest multiple of 5 that is ≤ 58.

Suppose we are given any pair of counting numbers, d and n. Why should such a q and r exist? The reason is straight forward. The list of counting numbers produced by skip-counting by d starting at 0 generates every counting number either as an exact multiple of d, or as a member of a gap. This is a direct consequence of the fundamental property that every counting number can be obtained using successor starting at 1, in other words, by the counting process.

The fact that quotients and remainders must exist does not help us find them. What is required is a method. We turn now to the algorithm for finding q and r that is taught to children.

12.4.1   The Division Algorithm: Long Division

The Division algorithm, commonly known as **long division** is a mechanical procedure that takes as input a dividend, n, and a divisor, d, and produces as output a quotient, q, and a remainder, r. We go over how this algorithm is implemented in practice. The reader should be aware that unlike our previous methods, **trial and error** will be a prominent feature of this algorithm. The reader should keep in mind the idea of skip-counting through multiples of d to find the largest multiple of d that is less than n. We will continually stress this in what follows.

Example 1

Our first example has $n = 785$ and $d = 8$. This example might appear on a student's problem sheet as: Find $785 \div 8$.

Students should understand that they have to find the largest multiple of 8 that is ≤ 785 and that the problem could be solved by skip-counting. However, skip counting is not efficient, so we find the multiple by using trial and error using **educated guesses**. This algorithm works **left to right**, as we will explain below.

To find this multiple, the problem is set up as shown:

$$8 \mid \overline{7 \quad 8 \quad 5}$$

Digits in the quotient will be written above the line, test multiplication data goes on lines underneath the 785.

The standard algorithm starts by asking: Does $d = 8$ divide 7 which is the leading digit in 785? The answer of course is: No, because $8 \nmid 7$. Equivalently, we ask: what is the largest multiple of 8 that is ≤ 7? Phrased this way, the answer is: 0.

But let's carefully analyze what this question means.

The lead digit in the divisor, 8, is a *ones* digit, whereas the lead digit in 785 is a *hundreds* digit. So, when we ask does 8 divide 7, we are really asking: Is there a positive multiple of 8 and 100 that is ≤ 785? The answer is no because the smallest multiple of 8 and 100 is 800 and $800 \nmid 785$. Thus, the answer to: Find the largest whole number[1] k such that:

$$8 \times (k \times 100) \leq 785$$

is 0, which tells us that the quotient must have 0 in the *hundreds* place. Since this digit is 0, we proceed to the next step.

The next step in the standard algorithm is to ask: Is there a positive multiple of 8 that is ≤ 78? Here the answer is yes; it is 9. Again the question can be rephrased to: What is the largest multiple of 8 that is ≤ 78? This time the answer is the same, namely, 9.

Again we analyze the result. Because the 8 in 78 is in the *tens* place, the question above is equivalent to asking: What is the largest single digit multiple of 8 and 10 that is ≤ 785? We find this multiple by trial and error; the largest multiple is 9 as shown by

$$8 \times (\boxed{9} \times 10) = 720 \leq 785 < 800 = 8 \times (10 \times 10),$$

so the issue now is: what to record, where? Following the standard algorithm, the 9 is recorded above the 8 in 78 as shown below.

$$
\begin{array}{r}
\boxed{9} \quad\ \\
\hline
8 \mid 7 \ \ 8 \ \ 5
\end{array}
$$

The analysis above tells us why the 9 is placed above the 8 in the dividend. The divisor is 8. The largest multiple ≤ 785 is

$$720 = 8 \times 90.$$

So the *tens* digit in the quotient is a 9. The quotient is recorded above the line and the *tens* place is determined by the *tens* place in the dividend as shown above.

The next step in the process is to record the 72, below the 78 in preparation for subtraction as shown:

---

[1]Recall that in Chapter 9 we defined the non-negative whole numbers as being the set of counting numbers together with 0. Since these are the only numbers available at this point, we will simply use the term whole numbers to describe this set.

$$
\begin{array}{r}
9 \\
\hline
\end{array}
$$

```
            9
          ─────────
  8  |   7   8   5
    −    7  [2]
         ──────
```

The 2 is placed in the *tens* column since, as discussed above, the 72 is actually 720.

The next step in the process is to subtract 72 from 78 using the standard method and record the answer below the line in the *tens* column. To see why we do this, recall our objective: find the largest multiple of 8 that is ≤ 785. What we have so far is the largest multiple of 8 and 10 that is ≤ 785. Only when we have performed the subtraction will we know whether this multiple also satisfies the criterion that it is the largest multiple of 8 that is ≤ 785.

```
            9
          ─────────
  8  |   7   8   5
    −    7   2
         ──────
            [6]
```

At this point, students must verify that the result of the subtraction, which we have put in a box for emphasis, is **strictly less than** the divisor, which is 8. If the divisor is less than or equal to the result of subtraction, the largest multiple of 8 has not been found.

Next, the student brings down the 5 directly in the *ones* column as shown:

```
            9
          ─────────
  8  |   7   8   5
    −    7   2
         ──────
            6  [5]
```

Students should understand that the 65 results from

$$785 - 720 = 65$$

and also that because $8 < 65$ we have not yet found the largest multiple of 8 that is ≤ 785. This means the next step is to find the largest multiple of 8 that is ≤ 65. From recall we have:

$$8 \times \boxed{8} = 64 < 65 < 8 \times (8+1) = 72.$$

Since the 5 in 65 is in the *ones* place, the student records the $\boxed{8}$ from $8 \times \boxed{8}$ in the *ones* column on the top line (next to the 9) and records the product 64 below the 65 as shown:

$$\begin{array}{r}
9 \quad \boxed{8} \\
8 \mid \overline{7 \quad 8 \quad 5} \\
- \quad 7 \quad 2 \\
\overline{\quad 6 \quad 5} \\
6 \quad 4
\end{array}$$

The last step is to compute the remainder $65 - 64$ and record the result as shown.

$$\begin{array}{r}
9 \quad 8 \\
8 \mid \overline{7 \quad 8 \quad 5} \\
- \quad 7 \quad 2 \\
\overline{\quad 6 \quad 5} \\
- \quad 6 \quad 4 \\
\overline{\quad 0 \quad \boxed{1}}
\end{array}$$

Again, children must check that the result of this subtraction is less than the divisor, that is, that

$$65 - 64 = 1 < 8.$$

Since the digit $\boxed{8}$ in the quotient is in the *ones* place, the process stops.

As students can now check, $98 \times 8 + 1 = 785$, and $r = 1 < 8$, so we have found $q = 98$ and r. Moreover, the steps just recreate the fact that

$$785 = 8 \times (9 \times 10) + 8 \times 8 + 1 = 8 \times (9 \times 10 + 8) + 1$$

where the RHS makes clear why each digit is in the place shown in the quotient.

This procedure will work for any pair of numbers. We perform one more computation, dividing a two digit number into a four-digit number, to illustrate the process. The instructions are intended to be suitable for students.

Example 2

Find q and r such that

$$27 \times q + r = 4316, \quad \text{and} \quad 0 \leq r < 27.$$

The actual problem might well be stated as $4316 \div 27$. Children should understand that the request is to find the largest multiple of the divisor, 27, that is less than or equal to the dividend, 4316, with the residual on subtraction being the remainder.

The set up is shown below:

$$27 \mid \overline{4 \quad 3 \quad 1 \quad 6}$$

Since the divisor, 27, has two digits and the dividend has four digits, the process starts by asking whether there is a multiple of 27 that is ≤ 43? The answer is yes since, $27 \times 1 < 43$. Also, 1 is the largest such multiple, since $43 < 2 \times 27 = 54$. We rewrite this as

$$27 \times (1 \times 100) < 4385,$$

which tells us that 1 is the *hundreds* digit in the quotient, q. So the 1 is placed directly above the 3 in 4385. The 27 is recorded below 43 making sure to place the 7 in the *hundreds* column directly below the 3 because as we know from the above discussion, the 27 is actually 2700. We show this below:

$$
\begin{array}{r}
1 \\
\hline
27\,|\ \ 4\ \ \boxed{3}\ \ 1\ \ 6 \\
2\ \ 7
\end{array}
$$

To complete the first step, we subtract $43 - 27 = 16$ and verify the result is less than the divisor, 27. Since it is, we record the result as shown.

$$
\begin{array}{r}
1 \\
\hline
27\,|\ \ \ 4\ \ 3\ \ 1\ \ 6 \\
-\ \ 2\ \ 7 \\
\hline
1\ \ 6
\end{array}
$$

Our next step is to determine the *tens* digit in the quotient. To begin the process, we bring down the 1 from the *tens* place (marked with a box), as shown:

$$
\begin{array}{r}
1 \\
\hline
27\,|\ \ \ 4\ \ 3\ \ \boxed{1}\ \ 6 \\
-\ \ 2\ \ 7 \\
\hline
1\ \ 6\ \ 1
\end{array}
$$

At this point we want the largest single digit multiple of 27 that is ≤ 161. We want to emphasize that asking the question this way is shorthand for the actual question: What is the largest single digit multiple of 27 and 10 that is $\leq 1616 = 4316 - 2700$? Finding the largest such multiple is trial and error. So, let's try 4. To do this, write 4 above the line in the the *tens* place which is the in same column as the 1 that was just brought down.

$$
\begin{array}{r}
1\ \ 4 \\
\hline
27\,|\ \ \ 4\ \ 3\ \ \boxed{1}\ \ 6 \\
-\ \ 2\ \ 7 \\
\hline
1\ \ 6\ \ 1
\end{array}
$$

Next we multiply 27×4 and record the result as shown. The setup actually facilitates the multiplication which is by a single digit, in this case 4.

```
          1   4
27 |  4   3  |1|  6
   -  2   7
      1   6   1
      1   0   8
```

The student performs the required subtraction as shown:

```
          1   4
27 |  4   3  |1|  6
   -  2   7
      1   6   1
   -  1   0   8
          5   3
```

The result, 53, is greater than 27, so the student should know the largest multiple of 27 that is ≤ 161 has not been found and tries a larger multiple, in this case 5. The result of 5×27 is recorded below

```
          1   5
27 |  4   3  |1|  6
   -  2   7
      1   6   1
      1   3   5
```

This time the subtraction produces $161 - 135 = 26$ which is less than the divisor 27:

```
          1   5
27 |  4   3   1   6
   -  2   7
      1   6   1
   -  1   3   5
          2   6
```

Thus the *tens* digit in the quotient is a 5. Note that if the student had tried 6 as the multiplier, the result would have been 162 which is larger than 161 and fails the condition that $q \times d \leq n$.

The last step is to find the *ones* digit in q. To begin this step, we bring down the 6 from the *ones* place as shown:

$$
\begin{array}{r}
\;1\;\;5\phantom{\;\;\boxed{6}}\\
\hline
27\,|\;\;4\;\;3\;\;1\;\;\boxed{6}\\
-\;\;2\;\;7\phantom{\;\;1\;\;\boxed{6}}\\
\hline
1\;\;6\;\;1\\
-\;\;1\;\;3\;\;5\\
\hline
2\;\;6\;\;6
\end{array}
$$

Again, the student is looking for the largest single digit multiple of 27 that is ≤ 266. Because the subtraction result in the previous step was so close to 27, a good guess for the required multiple would be 9. This results in:

$$
\begin{array}{r}
1\;\;5\;\;9\\
\hline
27\,|\;\;4\;\;3\;\;1\;\;\boxed{6}\\
-\;\;2\;\;7\\
\hline
1\;\;6\;\;1\\
-\;\;1\;\;3\;\;5\\
\hline
0\;\;2\;\;6\;\;6\\
-\;\;2\;\;4\;\;3\\
\hline
0\;\;2\;\;3
\end{array}
$$

The process stops because the 9 in the quotient, 159, is in the *ones* place directly above the 6. The remainder, 23, is less than the divisor, 27 and we now know that

$$27 \times 159 + 23 = 27 \times (1 \times 100 + 5 \times 10 + 9) + 23 = 4316.$$

12.5   What Your Child Needs to Know

The procedures for performing the arithmetic of whole numbers are complicated. Moreover the ability to carry them out accurately and with ease,[2] requires both knowledge at the level of recall and practice. A natural question is:

> Why is fluency with the procedures of arithmetic necessary in an age of calculators?

We take as given, that the reader accepts the necessity for their child to successfully complete Algebra II as a gateway to future success. So the question posed really becomes:

> Can a calculator replace fluency with the procedures of arithmetic?

------

[2]In curriculum guides this capacity is referred to as **fluency.**

Calculators have been available for computational support for at least 30 years. For the last 20 years of my teaching career, students had regular access to them throughout their schools years and their use was accepted by teachers. During the same time period, it appeared to me that students arriving for post-secondary training were less able to learn what we were trying to teach and had fundamental difficulties with basic arithmetic. Although my evidence is anecdotal, there is standardized test evidence consistent with my observations. For this reason, I conclude that replacing computational skill with the procedures of arithmetic by pushing buttons on a calculator is a major mistake. That is why I believe the fluidity requirement with the standard algorithms is essential.

12.5.1   Goals for Grade 3

By the end of Grade 3 it is expected your child will:

1. understand division of whole numbers as a solution to equations like $8 \times \boxed{?} = 48$, or equivalently, $48 \div 8 = \boxed{?}$;

2. understand division as a process for subdividing collections into equal groups, for example, $20 \div 5$ produces 5 groups of 4 from a collection of 20 objects;

3. understand division generally as a process for subdividing collections into a maximum number of equal groups and a remainder;

4. recognize where division should be applied to solve word problems: A school has 350 children to take on a field trip; how many buses that have 40 seats each are required to accommodate all the children and 7 teachers?

5. understand that division by 0 is impossible, and why this is so in terms of division arising out of multiplication, specifically, $\boxed{?} \times 0 = 0$ for all counting numbers, so equations like $\boxed{?} \times 0 = 18$ can not have a solution;

6. fluently multiply and divide within 100 by using relationship between multiplication and division, e.g., knowing that $8 \times 5 = 40$, one knows $40 \div 5 = 8$;

7. be able to find quotients and remainders using the standard procedure for problems involving four digit dividends and single digit divisors;

8. understand the role of place in the division procedure for single-digit divisors and why it works.

In respect to the above, by the middle of Grade 3, a child should immediately recall all the products of two one-digit numbers.

12.5.2 Goals for Grade 4

By the end of Grade 4 it is expected your child will:

1. be able to find all factor pairs for numbers ≤ 100, for example, 85 factors into 17 and 5;

2. be able to find quotients and remainders using the standard procedure for problems involving four digit dividends and two digit divisors;

3. understand the role of place in the division procedure for two-digit divisors and why it works;

4. solve multi-step word problems posed with whole numbers and having whole-number answers using the four operations, including problems in which remainders must be interpreted;

5. represent these problems using equations with a letter standing for the unknown quantity;

6. illustrate and explain multiplication and division calculations by using equations, rectangular arrays, and/or area models.

Chapter 13

Fractions: Topics for Grade 3

A+ curricula postpone work on fractions until at least Grade 3. Indeed, according to the Schmidt study (see §1.1) more than 30% of A+ countries do not start fractions until Grade 4. Contrast this with North America where curricula typically introduce the topic in Kindergarten. Our approach is to adhere to the A+ schedule.

In Chapters 13-15 we develop the arithmetic of **common fractions**, which as readers may recall are fractional forms having a whole number[1] in the numerator and a counting number in the denominator. The intent is that children in Grade 3 will gain an understanding of how fractions arise as numbers and the notation used to represent common fractions. In Grade 4, children will learn the rules for computing with common fractions. And in Grade 5 children will learn about decimal fractions and computing with decimals. The material is presented in a manner that will enable you to help your child succeed with fractions, even if fractions were a total mystery to you at the end of your school years.

The topic of fractions is considered the hardest in the arithmetic curriculum. It consists of two separate parts, namely,

1. What are fractions as numbers?

2. How do we perform calculations with fractions?

All children who have mastered the whole-number operations of arithmetic and can apply a formula should be able to master the computations with fractions with little difficulty.

In this chapter, we concentrate on what fractions are and the important properties that are associated with notation. The treatment emphasizes the ideas whose understanding is essential to understanding why the computations work the way they do.

---

[1]We remind readers that the whole numbers include 0 as well as the counting numbers.

13.1 Numbers That Are Not Whole

Let's think about numbers in terms of measurements of length. Children begin making such measurements as early as Grade 1. For our purposes here we want to think about a line as having a fixed starting point and extending in one direction in a manner similar to a tape measure. The critical fact that we state as a fundamental principle is that:

every identifiable place on this line is associated with a number.

This fact is illustrated in the following diagram:

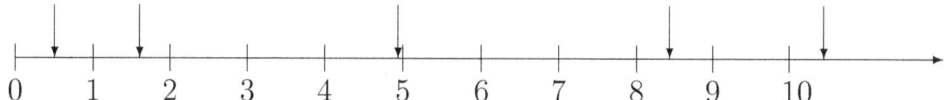

Vertical arrows that identify places on the line not associated with an exact unit of length. Each such place identifies a unique number.

That **every identifiable place on a line can be associated with a number** is a fact that your child must know. Making measurements of the length of real objects is a great way to study this idea. The website *softschools* has worksheets organized by grade level to help your child with this activity.

We can use places on the line to find numbers that are not whole numbers as follows. Let n be any counting number. We can create a number that is not a whole number by dividing one unit of length into n equal parts. Children are familiar with this idea in the context of dividing something into two parts, e.g., a candy bar.

13.2 Unit Fractions

Unit fractions are numbers that result by the process of dividing the number 1 into n equal parts. Unit fractions are the foundation on which the arithmetic of fractions is developed and are originally introduced using small values of n, for example 2 and 4.

To make this idea concrete, we need to start with a real-world example. Consider a portion of a line that corresponds to one unit of length. We refer to such an interval as: the **unit interval**, and view it as the portion of the number line having 0 at the left and 1, as shown:

Now to illustrate the meaning of unit fraction, we divide this interval into $n = 5$ parts of equal length.

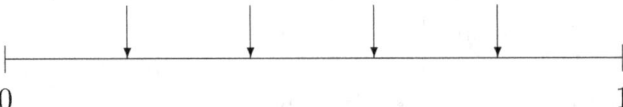

The vertical arrows subdivide the unit interval into five equal, non-overlapping parts. Each of the parts is identical to every other in respect to **length**.

The essential fact that each child must understand about this diagram is that the number associated with the **length** of each interval between the vertical arrows is the same. In respect to the attribute of length, **each of the five subintervals is identical to every other**.

We illustrate this idea again by subdividing the unit interval into $n = 8$ parts of equal length.

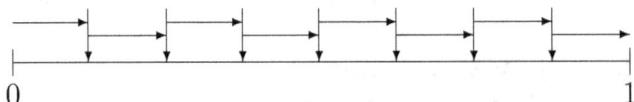

The vertical arrows subdivide the unit interval into eight equal, non-overlapping parts and each horizontal arrow has the same length.

The intrinsic idea expressed in these diagrams is simple: given one whole of something, we can divide that something into any counting number of equal parts. Since the parts all have the same size, **each one** of these parts can be described by the same number. We stress that the process of dividing results in parts that are **indistinguishable** in respect to some numerical attribute.

We illustrate this concept again using the numerical attribute of area. The box in the diagram has a unit area and we divide it into 16 equal parts, as shown.

A rectangle of unit area subdivided into 16 equal parts.

We stress that in respect to the numerical attribute of area, each part is identical to every other irrespective of position within the rectangle. They are indistinguishable.

13.2.1 Naming Unit Fractions

Recall our discovery that for numbers to be useful, they have to have names. Thus, for unit fractions to be useful, they must have names as well.

The fact that unit fractions arise by subdividing a unit whole into some counting number of parts must be captured by our notation. Our first two examples involved subdividing a unit interval into 5 and 8 parts, respectively, as shown simultaneously on the next diagram.

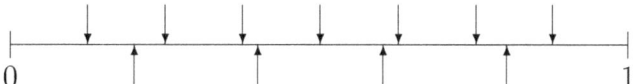

The down-pointing arrows subdivide the unit interval into eight equal, non-overlapping parts. The up-pointing arrows subdivide the unit interval into five equal, non-overlapping parts. In each case, the numerical attribute is length.

The length of each segment between adjacent up-pointing arrows is the same. For this reason, the number assigned as the length of each segment between adjacent up-pointing arrows must be the same. Similarly, the number assigned as the length of each segment between adjacent down-pointing arrows must also be the same.

The numbers used will not include the units of length. Rather, the numbers will be an abstraction in the same way that counting numbers are abstractions. But the names must not be arbitrary symbols, rather they must tell us about the process that created the segments and connect to our previous notation for counting numbers. For example, in the case of the up-pointing arrows, the line segment was divided into 5 equal parts and all this information should be built into the fractional notation.

The notational answer that achieves this objective is:

if we have one whole, and we divide that whole into 5 equal parts, we will use the combined symbols

$$\frac{1}{5}$$

as the notation for the number associated with the numerical measure of each of the identical parts resulting from this division process.

Thus, the notation consists of three parts:

- the 1 on top, called the **numerator**, tells us we are dealing with one whole entity;

- the 5 on the bottom, called the **denominator**, tells us how many equal parts this whole is divided into;

- the horizontal line tells us that the top is to be divided by the bottom.

Similarly, if we divide 1 whole into 8 equal parts, as we did with the unit length, we would use the combined symbols

$$\frac{1}{8}$$

where the 1 tells us that 1 is being subdivided and the 8 tells us that the 1 is divided into 8 equal parts.

Consider another rectangle having 1 unit of area. Suppose we subdivide the area into equal parts as shown:

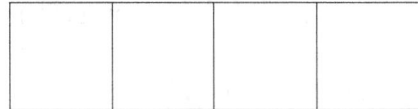

Find the number of parts by counting.

The unit of area determined by the rectangle has been divided into 4 equal parts. Therefore, the fractional part of the area in each of the 4 equal parts is assigned the numeral

$$\frac{1}{4}.$$

And in the case of the previous unit area diagram which was divided into $n = 16$ equal parts, we would use

$$\frac{1}{16}$$

as the numeral for any one of the fractional parts.

Further examples of the notation for such numbers are:

$$\frac{1}{2}, \ \frac{1}{3}, \ \frac{1}{7}, \ \frac{1}{17}, \ \frac{1}{25}, \ \frac{1}{132}, \ \frac{1}{845}, \ \frac{1}{9847}$$

and so forth. In general, for any **counting number** n, as in the numerical cases above, we will write

$$\frac{1}{n}$$

to signify the number that corresponds to the result of dividing 1 into n equal parts and refer generally to numbers of this type as **unit fractions**. Their distinguishing feature is a 1 in the numerator.

It is expected that children in Grade 3 will be fully conversant with the meaning and notation for unit fractions and be able to correctly use their names as in: *one fourth, one fifteenth* and *one one-hundred and twenty-fifth* to speak of $\frac{1}{4}$, $\frac{1}{15}$ and $\frac{1}{125}$, respectively.

Although it is not obvious why, the unit fraction concept is critical to all the development of arithmetic with fractions because each unit fraction can be counted in the same manner as buttons. In this way the arithmetic of fractions becomes just like the arithmetic of whole numbers all over again.

Again, and most importantly, children need to recognize that when a whole is divided into some counting number of equal parts, the resulting parts are **indistinguishable** in respect to the numerical attribute targeted by the division. For example, that attribute might be length, area, volume, weight, etc.

13.3   The Effect of Indistinguishable

Let's explore why it is important to recognize that the process of subdividing a whole into equal parts results in parts that are indistinguishable with respect to a target numerical attribute. Consider the following problem which is typical of many beginning work sheets on fractions:

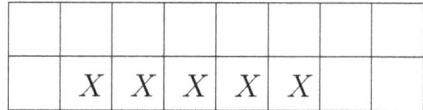

Write the fraction that represents the portion of the total area of the rectangle that is associated with squares marked with an X.

First, it is expected the child will recognize that the underlying attribute being considered is **area**. To solve the problem the child must determine the number of equal parts into which the whole is being divided. This number is found by counting and, as the reader can check, is 16. Counting also informs us that a total of 5 parts are marked with an X.[2] At this point the child can mechanically write an answer, namely, $\frac{5}{16}$, which is the standard notation for this common fraction.

Now consider the alternative problem shown below:

Write the fraction that represents the portion of the total area of the rectangle that is associated with squares marked with an Y. Understand

---

[2]Counting the squares marked with an X is just like counting buttons in a jar.

190

that this fraction must be the same as the fraction identified by squares marked with an X in the diagram above.

Children must understand that because each of the squares (parts) has the same area, the total area associated with 5 squares will be the same however these squares are chosen. So in the diagram above, the squares marked with Y have the same total area as the squares marked with X in the previous diagram. Thus, the numerical measure of the portion of the area associated with Y—squares is the same as for X—squares. Indeed, any way we choose 5 squares from the 16 will result in this same portion of the area, albeit from a different physical part. This is the essential consequence of indistinguishability.

But there is another step that is even more critical, namely that counting corresponds to addition. In this case we are counting unit fractions. So the total has to be given by the following sum:

$$\frac{1}{16} + \frac{1}{16} + \frac{1}{16} + \frac{1}{16} + \frac{1}{16}.$$

Let's consider a second example using a different numerical attribute, length. In the physical example of unit fractions given in the last section, we asserted that when we divide the unit interval into equal parts, each part will have the same length. In the diagram that follows the unit interval is divided into six equal parts, 3 of which are marked with horizontal arrows. We know that each of these horizontal arrows that starts at a vertical arrow and ends at the next vertical arrow to the right will have length $\frac{1}{6}$ when measured in the same units as the unit interval. Using counting as above, we also know that the fractional part of the length of the unit interval associated with the 3 horizontal arrows is $\frac{3}{6}$ in standard notation. This time we frame a different question:

> Is there any way to choose 3 of the subintervals that will produce a different fractional part of the total length?

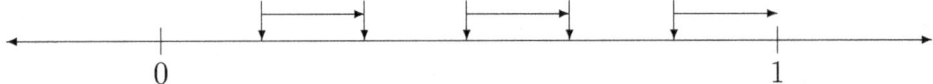

> The vertical arrows subdivide the unit interval into six equal, non-overlapping parts. Each horizontal arrow has length $\frac{1}{6}$.

The answer is: No. And the reason is that each of the subintervals is indistinguishable in respect to its length. Further, because segments of fixed length can be used to represent addition (see §10.3), the resulting fractional part of the total length must be the sum of the unit fractions measuring length of each subinterval identified by a horizontal arrow, namely,

$$\frac{1}{6} + \frac{1}{6} + \frac{1}{6}.$$

We will explore this idea in the next several sections.

13.4 The Fundamental Equation

Forming a unit fraction in the context of the real world is intended to divide a whole into some number of indistinguishable parts. When these parts are brought back together it should restore the whole. For example, if we divide a pie into two indistinguishable parts, when we bring those parts back together we should end up with a whole pie.

In respect to subdividing the interval into eight parts of equal length, we have the following diagram.

The vertical lines subdivide the unit interval into eight equal, non-overlapping parts.

Bringing the equal parts together in this context should amount to nothing more than removing the lines of subdivision. When we do this we end up with a diagram of the unit interval.

The vertical lines subdividing the unit interval into eight equal, non-overlapping parts have been removed restoring the unit interval.

We want to think about how to capture this idea with our arithmetic model. What we know is that when we bring all the subdivided parts of the whole back together, it restores the whole. This is an instance of **conservation**.

We can express this idea with an addition equation. For example, if we subdivide a pie into 2 equal parts, each described by the unit fraction $\frac{1}{2}$, the equation

$$1 = \frac{1}{2} + \frac{1}{2}$$

expresses the result of bringing those two parts back together.

If we think instead of the unit interval divided into 8 parts, the relevant equation would be

$$1 = \frac{1}{8} + \frac{1}{8} + \frac{1}{8} + \frac{1}{8} + \frac{1}{8} + \frac{1}{8} + \frac{1}{8} + \frac{1}{8}.$$

This equation is visually represented by the following diagram.

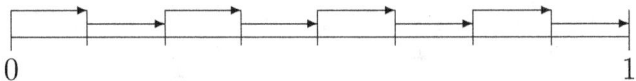

0   1

The vertical lines subdivide the unit interval into eight equal, non-overlapping parts. Each horizontal arrow has length $\frac{1}{8}$. A total of 8 arrows placed end-to-end and directed left-to-right beginning at 0 will terminate at 1. In the context of the last section, we could ask: What fraction of the length of the unit interval is associated with all subintervals identified by a horizontal arrow?

The last diagram is completely consistent with the interpretation of addition and the number line presented in §10.3. In answer to the question posed, we can say when all the subunits are identified as in the diagram above, we must have the whole and this should be evident to children who have experience with the sorts of worksheets on fractions based on counting that are typical of K-Grade 2.[3]

The deeper point comes by realizing that counting unit fractions having the same denominator corresponds to addition. Thus, given unit fractions, like $\frac{1}{2}$, $\frac{1}{3}$, $\frac{1}{4}$, $\frac{1}{5}$, and so forth, each satisfies an equation of the form

$$1 = \frac{1}{2} + \frac{1}{2}$$
$$1 = \frac{1}{3} + \frac{1}{3} + \frac{1}{3}$$
$$1 = \frac{1}{4} + \frac{1}{4} + \frac{1}{4} + \frac{1}{4}$$
$$1 = \frac{1}{5} + \frac{1}{5} + \frac{1}{5} + \frac{1}{5} + \frac{1}{5}$$

where the number of unit fractions contributing to the sum on the RHS is exactly the number given in the denominator of the fraction. **Every child should know and understand why these equations are true by the end of Grade 3.** This understanding should be evidenced by an ability to produce a diagram, like the one above for *eighths*, that corresponds to each equation.

---

[3] Although such sheets are available, the A+ curriculum does not start children working on them before Grade 3.

13.5 Repetitive Addition is Multiplication

In §11.1 multiplication of counting numbers was defined as repetitive addition. In §11.4.2 we discussed how this definition gives rise to the Distributive Law. It is critical that our readers remember that when we multiply any number by a counting number, the effect is adding that number to itself however many times are specified by the counting number. Thus,

$$4 \times x = x + x + x + x$$

where x stands for any number whatsoever. To see why this must be so, we apply the Distributive Law:

$$
\begin{aligned}
4 \times x &= (1 + 1 + 1 + 1) \times x \\
&= 1 \times x + 1 \times x + 1 \times x + 1 \times x \\
&= x + x + x + x,
\end{aligned}
$$

since $1 \times x = x$. Thus, $4 \times x$ is simply x added to itself 4 times. But this equation can be read the other way. It also says that

$$x + x + x + x = 4 \times x,$$

that is, whenever we add the same number to itself four times, we get 4 times that number.

The sum on the RHS of an equation like

$$1 = \frac{1}{7} + \frac{1}{7} + \frac{1}{7} + \frac{1}{7} + \frac{1}{7} + \frac{1}{7} + \frac{1}{7}$$

is repetitive addition. The same number, in this case $\frac{1}{7}$ is being added to itself 7 times. So this equation translates to

$$1 = \frac{1}{7} + \frac{1}{7} + \frac{1}{7} + \frac{1}{7} + \frac{1}{7} + \frac{1}{7} + \frac{1}{7} = 7 \times \frac{1}{7}.$$

We have made the point previously that what enforces the fact that multiplication is repetitive addition is the Distributive Law.

Every child in Grade 3 should understand these computations in their numerical form, so we write out the equations involving $\frac{1}{3}$. Recall that $3 \equiv 1 + 1 + 1$[4] and

---

[4]The symbol $\equiv$ is read: is defined to be. So $3 = 1 + 1 + 1$ by definition.

$1 \times \frac{1}{3} = \frac{1}{3}$. Using these facts we have,

$$
\begin{aligned}
3 \times \frac{1}{3} &= (1 + 1 + 1) \times \frac{1}{3} \\
&= 1 \times \frac{1}{3} + 1 \times \frac{1}{3} + 1 \times \frac{1}{3} \\
&= \frac{1}{3} + \frac{1}{3} + \frac{1}{3} = 1
\end{aligned}
$$

The second line results from the Distributive Law, and it is essential that children understand that we do not need parentheses in this line because multiplication must be performed before addition. The last line is obtained because 1 is the multiplicative identity (see §11.4.4), and all children should know this fact. That the last sum must be 1 is conservation.

We express this fact about unit fractions and counting numbers from which they arise as the **Fundamental Equation**:

$$
1 = n \times \frac{1}{n} = \frac{1}{n} \times n,
$$

where n is any counting number. Any pair of numbers that has the property that their product is 1 are said to be **multiplicative inverses** for one another (see §A.6.1). This is an important concept. In the context of a counting number n and a unit fraction $\frac{1}{n}$, it is clear why this product is 1. The product at the far right is also 1 due to the Commutative Law and is included to emphasize that the product of a number and its multiplicative inverse computed in either order results in 1.

Lastly, we are clearly thinking of $\frac{1}{n}$ as arising by the process of division and would normally express this as $1 \div n$ (see §A.6.3). If we think in this manner, it suggests that $\frac{1}{n}$ is the notation we use for the **quotient** resulting from dividing the **dividend**, 1, by the **divisor**, n, and that the Fundamental Equation merely expresses the fact that

> **the product of the quotient times the divisor should be the dividend.**

13.5.1   Unit Fractions and Repetitive Addition in General

Let us fix our attention on a single unit fraction. Rather than take a specific numerical value for this fraction, we can simply identify it by its denominator which we take to be the counting number n. Suppose we have some number of these unit fractions and we want to add them up. As discussed in the previous section in respect to *sevenths* and *thirds*, adding the fraction $\frac{1}{n}$ to itself repetitively comes down to multiplication by a counting number. The number that we multiply by is simply the number of times $\frac{1}{n}$ is added to itself and could be any counting number whatsoever. The fact that repetitive addition is multiplication applies in this situation equally well. Specifically, for counting numbers from 2 to 7, using the fact that repetitive addition is multiplication gives us the following equations

$$2 \times \frac{1}{n} = \frac{1}{n} + \frac{1}{n}$$

$$3 \times \frac{1}{n} = \frac{1}{n} + \frac{1}{n} + \frac{1}{n}$$

$$4 \times \frac{1}{n} = \frac{1}{n} + \frac{1}{n} + \frac{1}{n} + \frac{1}{n}$$

$$5 \times \frac{1}{n} = \frac{1}{n} + \frac{1}{n} + \frac{1}{n} + \frac{1}{n} + \frac{1}{n}$$

$$6 \times \frac{1}{n} = \frac{1}{n} + \frac{1}{n} + \frac{1}{n} + \frac{1}{n} + \frac{1}{n} + \frac{1}{n}$$

$$7 \times \frac{1}{n} = \frac{1}{n} + \frac{1}{n} + \frac{1}{n} + \frac{1}{n} + \frac{1}{n} + \frac{1}{n} + \frac{1}{n}.$$

What is critical for children to absorb from this sequence of equations is that

> every sum of **unit fractions having the same denominator** can be written as a product and determining the multiplier is simply a matter of counting!

13.6 Counting and the Notation for Fractions

To really understand the effect of the equations in the last section, let's consider the typical worksheet problem discussed in §13.3 in relation to the notation for common fractions. Recall, a common fraction looks like:

$$\frac{m}{n}$$

where m is a whole number and n is a counting number. Thus,

$$\frac{4}{7}, \quad \frac{3}{5}, \quad \frac{2}{8}, \quad \frac{6}{15}, \quad \frac{8}{10}, \quad \frac{5}{24},$$

are all numerals for common fractions. Now the worksheet problem requests that a child assign a numeral for a common fraction to a diagram:

196

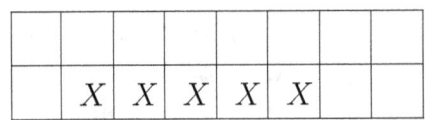

Write the common fraction that represents the portion of the total area of the rectangle that is associated with squares marked with an X.

The required numeral is obtained by simply counting the squares marked with an X to determine the numerator, counting the total number of squares to determine the denominator, and writing the result in standard form. So the child would write the result as $\frac{5}{16}$ as a direct result of counting, just like counting identical buttons. This process is a **mechanical** method to assign a numeral to a diagram. But the method tells us nothing about the arithmetic properties of the number the numeral represents, any more than the symbol 2 tells us anything about the number that it identifies. In the case of 2 it is only when we say that 2 names the successor of 1, or write the equation $2 = 1 + 1$ that we know what number is being named by 2.

13.6.1 Repetitive Addition and the Notation for Fractions

In the case of common fractions, it is only when we connect counting to addition at the next level of abstraction that the fraction $\frac{5}{16}$ begins to gain arithmetic meaning. The equation that provides this meaning is:

$$\frac{5}{16} = \frac{1}{16} + \frac{1}{16} + \frac{1}{16} + \frac{1}{16} + \frac{1}{16}$$

and informs us we should think about $\frac{5}{16}$ as a sum of unit fractions all of which have the same denominator, namely, 16. This idea is reinforced by saying the name: *five sixteenths* and thinking to oneself that we have 5 unit fractions each of which is $\frac{1}{16}$. Thus, the equation:

$$\frac{5}{16} = \frac{1}{16} + \frac{1}{16} + \frac{1}{16} + \frac{1}{16} + \frac{1}{16}$$

expresses an exact arithmetic relationship between the two numbers, $\frac{1}{16}$ and $\frac{5}{16}$ and this relationship is completely analogous to that between 1 and $2 = 1 + 1$. Both equations act as definitions for the notation on the LHS.

In a similar manner we conclude:

$$\frac{6}{15} = \frac{1}{15} + \frac{1}{15} + \frac{1}{15} + \frac{1}{15} + \frac{1}{15} + \frac{1}{15}$$

or the equivalent equation using the numeral for any other common fraction. Again, this and similar equations express a relationship between a particular unit fraction

197

and common fractions having the **same denominator**. But they do not tell us anything about the numerical relations between the numbers denoted by the various unit fractions. To get that information, we will have to look elsewhere.

13.6.2 Multiplication and the Notation Equation

There is an even higher level of abstraction that arises when we recognize repetitive addition as multiplication. With this recognition we can write what we refer to as the **Notation Equation**:

$$\frac{5}{16} = 5 \times \frac{1}{16}$$

which instructs us to think of $\frac{5}{16}$ as a product. Generally stated, the Notation Equation asserts:

$$\frac{m}{n} = m \times \frac{1}{n}$$

for any common fraction.

Every child should know the two equations in this section:

$$\frac{5}{16} = \frac{1}{16} + \frac{1}{16} + \frac{1}{16} + \frac{1}{16} + \frac{1}{16},$$

and

$$\frac{5}{16} = 5 \times \frac{1}{16}$$

and that they can be expressed simply as:

$$\frac{5}{16} = \frac{1}{16} + \frac{1}{16} + \frac{1}{16} + \frac{1}{16} + \frac{1}{16} = 5 \times \frac{1}{16}.$$

Knowledge is critical to the child's success with common fractions because this equation explains the relationship between unit fractions and common fractions having the same denominator. To come to terms with equations like these, the child must completely understand that

1. unit fractions represent distinct numbers determined by their denominator;

2. combining unit fractions having the same denominator is like counting identical buttons;

3. counting unit fractions having the same denominator amounts to repetitive addition of the unit fraction;

4. that because repetitive addition corresponds to multiplication, adding the unit fraction $\frac{1}{n}$ to itself m times is the same as multiplying the unit fraction by m.

One way to teach these ideas to young children is to get them to think of the notation $\frac{7}{8}$ in terms of a jar which contains seven items, each of which is the unit fraction $\frac{1}{8}$.

13.6.3   Language for Common Fractions

Common fractions can have any integer in the numerator. Since negative numbers are not introduced until Grade 6, most fractions discussed in elementary school are composed of two counting numbers as in

$$\frac{4}{7}, \quad \frac{3}{5}, \quad \frac{2}{8}, \quad \frac{6}{15}, \quad \frac{8}{10}, \quad \frac{5}{24},$$

and so forth. It is expected that children in Grade 3 will know that equations like

$$\frac{4}{5} = 4 \times \frac{1}{5}$$

are valid and know that it arises from viewing $\frac{4}{5}$ as resulting from

$$\frac{4}{5} = \frac{1}{5} + \frac{1}{5} + \frac{1}{5} + \frac{1}{5}.$$

Early in Grade 4 children should generally come to recognize every fractional form as a product as in

$$\frac{m}{n} = m \times \frac{1}{n}$$

because this recognition is critical to correctly performing and understanding computations. This recognition is so important that we give this equation a special name: **Notation Equation**. The reason we stress this equation's importance is that it is only when fractions are written as products that we can apply the full power of the Associative, Commutative and Distributive Laws to make computations with common fractions easy. So this idea needs to be carefully explained to children.

Referring to $\frac{1}{n}$ as the **reciprocal** of n is appropriate in Grade 3, but the introduction of negative exponents in the form n^{-1} has to wait until Grade 6.

By the end of Grade 3, children should be comfortable with correctly using all the standard language for common fractions. Thus, In a fraction, for example $\frac{m}{n}$, the number appearing below the line, in this case, n, is called the **denominator**. The number appearing above the line, in this case, a m, is called the **numerator**. Children should know that the terms denominator and numerator apply to all fractional forms, not merely common fractions. As already noted, given specific numerical values for n, we have standard names for unit fractions. Thus, we write $1/2$, and say, one half; we write $1/3$, and say, one third; we write $1/4$, and say, one fourth; we write $1/5$, and say, one fifth; skipping on, we write $1/10$, and say, one tenth; we write $1/27$, and say, one twenty-seventh; and so forth.

13.7 Common Fractions from Division

In §13.2 unit fractions were developed by dividing one unit whole into n equal parts. In §13.5 we described how to think about $\frac{1}{n}$ as the quotient resulting from $1 \div n$.

In §13.6, we obtained common fractions having numerators larger than 1 as sums of unit fractions and these considerations led to the Notation Equation:

$$\frac{m}{n} = m \times \frac{1}{n}.$$

Here, we want to think about obtaining the fraction $\frac{m}{n}$ from the division process: $m \div n$.

Since this process is conceptually very different from the Notation Equation approach, we start with a physical example using length.

Take a line segment 3 units long and divide it into two equal parts as shown:

The down pointing arrow divides the line segment into two equal parts.

Next ask the question: How many segments of length $\frac{1}{2}$ are required to divide the interval of length 3 into two equal parts? The following figure provides the answer.

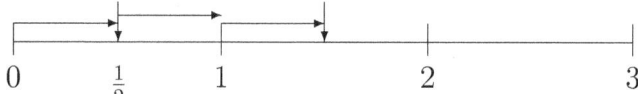

The down pointing arrow at the left identifies $\frac{1}{2}$ and each horizontal arrow has a length of $\frac{1}{2}$. Three of these arrows placed end-to-end identify the mid-point of the line segment having length 3.

The conclusion is that dividing a line segment having length 3 into two equal parts identifies the same point on the line as the fraction $\frac{3}{2}$, so that

$$3 \div 2 = \frac{3}{2} = 3 \times \frac{1}{2} = \frac{1}{2} + \frac{1}{2} + \frac{1}{2}.$$

In Appendix A we will deepen our understanding of this fact when we show that the division process is really just multiplication by a multiplicative inverse (see also §13.5).

13.8   Placing Proper Fractions in the Unit Interval

The term **proper** fraction applies to those common fractions in which the denominator is larger than the numerator. It turns out that the relations between common fractions in any other interval between consecutive whole numbers simply repeat the relations for proper fractions. For example, the common fractions in the interval between 4 and 5 have this property. So we want to determine the relationships that exist in the unit interval and the best place to start is to look at where the number associated with a given numeral for a proper fraction can be found.

Examining this problem has two parts:

1. Where are proper fractions having the same denominator found in the unit interval and what is their relationship to each other?

2. Where are unit fractions having different denominators found in the unit interval and what is their relationship to each other?

13.8.1   Proper Fractions with the Same Denominator

In §10.3 we described how to represent addition using line segments of various lengths placed end-to-end. We will use that representation together with the Notation Equation to determine how the proper fractions sit in the unit interval. Our unit interval has a 0 for its left end-point and a 1 for its right end-point, but otherwise, the units are arbitrary.

To make things concrete, we will work with the unit interval subdivided into five parts of equal length. We know that each of the subintervals has a length of $\frac{1}{5}$ and also that

$$1 = \frac{1}{5} + \frac{1}{5} + \frac{1}{5} + \frac{1}{5} + \frac{1}{5}.$$

Using the representation for addition in §10.3, this sum physically corresponds to placing five horizontal arrows end-to-end starting at 0 and terminating at 1. This last relation, between the arrows and the equation, forces us to identify as $\frac{1}{5}$ the number corresponding to a horizontal arrow starting at 0 and terminating at the first vertical arrow as shown in the following diagram.

The vertical arrows subdivide the unit interval into five equal, non-overlapping parts. The horizontal arrow identifies the position of $\frac{1}{5}$ as shown.

The relation between addition and length described in §10.3 asserts that the sum

$$\frac{1}{5} + \frac{1}{5}$$

corresponds to two horizontal arrows of length $\frac{1}{5}$ placed end-to-end and two arrows starting from 0 would clearly extend beyond the left-most vertical arrow in the diagram above. Thus length considerations alone force us to assign $\frac{1}{5}$ to the left-most vertical arrow.

Applying the Notation Equation to the last sum, we obtain:

$$\frac{1}{5} + \frac{1}{5} = 2 \times \frac{1}{5} = \frac{2}{5}.$$

Since addition can be represented on the line by placing arrows end-to-end starting at 0, we represent this sum as follows.

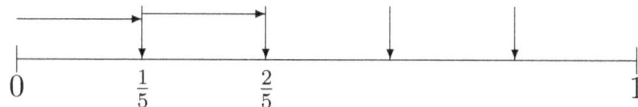

The horizontal arrows each have length $\frac{1}{5}$. As discussed in the text and §10.3, $\frac{2}{5} = 2 \times \frac{1}{5} = \frac{1}{5} + \frac{1}{5}$ must be the number associated with the second vertical arrow to the right of 0.

In a similar manner, we conclude that the remaining places on the line must identify the following numbers as shown below.

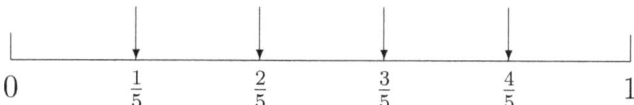

We have now found a number to go with each place identified by vertical arrows dividing the unit interval into 5 equal parts.

Most importantly, we understand that the determination of which number is associated with a particular place on the line is determined by the arithmetic relationships:

$$\frac{k}{5} = k \times \frac{1}{5} \quad \text{and} \quad 1 = 5 \times \frac{1}{5}.$$

There is nothing special about the unit fraction $\frac{1}{5}$. Given any fixed counting number, n, we know that it can be used to subdivide the unit interval into n non-overlapping subintervals and that the numbers marking the points separating the subdivisions will be labeled consecutively:

$$\frac{1}{n}, \ \frac{2}{n}, \ \frac{3}{n}, \ \frac{4}{n}, \ \frac{5}{n}, \ \frac{6}{n}, \ \ldots, \ \frac{n-1}{n}.$$

This instruction gives us all the relations between common fractions having the same denominator, namely:

$$0 < \frac{1}{n} < \frac{2}{n} < \frac{3}{n} < \frac{4}{n} < \frac{5}{n} < \frac{6}{n} < \ldots < \frac{n-1}{n} < \frac{n}{n} = 1.$$

13.8.2 Unit Fractions with Different Denominators

Answering the second question, where are the unit fractions and what is their relation to one another is reasonably straight forward. The first part of the answer is determined from dividing the unit interval into equal parts. Once this is accomplished, it is simply a matter of finding the point on the unit interval that is one fractional unit of distance to the right of 0. The process for finding that point was discussed above.

What is left is determining the relations between the unit fractions with different denominators. It is a fact learned by children early in life that given a fixed whole,

> **the more equal parts that whole is divided into, the smaller each of the parts will be.**

This fact applies to all objects in the world and hence to all numerical attributes, e.g., length. Thus, the line segment measured by $\frac{1}{4}$ will be shorter than the line segment measured by $\frac{1}{3}$, and so forth. We can easily represent these relations using a diagram:

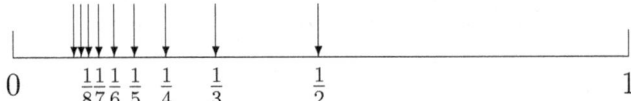

> The positions of the first ten unit fractions. As the denominators get larger the fractions get smaller and their positions closer together on the line. The fractions $\frac{1}{9}$ and $\frac{1}{10}$ are not labelled due to crowding.

As the reader can see, the unit fractions are positioned in order, each closer to 0 than its predecessor. The distance between successive unit fractions is not a constant and there is no simple arithmetic relationship between them the way there is between the fractions having a fixed denominator. It is this complexity that makes addition of fractions having different denominators complicated.

13.9 What Your Child Needs to Know

The material in this chapter is spread from Kindergarten to Grade 3 in the CCSS-M. In an A+ curriculum, it is not begun until Grade 3 so as to ensure that children master most of whole number arithmetic before being confronted by the fraction concept.

13.9.1 Goals for Grade 3

In respect to visual understanding of the unit fraction concept, children should be able to

1. partition circles and rectangles into two and/or four equal shares;[5]

2. correctly use the words *half, quarter, fourths* to describe the portions;

3. partition rectangles into rows and columns of equal size squares and count to find the total numbers;

4. partition rectangles into three equal parts by measuring and correctly use *thirds, third of;*

5. correctly use the phrases *half of, quarter of, fourth of* to describe the portions in respect to the whole;

6. understand that combining two halves (four quarters) physically restores the whole;

7. understand that dividing a whole into fourths, as opposed to halves, produces smaller shares;

8. divide rectangles into ≤ 16 equal parts and correctly use language to describe the resulting unit fraction.

In respect to numerical understanding of fractions:

1. understand the fraction $\frac{1}{n}$ partitions a whole into n equal parts, e.g., $\frac{1}{15}$ partitions a whole into fifteen equal parts called *fifteenths*;

2. understand that when something is divided into equal parts, all of the parts will have the same numerical measure;

3. understand that when the n equal parts partitioning a whole are brought back together, the whole is restored and recognize this as an instance of the principle expressed by the Fundamental Equation:

$$n \times \frac{1}{n} = \frac{1}{n} \times n = 1;$$

---

[5]The **math-aids** website has visual material to help children conceptually. The **softschools** website has interactive fraction games. The **superkids** website has worksheets at all levels.

4. understand the fraction $\frac{m}{n}$ as being the quantity obtained in either of two distinct ways:

(a) using unit fraction and the Notation Equation: by taking m parts of size $\frac{1}{n}$, as in:

$$\frac{3}{5} = 3 \times \frac{1}{5} = \frac{1}{5} + \frac{1}{5} + \frac{1}{5};$$

(b) using division by starting with a whole that has a numerical measure m greater than 1 that is then divided into n equal parts, e.g., when the interval from 0 to 3 is divided into 5 equal parts, each part has numerical measure $\frac{3}{5}$ (see §13.7);

(c) know, and be able to demonstrate with a simple example that both these processes yield the same result; for example, using a line segment 4 units long divided into 8 equal parts, show by counting that each part will have the same total length as summing 4 unit fraction pieces of length $\frac{1}{8}$;

In respect to measurement and fractions, children should be able to

1. measure the length of an object;[6]

(a) understand that the result depends on the units, e.g., centimeters vs inches vs feet vs meters;

(b) measure to determine how much longer one object is than another;

(c) use addition and subtraction to relate (compare) lengths of objects in common units;

2. understand the concept of a **unit of area** and partition geometric shapes into equal parts based on equal areas;

3. recognize that equal portions of identical wholes need not have the same shape; for example, in a 3 by 2 partition of a rectangle into squares, any three adjacent squares have the same share of the area, although not necessarily the same shape.

4. understand a fraction as a number on the number line;

5. represent the fraction $\frac{1}{n}$ by dividing a unit length into n equal parts;

6. represent fractions on a number line diagram;

---

[6]The **softschools** website has measurement worksheets by grade level. These can be helpful in learning about the real line, fractions and decimals.

(a) represent a fraction $\frac{1}{n}$ on a number line diagram by defining the interval from 0 to 1 as the whole and partitioning it into n equal parts;

(b) recognize that each part has size $\frac{1}{n}$, and that the right-hand endpoint of the segment having left-hand endpoint 0 locates the number $\frac{1}{n}$ on the number line;

(c) represent a fraction $\frac{m}{n}$ on a number line diagram by marking off m lengths $\frac{1}{n}$ from 0;

(d) recognize that the resulting interval has size $\frac{m}{n}$ and that its endpoint locates the number $\frac{m}{n}$ on the number line.

Chapter 14

Fractions for Grade 4

The last chapter focussed on the question:

What are fractions?

In this chapter our focus will be on how to compute with common fractions. Our focus will be on straightforward methods that are easily taught to children. In addition, these methods should **always work** and it should be possible for children to understand why they work.

The rules governing the arithmetic of common fractions are actually the same as the rules governing the arithmetic of structures that mathematicians refer to as **fields**. These structures are defined by a set of axioms and the rules governing computations are then derived and proved as theorems. The main ideas that drive all computations in arithmetic are the Associative, Commutative and Distributive Laws. These laws together with the Fundamental and Notation Equations produce the rules for fractions. Readers who want to see a mathematical treatment of the underlying structures are referred to the two Appendices on the integers and real numbers. In what follows, we try to limit the scope of discussion to what is relevant to teaching children. But we note that the deeper the understanding on the part of the teacher, the more it becomes possible to construct explanatory material for children.

14.1   The Key Equations

In Chapter 13 two essential equations were developed, the **Fundamental Equation**:

$$n \times \frac{1}{n} = \frac{1}{n} \times n = 1$$

and the **Notation Equation**:

$$\frac{m}{n} = m \times \frac{1}{n}.$$

The Fundamental Equation simply asserts that when the counting number n and the unit fraction $\frac{1}{n}$ are multiplied together, the product is 1. Because multiplication by a counting number is defined by repetitive addition, this fact can be restated as:

> when n unit fractions having the form $\frac{1}{n}$ are added together, the sum is 1.

Expressed as a sum, the Fundamental Equation is clearly about restoring the whole (see §13.4) and this is the understanding that children should bring with them from Grade 3.

The **Notation Equation** relates the fraction notation, $\frac{m}{n}$, to the arithmetic operations of multiplication and addition. Explanatory material for the Notation Equation in terms of counting unit fractions is given in §13.6. Again, this is the understanding that children should bring with them from Grade 3.

These two equations connect unit fractions to the abstract operations of arithmetic. If they are understood, they make computations with common fractions simple! Our development of the computational procedures will also use the Commutative, Associative and Distributive Laws which should be covered in numerical form in Grade 3.

To illustrate the use of these ideas, we begin with the problem of multiplying a whole number times a fraction. We note that children leaving Grade 3 should have become comfortable with numerical equations of the type appearing in the next section.

14.1.1   Applying the Notation Equation: Multiplying a Whole Number by a Common Fraction

We start with the simplest type of numerical example, a counting number, say 6, times a common fraction, say $\frac{2}{9}$.

The Notation Equation asserts that $\frac{2}{9} = 2 \times \frac{1}{9}$, so that

$$6 \times \frac{2}{9} = 6 \times \left(2 \times \frac{1}{9}\right).$$

Applying the Associative Law for multiplication to the RHS and performing the indicated multiplication gives:

$$6 \times \frac{2}{9} = (6 \times 2) \times \frac{1}{9} = 12 \times \frac{1}{9}.$$

The computation is completed by again applying the Notation Equation to the expression at the far right:

$$6 \times \frac{2}{9} = \frac{12}{9}.$$

This computation illustrates the importance of the Notation Equation as a tool for computation with fractions. While we have given a numerical example, it is clear that the computation results from the general rule:

$$m \times \frac{p}{q} = \frac{m \times p}{q}.$$

It is expected that children will be completely comfortable with the use of this computational rule starting in Grade 4 and that by the end of Grade 4 they will be comfortable with its expression using letters as opposed to numbers. It is also expected that children will understand each step in the process (see Goals for Grade 4).

14.2   Multiplication of Common Fractions

Suppose we want to multiply two common fractions $\frac{m}{n}$ and $\frac{p}{q}$, or, in numerical form $\frac{2}{5}$ and $\frac{3}{7}$. The Notation Equation tells us that each of these fractions is actually the product of a counting number and a unit fraction so that

$$\frac{m}{n} \times \frac{p}{q} = \left(m \times \frac{1}{n} \right) \times \left(p \times \frac{1}{q} \right).$$

The numerical example illustrates this fact:

$$\frac{2}{5} \times \frac{3}{7} = \left(2 \times \frac{1}{5} \right) \times \left(3 \times \frac{1}{7} \right).$$

The Commutative and Associative Laws for multiplication tell us we can write the four products on the RHS in any order we choose, so that:

$$\frac{m}{n} \times \frac{p}{q} = (m \times p) \times \left(\frac{1}{n} \times \frac{1}{q} \right)$$

or numerically,

$$\frac{2}{5} \times \frac{3}{7} = (2 \times 3) \times \left(\frac{1}{5} \times \frac{1}{7} \right).$$

Since m and p are counting numbers, we know how to form their product, in this case $2 \times 3 = 6$. What we **need to know** is how to compute the product of the two unit fractions which we explore next.

14.2.1 Multiplying Unit Fractions

For the counting numbers n and q, R12 (see Appendix B.9.1)[1] asserts

$$\frac{1}{n} \times \frac{1}{q} = \frac{1}{n \times q} \quad \text{or} \quad \frac{1}{5} \times \frac{1}{7} = \frac{1}{5 \times 7}.$$

So R12 tells us what we want to know, but not why it is so. To answer that question, we turn to the Fundamental Equation.

Consider the unit fractions $\frac{1}{5}$ and $\frac{1}{7}$. The Fundamental Equation asserts that

$$5 \times \frac{1}{5} = 1 \quad \text{and} \quad 7 \times \frac{1}{7} = 1.$$

The Fundamental Equation also tells us that

$$(5 \times 7) \times \left(\frac{1}{5 \times 7}\right) = 35 \times \frac{1}{35} = 1.$$

Now multiplying the first two instances of the Fundamental equation together produces

$$\left(5 \times \frac{1}{5}\right) \times \left(7 \times \frac{1}{7}\right) = 1 \times 1 = 1.$$

Applying the Associative and Commutative Laws to the LHS we can obtain:

$$(5 \times 7) \times \left(\frac{1}{5} \times \frac{1}{7}\right) = 1$$

whence by Transitivity of Equality

$$(5 \times 7) \times \left(\frac{1}{5 \times 7}\right) = 1 = (5 \times 7) \times \left(\frac{1}{5} \times \frac{1}{7}\right).$$

Since 5×7 is a counting number, hence not 0, we can apply the Cancellation Law, R17 of B.9.1, to obtain:

$$\frac{1}{5 \times 7} = \frac{1}{5} \times \frac{1}{7}.$$

The numerical essence of this is that

$$\cancel{35} \times \frac{1}{35} = \cancel{35} \times \left(\frac{1}{5} \times \frac{1}{7}\right)$$

---

[1]Appendix A presents the basic theory of the integers. Appendix B presents the theory for the field of real numbers and concludes with a list of 17 computational rules that all real numbers obey. These rules are used in the course of this chapter and referenced as R(number), e.g., R15.

whence

$$\frac{1}{35} = \frac{1}{5} \times \frac{1}{7}$$

by an application of the Cancellation Law for multiplication, R17 of B.9.1.

The computation can be repeated for any pair of unit fractions, and your child should become adept at applying this formula,

$$\frac{1}{n} \times \frac{1}{q} = \frac{1}{n \times q},$$

by the start of Grade 5.

14.2.2   Multiplying Common Fractions

The rule for multiplying the two common fractions, $\frac{m}{n}$ and $\frac{p}{q}$, is easily stated in words (see R13, Appendix B.9.1):

> **the product of two common fractions is the product of the numerators over the product of the denominators.**

To obtain this rule, we apply the Notation Equation to the two fractions forming the product:

$$\frac{m}{n} \times \frac{p}{q} = \left(m \times \frac{1}{n}\right) \times \left(p \times \frac{1}{q}\right).$$

The Commutative and Associative Laws for multiplication turn the RHS into:

$$\frac{m}{n} \times \frac{p}{q} = (m \times p) \times \left(\frac{1}{n} \times \frac{1}{q}\right).$$

Applying R12 to the RHS takes us a further step:

$$\frac{m}{n} \times \frac{p}{q} = (m \times p) \times \frac{1}{n \times q},$$

and one more application of the Notation Equation to the RHS gives:

$$\frac{m}{n} \times \frac{p}{q} = \frac{m \times p}{n \times q}.$$

Since m, n, p and q are all whole numbers, a child knows how to do the required multiplications on the RHS and that the results of these multiplications are whole numbers. Thus, finding the product of two fractions comes down to finding the product of two pairs of counting numbers (the product of the numerators and the product of the denominators). You may have been wondering why we demand that your child should know and understand the Notation Equation:

$$\frac{m}{n} = m \times \frac{1}{n}.$$

As you can see, the Notation Equation is an essential relation that enables computations with common fractions.

Let's review these steps for our numerical example: $\frac{2}{5} \times \frac{3}{7}$. The Notation Equation tells us:

$$\frac{2}{5} \times \frac{3}{7} = \left(2 \times \frac{1}{5}\right) \times \left(3 \times \frac{1}{7}\right).$$

The RHS involves only multiplications which can be manipulated using only the **Commutative and Associative Laws** to obtain

$$\left(2 \times \frac{1}{5}\right) \times \left(3 \times \frac{1}{7}\right) = (2 \times 3) \times \left(\frac{1}{5} \times \frac{1}{7}\right).$$

Applying R12 we have $\frac{1}{5} \times \frac{1}{7} = \frac{1}{5 \times 7} = \frac{1}{35}$ followed by the Notation Equation gives

$$\frac{2}{5} \times \frac{3}{7} = \frac{2 \times 3}{5 \times 7} = \frac{6}{35}$$

It is expected that children will understand all aspects of multiplication of fractions in Grade 4.

14.2.3 Concrete Meaning for Products of Unit Fractions

We have made much of the idea that multiplication was repetitive addition. Indeed, this was the definition of multiplication for counting numbers. However, the reader may be wondering, how can we interpret the computation

$$\frac{1}{4} \times \frac{1}{3} = \frac{1}{12}$$

as repetitive addition. If you have been thinking about this and find yourself stumped, there is a reason. There is no sensible way to interpret this product as repetitive addition because for this interpretation to be valid, one factor in the product must to be a whole number.

There is, however, an alternative interpretation of multiplication that is universally valid: it is **scaling**. Consider the following list of numbers

$$2, 10, 50.$$

Speaking in comparative terms, we would say that 10 is five times as big as two, but only one fifth as big as 50. We would also say that 2 is one fifth as big as 10 and one twenty-fifth as big as 50. The factor that we multiply one number by to obtain another number is often referred to as a **scale factor**. Thus, in the equation

$$2 = \frac{1}{25} \times 50,$$

we would say *the scale factor is one twenty-fifth*.

To get the idea of scaling in respect to products, think of a unit square. If we reduce the width by a scale factor of one third, we obtain a rectangle having an area that is $\frac{1}{3}$ as big as the original. On the other hand, if we reduce the height by a factor of one fourth, the resulting rectangle has an area $\frac{1}{4}$ as large as the original. Scaling the width by one third and the height by one fourth results in the rectangle on the far right which has its area scaled by $\frac{1}{12}$.

Concrete meaning is given to the product $\frac{1}{3} \times \frac{1}{4}$ by thinking of the product as successive acts of scaling. Since the area of the final figure, $\frac{1}{12}$ sq units, is independent of the orientation of the figure, the scaling can be done in either order without changing the final result as shown next.

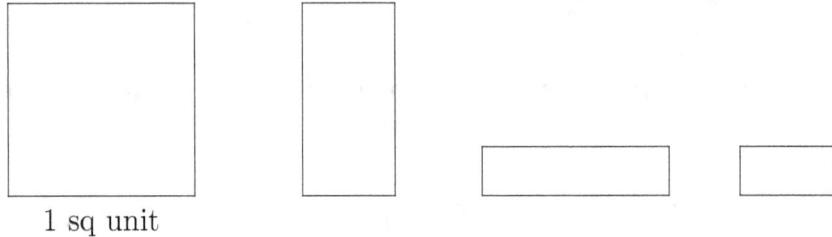

1 sq unit

> Start with a unit square. Scale the width by one third, or scale the height by one quarter to get the middle two rectangles. Scale by both to get the rectangle on right. Its area is scaled by one twelfth.

14.3 Adding Common Fractions

14.3.1 Adding Fractions with the Same Denominator

The Notation Equation,

$$\frac{m}{n} = m \times \frac{1}{n},$$

combined with the **Distributive Law** makes adding fractions with the same denominator easy. The straightforward rule for computation is (see R15, Appendix B.9.1):

To add fractions having the same denominator, put the sum of the numerators over the common denominator.

For example, suppose we want to add $\frac{9}{17}$ and $\frac{4}{17}$. The computation, which uses the Distributive Law to factor out $\frac{1}{17}$, in full detail looks like

$$\begin{aligned} \frac{9}{17} + \frac{4}{17} &= 9 \times \frac{1}{17} + 4 \times \frac{1}{17} = (9+4) \times \frac{1}{17} \\ &= \frac{9+4}{17} = \frac{13}{17}. \end{aligned}$$

We remind the reader that in the middle sum of the first line, the requirement to perform multiplication before addition avoids the use of parentheses.

R15 of B.9.1 asserts for the common fractions, $\frac{m}{n}$ and $\frac{k}{n}$ that

$$\frac{m}{n} + \frac{k}{n} = \frac{m+k}{n}$$

and arises through exactly the same sequence of manipulations starting with the Notation Equation applied to the LHS (see Appendix B).

It is expected that children in Grade 4 will be able to understand and apply the Notation Equation and the Distributive Law to follow the numerical sequence leading to the sum. Again, the Notation Equation is essential reason why this formula is valid.

Children need to know the rule. And they also should know the companion rule:

Do not try to add fractions by adding the numerators, unless the fractions have the same denominator!

It is always helpful to have easy ways to remember things. In my own mind, I think of these two rules as examples of the rule:

Always add apples to apples.

To see what I mean by *adding apples to apples*, if we want to add

$$\frac{4}{15} + \frac{7}{15},$$

we would think of adding 4 *fifteenths* to 7 *fifteenths*. Here we are regarding each *fifteenth* as an **apple**. So we are adding four apples to seven apples, and we end up with eleven apples or more precisely 11 *fifteenths*:

$$\frac{4}{15} + \frac{7}{15} = \frac{4+7}{15} = \frac{11}{15}.$$

214

Thought of in this way, we realize the computation merely amounts to counting up the total number of *fifteenths*. Thus, adding the numerators and leaving the denominator alone really is the correct procedure. Moreover, the actual computation, in this case $4 + 7$, is mere recall, provided the child knows the addition table.

On the other hand, if we want to add

$$\frac{3}{7} + \frac{2}{9},$$

we don't satisfy the *apples to apples* rule, since when we put this in words, we have to add 3 *sevenths* and 2 *ninths*. If the *sevenths* are apples, then the *ninths* have to be oranges, or fish, or whatever, but not apples.

14.3.2 Adding Fractions with Different Denominators

Suppose we want to add fractions having different denominators, as in:

$$\frac{3}{7} + \frac{2}{9}.$$

The companion rule tells us not to try to perform the computation unless the two fractions have the same denominator. The obvious question is:

Can we make the denominators the same?

When we ask this question what we really want to know is: Can we make the denominators the same without changing the problem in any arithmetic way?

There are two ways to change an arithmetic problem without affecting the answer: add 0, or multiply by 1. In this case, we multiply by 1, as follows:

$$\frac{3}{7} \times 1 + \frac{2}{9} \times 1.$$

As the reader can see, the problem is unchanged, but we are still no better off. This is where someone way back when got clever by observing that the Fundamental Equations tells us that

$$\frac{7}{7} = \frac{9}{9} = 1,$$

so that

$$\frac{3}{7} \times 1 = \frac{3}{7} \times \frac{9}{9} = \frac{27}{63} \quad \text{and} \quad \frac{2}{9} \times 1 = \frac{2}{9} \times \frac{7}{7} = \frac{14}{63}$$

215

where the resulting fractions now have the same denominator, called a **common denominator**. The **common denominator**, 63, is the product of the two starting denominators

$$9 \times 7 = 63.$$

The product of the two denominators can **always** be used as the common denominator in problems where the starting denominators are different. Children need to realize that each fraction is multiplied by:

$$\frac{\text{missing factor}}{\text{missing factor}}.$$

This is why $\frac{3}{7}$ is multiplied by $\frac{9}{9}$; similarly, $\frac{2}{9}$ is multiplied by $\frac{7}{7}$. Using the product, 63, we have

$$\frac{3}{7} + \frac{2}{9} = \frac{27}{63} + \frac{14}{63}$$

where the RHS now satisfies the **apples to apples** rule, where **apples** are *sixty thirds*. The computation is now straight forward using the rule for adding fractions with the same denominator:

$$\frac{3}{7} + \frac{2}{9} = \frac{27}{63} + \frac{14}{63} = \frac{27 + 14}{63} = \frac{41}{63}.$$

The example just given merely implements R16 of B.9.1 for finding the sum of two fractions.

To generate R16, we first apply the Fundamental Equation to multiply each of the fractions by the two forms of 1, $\frac{k}{k}$ and $\frac{n}{n}$, to obtain:

$$\frac{m}{n} + \frac{p}{k} = \left(\frac{m}{n} \times \frac{k}{k} \right) + \left(\frac{p}{k} \times \frac{n}{n} \right).$$

Applying the rule for multiplying fractions followed by the rule for adding fractions having the same denominator gives:

$$\frac{m}{n} + \frac{p}{k} = \frac{m \times k + p \times n}{n \times k}$$

(see Appendix B for details).

14.3.3 Adding Fractions: A General Rule for Children that Always Works

The content of the preceding material on addition can be turned into a three-step procedure which **always works**, even for algebraic fractions!

Given the problem:

$$\frac{m}{n} + \frac{p}{k} = ?$$

the three steps are:

Step 1: **Always** check whether the two denominators, n and k, are equal;

Step 2: If the two denominators are equal, that is, $n = k$, find $m + p$ and the sum is:

$$\frac{m}{n} + \frac{p}{n} = \frac{m+p}{n};$$

Step 3: If the denominators are not equal, that is, $n \neq k$, find the products $n \times k$, $m \times k$ and $p \times n$, then the sum is:

$$\frac{m}{n} + \frac{p}{k} = \frac{m \times k + p \times n}{n \times k}.$$

In doing examples requiring Step 3, one would want to remind children of the order of precedence rule that requires products to be computed before sums (see §11.4.2).

The key to success is that a child faced with a fraction addition problem always starts by asking:

Are the denominators the same?

The answer then sets the child on the right track computationally.

In explaining this procedure to children, one would start with numerical examples in which the answer to the initial question is: 'Yes, the denominator is same.', as in

$$\frac{2}{9} + \frac{4}{9} = \frac{2+4}{9}, \quad \frac{3}{19} + \frac{7}{19} = \frac{3+7}{19}, \quad \frac{11}{15} + \frac{4}{15} = \frac{11+4}{15}.$$

and so forth. After facility with this computation is achieved, one would proceed to examples for which the answer is: 'No, the denominators are not the same.', and that require the third step as in

$$\frac{2}{3} + \frac{4}{9} = \frac{18+12}{27}, \quad \frac{3}{7} + \frac{1}{3} = \frac{9+7}{21}, \quad \frac{4}{15} + \frac{1}{5} = \frac{20+15}{75}.$$

It is expected that children will be able to master calculations with common fractions using this rule in Grade 4.

If the procedures for addition and multiplication were limited to those above, fractions would not be the hardest topic in the elementary school curriculum. Much to the chagrin of many, there are topics yet to consider. However, if your child understands and can perform fluently the procedures described to this point, they have the knowledge to complete the task. The point is, first mastering these procedures at the level of recall is the key to future success because they apply to all quantities that are real numbers.

14.4 What Your Child Needs to Know

14.4.1 Goals for Grade 4

The intention is that on completion of Grade 4, children will have mastered the operations of multiplication and addition for common fractions. Because the procedure is simpler, it is recommended that multiplication be covered before addition. It is essential that children know both the procedures and why these procedures work. Division and subtraction are treated is advanced topics in §15.1-15.2.

As prerequisite knowledge for learning to compute with fractions, children need to know the basic properties of addition and multiplication and how to use them numerically. Specifically, they need to be comfortable with the associative, commutative and distributive properties and that 1 is the multiplicative identity. They should be able to express and interpret these laws in algebraic or numeric form:

$$2 + (7 + 4) = (2 + 7) + 4 \quad \text{is an example of} \quad n + (m + p) = (n + m) + p.$$

To understand why the numerical example is true, children must know how to substitute numerical values into the general formula on the right.

Children should

1. know and use the Notation Equation: for $m > 1$, the fraction $\frac{m}{n}$ is the sum of m unit fractions $\frac{1}{n}$;

2. understand that an alternative correct interpretation of a fraction is as division of the numerator by the denominator, i.e., $\frac{m}{n} = m \div n$;

   (a) use this interpretation to solve problems like: (A 20 kg sack of sugar is to be equally shared by 25 people. Express each person's share as a fraction.);

3. apply the multiplicative form of the Notation Equation $\frac{m}{n} = m \times \frac{1}{n}$, to multiply fractions by whole numbers;

   (a) fully understand $p \times \frac{m}{n} = \frac{p \times m}{n}$, computationally;

   (b) fully understand $p \times \frac{m}{n} = \frac{p \times m}{n}$, visually, e.g., $3 \times \frac{2}{5}$ is equivalent to $(3 \times 2) \times \frac{1}{5}$;

   (c) solve word problems involving multiplication of fractions by whole numbers using equations and visual models. e.g., If each child at a party drinks $\frac{2}{3}$ of a liter of lemonade and five children are attending the party, how much lemonade must be made?;

4. understand the equations

$$\frac{m}{n} \times q = \frac{m \times q}{n} = m \times \frac{q}{n}$$

in all the aspects that follow:

(a) interpret the product $\frac{m}{n} \times q$ as m parts of a partition of q into n equal parts by using the Notation Equation;

(b) obtain this partition as the result of a sequence of operations leading to $m \times \frac{q}{n}$;

(c) understand this is equivalent to partitioning $m \times q$ into n equal parts;

(d) for example, use a visual fraction model to show $\frac{2}{3} \times 4 = \frac{8}{3}$, and create a story context for this equation;

(e) do the same with $\frac{2}{3} \times \frac{4}{5} = \frac{8}{15}$;

(f) more generally, understand and interpret $\frac{m}{n} \times \frac{q}{p} = \frac{m \times q}{n \times p}$;

5. for fractions having the same denominator:

(a) understand addition and subtraction of fractions as joining and separating parts in respect to the same whole;

(b) decompose a fraction into a sum in multiple ways, e.g., $\frac{3}{8} = \frac{1}{8} + \frac{1}{8} + \frac{1}{8}$ or $\frac{3}{8} = \frac{1}{8} + \frac{2}{8}$;

(c) decompose a mixed number into a sum in multiple ways, e.g., $2\frac{3}{8} = 2\frac{1}{8} + \frac{1}{8} + \frac{1}{8}$ or $2\frac{3}{8} = 1\frac{1}{8} + 1\frac{2}{8}$, or $2\frac{3}{8} = 2 + \frac{1}{8} + \frac{2}{8}$, etc.;

(d) add and subtract mixed numbers involving like denominators by replacing the mixed numbers with equivalent fractions;

(e) solve word problems involving addition and subtraction of fractions having like denominators by using equations to represent the problem;

6. explain computationally using the Fundamental Equation why $\frac{n}{m} = \frac{n \times p}{m \times p}$ (see §15.3);

7. use the three-step procedure as a means to add fractions with unlike denominators (see §14.3);

8. recognize the formula

$$\frac{m}{n} + \frac{p}{q} = \frac{mq + pn}{nq};\,^{2}$$

---

[2]The reader should be aware that two letters written together as in nq is shorthand for $n \times q$ and such expressions appear in general usage and in the school curriculum. We have retained the use of $\times$ for clarity.

as an implementation of Step 3 (see §14.3.3);

9. interpret multiplication as scaling (resizing), by:

   (a) comparing the size of a product to the size of one factor on the basis of the size of the other factor, without performing the indicated multiplication;

   (b) explaining why multiplying a given number by a fraction greater than 1 results in a product greater than the given number (recognizing multiplication by whole numbers greater than 1 as a familiar case); explaining why multiplying a given number by a fraction less than 1 results in a product smaller than the given number; and relating the principle of fraction equivalence $\frac{m}{k} = \frac{nm}{nk}$ to the effect of multiplying $\frac{m}{k}$ by 1;

10. find the area of a rectangle with fractional side lengths by tiling it with unit squares of the appropriate unit fraction side lengths, and show that the area is the same as would be found by multiplying the side lengths, in other words, a visual demonstration of $A = L \times W$ when the length and width are fractions;

11. use multiplication of fractions to find the area of rectangles having sides with fractional lengths and represent fraction products as rectangular areas;

12. solve real world problems involving multiplication of fractions and mixed numbers.

Chapter 15

Common Fractions: Advanced Topics

Subtraction and division of fractions were not discussed in Chapter 14 because the computations are more difficult. For example, a complete treatment of subtraction involves negative numbers and they are not introduced until Grade 6. That said, the basic formulae developed in Chapter 14 for addition will still apply so that the development below is merely applying previous rules.

15.1   Subtracting Common Fractions

For the moment let's just act as though the procedure for addition works for subtraction. In fact, it seems as though it ought to because subtraction and addition are so closely related (see Chapter 9), so let's try it and see what happens.

Three Step Procedure for Subtraction of Fractions

Given the problem:

$$\frac{m}{n} - \frac{p}{k} = ?$$

where $\frac{p}{k}$ is smaller than $\frac{m}{n}$, the three steps are:

Step 1: **Always** check whether the two denominators, n and k, are equal;

Step 2: If the two denominators are equal, that is, $n = k$, find $m - p$ and the difference is:

$$\frac{m}{n} - \frac{p}{n} = \frac{m - p}{n};$$

221

Step 3: If the denominators are not equal, that is, $n \neq k$, find the products $n \times k$, $m \times k$ and $p \times n$, then the difference is:

$$\frac{m}{n} - \frac{p}{k} = \frac{m \times k - p \times n}{n \times k}.$$

In doing examples requiring Step 3, one would want to remind children of the order of precedence rule that requires products to be computed before sums (see §11.4.2).

Now consider finding $\frac{7}{11} - \frac{3}{11}$. Since the denominators are the same, the computation proceeds by Step 2:

$$\frac{7}{11} - \frac{3}{11} = \frac{7-3}{11} = \frac{4}{11}.$$

Clearly we recognize this process as *take away* where we start with 7 of the unit fractions $\frac{1}{11}$ and take away 3 of them, leaving 4 of the unit fractions $\frac{1}{11}$.

Consider an example that requires Step 3:

$$\frac{6}{7} - \frac{2}{3} = \frac{18 - 14}{21} = \frac{4}{21}.$$

The common denominator is 21, so the computation is recognizable as *take away*, where the units are: $\frac{1}{21}$.

Readers will recall from Chapter 9 that the minuend must be at least as large as the subtrahend. Knowing whether this requirement is satisfied is obvious in the case of counting numbers. It is not at all obvious when it comes to fractions as the following example shows:

$$\frac{1}{2} - \frac{5}{9} = \frac{1 \times 9 - 2 \times 5}{2 \times 9} = \frac{9 - 10}{18}.$$

The point is that great care needs to be taken when constructing exercises for children before Grade 6 to ensure the minuend exceeds the subtrahend in subtraction problems.

15.1.1 Advanced Subtraction of Common Fractions

Negative numbers are not introduced until Grade 6 and for this reason readers may want to skip this section for the time being. Negative numbers have an immediate effect on the treatment of subtraction because the requirement that the minuend exceeds the subtrahend disappears. There are two essential facts we need from the Appendices. These are that subtraction is defined by the equation (see Appendix A.2.5)

$$x - y \equiv x + (-y)$$

where $-y$ denotes the **additive inverse**[1] of y (see Appendix A.1.1) and R10 (see Appendix B.9.1) which asserts

$$(-1) \times x = -x.$$

Thus, the operation of subtraction is replaced by addition of the additive inverse, and additive inverses are found by multiplying a number by -1, which is the additive inverse of 1. Consider then that we have a common fraction: $\frac{m}{n}$. For fractional forms like this, R10 tells us that

$$(-1) \times \frac{m}{n} = -\frac{m}{n} = \frac{-m}{n}$$

where the expression in the center is the standard centered dash notation for the additive inverse of $\frac{m}{n}$. This fact leads to

$$\frac{p}{q} - \frac{m}{n} = \frac{p \times n - q \times m}{q \times n}$$

as a **computational formula** for subtracting common fractions (see R16 of Appendix B.9.1).

At the point students are introduced to negative numbers in Grade 6, they should learn that the three-step procedure they already know now works with no restrictions, but that the end result may be a negative number as in:

$$\frac{1}{2} - \frac{5}{9} = \frac{1 \times 9 - 2 \times 5}{2 \times 9} = \frac{9 - 10}{18} = -\frac{1}{18}.$$

15.2 Dividing Common Fractions

In Grade 3, it is expected that children will come to understand the relationship between multiplication and division as expressed by the relationship between the whole numbers d and k, where d is a factor of k:

$$\text{if } k \div d = q \text{ then } k = q \times d.$$

This is the key idea underlying division (see §12.1) and can be restated as:

quotient times the divisor should be equal to the dividend.

---

[1]Two numbers x and y are additive inverses of one another if their sum is 0.

We want to apply this idea in the context of common fractions.

Recall the Fundamental Equation which tells us that

$$n \times \frac{1}{n} = 1$$

for any counting number n. In Appendix B.6 it is shown that for any non-zero number, x,

$$x \times \frac{1}{x} = 1.$$

For this reason, we take as a requirement that the quantity

$$\frac{1}{\frac{p}{q}}$$

must satisfy:

$$\frac{p}{q} \times \left(\frac{1}{\frac{p}{q}}\right) = 1.$$

This leads to the conclusion that:

$$\frac{1}{\frac{p}{q}} = \frac{q}{p}$$

since

$$\frac{p}{q} \times \frac{q}{p} = \frac{p \times q}{q \times p} = 1.$$

Recall also that the unit fraction $\frac{1}{n}$ results from the process of dividing a unit whole into n equal parts. In other words:

$$1 \div n \equiv \frac{1}{n} = 1 \times \frac{1}{n}.$$

Similarly, we may consider that the common fraction $\frac{m}{n}$ results from dividing m units into n equal parts, so that:

$$m \div n \equiv \frac{m}{n} = m \times \frac{1}{n}$$

where the last step results from the Notation Equation. Notice that thinking about $\frac{m}{n}$ as a division process is completely different than thinking about it as repetitive addition, hence multiplication. Never the less, both are equally valid ways of thinking about the common fraction, $\frac{m}{n}$.

So the reader has a numerical example in mind, think of dividing 2 candy bars between 3 children. Each child should end up with $\frac{2}{3}$ of a candy bar. And at least one child must be given two pieces corresponding to one-third of a candy bar while the other two children could be given one piece corresponding to two-thirds of a candy bar.

The point is that like subtraction, division is a defined operation and in Appendix B.6 division is properly defined for all numbers x and $y \neq 0$ by:

$$x \div y \equiv x \times \frac{1}{y}.$$

Let us now suppose we have the two common fractions $\frac{m}{n}$ and $\frac{p}{q}$, where m, n, p, and q are all counting numbers. Applying the formula above for division with $x = \frac{m}{n}$ and $y = \frac{p}{q}$ gives

$$\frac{m}{n} \div \frac{p}{q} \equiv \frac{m}{n} \times \left(\frac{1}{\frac{p}{q}} \right)$$

and hence to the formula for dividing common fractions:

$$\frac{m}{n} \div \frac{p}{q} \equiv \frac{m}{n} \times \left(\frac{1}{\frac{p}{q}} \right) = \frac{m}{n} \times \frac{q}{p}.$$

Applying the rule for finding products gives:

$$\frac{m}{n} \div \frac{p}{q} = \frac{m}{n} \times \frac{q}{p} = \frac{m \times q}{n \times p}.$$

The rule for dividing common fractions is often described as **invert and multiply** because the divisor, $\frac{p}{q}$, is inverted to become $\frac{q}{p}$ and then the product with $\frac{m}{n}$ is taken. A careful development of division in a general setting is presented in Appendix B.6. Children should know that the procedure above always works.

A simple numerical example of division would be computing $\frac{2}{3} \div \frac{3}{4}$. Using the procedure above,

$$\frac{2}{3} \div \frac{3}{4} \;=\; \frac{2}{3} \times \left(\frac{1}{\frac{3}{4}} \right) \;=\; \frac{2}{3} \times \frac{4}{3} \;=\; \frac{2 \times 4}{3 \times 3} \;=\; \frac{8}{9}.$$

In the problem of dividing one common fraction by another,

$$\frac{m}{n} \div \frac{p}{q},$$

if $\frac{m}{n}$ is the dividend and $\frac{p}{q}$ is the divisor, the the quotient has to be given by:

$$\frac{m \times q}{n \times p},$$

exactly as specified in: $x \div y = x \times \frac{1}{y}$. To see that this is in fact the case, we compute the product of the proposed quotient times the divisor $\frac{p}{q}$ to obtain:

$$\frac{m \times q}{n \times p} \times \frac{p}{q} = \frac{m}{n} \times \frac{q \times p}{p \times q} = \frac{m}{n}$$

which is the dividend.

Children should come to terms with this equation in Grade 4 in the context of dividing fractions by whole numbers. Thus,

$$\frac{1}{8} \div 4 = \frac{1}{32}$$

exactly because

$$4 \times \frac{1}{32} = \frac{1}{8}$$

when reduced to lowest terms (see next section). In Grade 5 it is expected that children will become fully fluent with division of common fractions.

15.3 Equivalent Fractions

The reader may well recall from their own school years that each common fraction has many representations, in fact, infinitely many. For example, the numerals in the following list all identify the same number:

$$\frac{1}{2}, \quad \frac{2}{4}, \quad \frac{3}{6}, \quad \frac{4}{8}, \quad \frac{5}{10}, \quad \frac{6}{12}, \quad \cdots$$

and so forth. Equivalent fractions are useful because given a pair of fractions having different denominators, we can always find a pair of fractions having the same denominator that are equivalent to the original pair. The process by which the equivalent fractions are produced generates the common denominator and is discussed in §14.3.2.

Consider two counting numbers m and n and the fraction, $\frac{m}{n}$ where m and n have no common factor.[2] The critical point is that all fractions equivalent to $\frac{m}{n}$ by simply listing values on the RHS of

$$\frac{m}{n} = \frac{m}{n} \times 1 = \frac{m}{n} \times \frac{k}{k} = \frac{m \times k}{n \times k}$$

---

[2]This requirement is essential because what follows is not true about the pair $\frac{3}{6}$ and $\frac{2}{4}$ even though they are equal and hence, equivalent.

for $k = 1,\ 2,\ 3,\ \ldots$ The important fact here is that equivalent fractions are simply two numerals for the same number and the only way this can be so is if we can write an equality symbol between them.

Children in Grade 4 should understand the various ways that equivalent fractions can arise computationally and know that fractions can be equivalent only in the case that we can demonstrate they are equal.

15.3.1   Visualizing Equivalent Fractions

Fractions as Point on the Line

There are two easily understood visual interpretations of equivalent fractions. The first uses the unit interval. In this representation, we simply notice that the positions of certain common fractions coincide on the line. For example, the positions of $\frac{1}{3}$ and $\frac{2}{6}$ coincide. similarly, the positions of $\frac{3}{6}$ and $\frac{1}{2}$ coincide.

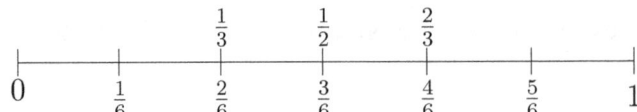

Equivalent fractions identify the same points on the line.

Computationally, it is clear why the diagram must be as shown. For example, starting at $\frac{1}{6}$,

$$\frac{1}{6} + \frac{1}{6} + \frac{1}{6} = \frac{3}{6} = \frac{1 \times 3}{2 \times 3} = \frac{1}{2} \times \frac{3}{3} = \frac{1}{2} \times 1 = \frac{1}{2}.$$

Since $\frac{3}{6}$ and $\frac{1}{2}$ are numerals for the same number, they must identify the same point on the line. This is a fundamental component of a child's understanding of what it means for things to be equal.

Fractions as Area

Recall again the computation to produce equivalent fractions:

$$\frac{m}{n} = \frac{m \times k}{n \times k}.$$

The essential feature of the RHS is that the denominator is a product. For this reason, a visual interpretation must begin with a unit fraction having a denominator that factors, for example, $\frac{1}{6} = \frac{1}{2 \times 3}$. In the diagram below, Box #1 is divided into six equal parts. What we want to think about is:

What fractions can be formed from *sixths*?

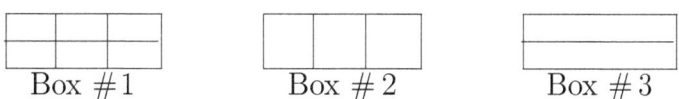

Box #1   Box #2   Box #3

The three boxes have the same area which we take to be 1 unit. The area of Box #1 is divided into *sixths*, the area of Box #2 is divided into *thirds*, and the area of Box #3 is divided into *halves*.

One way to answer this question is simply to start adding $\frac{1}{6}$ to itself and see if any other fractions turn up. Thus for example we could form the sum

$$\frac{1}{6} + \frac{1}{6} = 2 \times \frac{1}{6}.$$

In our diagram, forming this sum corresponds to removing the horizontal line from Box #1 to form Box #2. Box #2 is divided into 3 equal parts each of which is composed of 2 equal parts as shown in Box #1. Since each one of the 3 equal parts of Box #2 has area $\frac{1}{3}$, we know that

$$2 \times \frac{1}{6} = \frac{1}{3}.$$

Visually starting with the diagram of Box #2 and inserting the horizontal line produces Box #1 and illustrates why R10 tells us that $\frac{1}{3} = \frac{2}{6}$ through the computation

$$\frac{1}{3} = \frac{2}{2} \times \frac{1}{3} = \frac{2}{2 \times 3} = \frac{2}{6}.$$

Continuing in this way using Box #1 and #2, we see that

$$\frac{1}{6} + \frac{1}{6} + \frac{1}{6} + \frac{1}{6} = 4 \times \frac{1}{6}.$$

Combining the four subunits on the left side of Box #1, we see that the result must be $\frac{2}{3}$ which visually illustrates R10 for the case

$$\frac{2}{3} = \frac{2}{2} \times \frac{2}{3} = \frac{2 \times 2}{2 \times 3} = \frac{4}{6}.$$

Similarly, if we remove the vertical lines from Box #1 to form Box #3, we know the area of either the top or bottom will given by

$$\frac{1}{6} + \frac{1}{6} + \frac{1}{6} = 3 \times \frac{1}{6}.$$

Either way, we know the total area is

$$3 \times \frac{1}{6} = \frac{1}{2}.$$

The visual representations above can be reinterpreted numerically as

$$\frac{1}{2} = \frac{3}{3} \times \frac{1}{2} = \frac{3 \times 1}{3 \times 2} = \frac{3}{6}.$$

The reader might want to construct a similar diagram involving *tenths*, *fifths* and *halves* to better understand these ideas.

Visual results like those above definitely enhance understanding. But they are not a substitute for computational knowledge. Thus, every child is expected to know and understand that these results all follow from the application of the Associative Law and R6 as in:

$$
\begin{aligned}
4 \times \frac{1}{6} &= (2 \times 2) \times \frac{1}{2 \times 3} = (2 \times 2) \times \left(\frac{1}{2} \times \frac{1}{3} \right) \\
&= 2 \times \left(\left(2 \times \frac{1}{2} \right) \times \frac{1}{3} \right) = 2 \times \left(1 \times \frac{1}{3} \right) \\
&= 2 \times \frac{1}{3} = \frac{2}{3}.
\end{aligned}
$$

In the next section we introduce a simplified procedure referred to as **cancelling**. But we stress that these are the computations that underlie the cancelling process.

15.4 Lowest Terms and Common Factors

The fact that every common fraction like $\frac{3}{8}$ has an infinite number of notations that all give rise to the same number and identify the same place on the number line make the arithmetic of fractions complicated. For example:

$$\frac{3}{8}, \ \frac{6}{16}, \ \frac{9}{24}, \ \frac{15}{40}, \ \frac{-3}{-8}, \ \frac{-21}{-56}, \ \frac{30}{80}, \ \text{and} \ \frac{-2400}{-6400}$$

are all valid representations of the same rational number, $\frac{3}{8}$. However, there is a preferred numeral for each such number and that numeral is the one having the least positive denominator. In the above list, we see that the preferred numeral is $\frac{3}{8}$.

Let us recall how these multiple expressions arise. They all come from the application of R14:

$$\frac{m}{n} = \frac{m \times k}{n \times k}$$

229

Notice on the RHS of R14, the numerator is $m \times k$, and the denominator is $n \times k$. Thus the numerator is a product, the denominator is a product, and both contain the factor k. The factor k is referred to as a **common factor** of the numerator and denominator. In this situation, R14 is reinterpreted as follows:

$$\frac{m \times k}{n \times k} = \frac{m \times \cancel{k}}{n \times \cancel{k}} = \frac{m}{n},$$

where the overstrike denotes **cancelling**, and therefore eliminating, the common factor k from both denominator and numerator. This process preserves the value of the fraction, but **reduces** the fraction to a more compact form. Indeed, this process can be repeated until the numerator and denominator have no common factor, in which case, we say the fraction is in **lowest terms**.

In our visual example with *sixths*, we would reduce $\frac{4}{6}$ as follows

$$\frac{4}{6} = \frac{2 \times 2}{2 \times 3} = \frac{\cancel{2} \times 2}{\cancel{2} \times 3} = \frac{2}{3}.$$

The concept *lowest terms* applies more generally. However, our focus is on common fractions where both the numerator and denominator are integers and on determining how to ensure a common fraction is in lowest terms.

15.5 Factoring Whole Numbers

Fractions are not in lowest terms if there is a common factor in both the numerator and the denominator. Since what we are trying to determine is whether a given denominator and a given numerator have a common factor, the first thing to ask is whether the numerator and denominator factor at all. If a number, n, can be factored, we say it is **composite** which means that $n = k \times p$, where neither k nor p is 1.

Since for the most part the calculations required of children in elementary school deal with fractional forms in which both the denominator and the numerator are one or two-digit numbers, much of what children need to know can be at the level of recall provided they know the multiplication table well. The body of the multiplication table contains all single digit products. As such, it provides answers to the question: $n \times m =?$. This is the initial way in which the table is used. However, it can also be used in the opposite way as follows:

given n in the body of the multiplication table, what are its factors: $n =? \times ??$

For example, for entries like 45, 12 and 72 in the body of the multiplication table, a child needs to immediately recognize that

$$45 = 9 \times 5,$$
$$12 = 4 \times 3 = 2 \times 6,$$
$$72 = 9 \times 8,$$

and so forth.

Notice that from the first example, using the fact that $9 = 3 \times 3$, we get

$$45 = (3 \times 3) \times 5 = 3 \times (3 \times 5) = 3 \times 15.$$

Although the last product, 3×15, isn't in the multiplication table, but the factorization quickly follows from $(3 \times 3) \times 5$.

For 72, we can obtain the following factorizations not in the table

$$72 = 18 \times 4 = 36 \times 2 = 3 \times 24 = 6 \times 12,$$

by starting with $9 = 3 \times 3$ and $8 = 2 \times 2 \times 2$ which are in the table and using the Associative and Commutative Laws.

Clearly, immediate knowledge of the multiplication table at the level of recall is an essential prerequisite for success with factoring.

15.5.1   Prime and Composite Numbers

Once you are thoroughly familiar with the multiplication table, you notice that not every counting number, n, can be factored into a product of counting numbers in which neither of the factors is 1.[3]   There are some numbers which can only be expressed as a product of 1 and themselves, for example 2 and 3. Counting numbers, other than 1, having this property, are called **prime** numbers. Counting numbers having a proper factorization, that is, that can be expressed as $n = k \times p$, where $k,\ p \in \mathcal{N}$ and $k,\ p \neq 1$, are called **composite**. In the event n is composite and $n = k \times p$, we say p **divides** n, or equivalently, p **is a divisor of** n.

A list of the first several prime numbers is:

$$2,\ 3,\ 5,\ 7,\ 11,\ 13,\ 17,\ 19,\ 23,\ \dots$$

Integers between 2 and 23 not appearing in this list are composite, for example $22 = 2 \times 11$.

There are lots of fun facts about primes. For example, there is no largest prime. However, these are not relevant here. What is relevant is that using this short list of primes will enable you to factor any number less than $23 \times 23 = 529$.

---

[3]Expressing $n = 1 \times n$ is called the **trivial** factorization.

15.5.2   Procedure for Finding Factors of a Counting Number

Given a counting number, n, that is less than 529, we want to determine whether n is prime or composite, and, if it is composite, to completely factor n as a product of primes. We use the following procedure to factor n:

Step 1: make a list of primes in increasing order up to $\sqrt{529} = 23$, as shown;

$$2,\ 3,\ 5,\ 7,\ 11,\ 13,\ 17,\ 19,\ 23;$$

Step 2: starting with 2, successively test divide n by the primes in the list in increasing order until either, a divisor is found, or 23 is found not to be a divisor of n;

Step 3: if a divisor is found, so that $n = p_1 \times q$, where p_1 is the least prime divisor of n, record p_1 and q as factors of n;

Step 4 starting with p_1 (the previously found least prime divisor of n), successively test divide q by primes in the list in increasing order until either, a divisor is found, or 23 is found not to be a divisor of q;

Step 5: if a divisor is found, so that $q = p_2 \times q_1$, where p_2 is the least prime divisor of q, record p_2 and q_1 as factors of n and repeat Step 4 with q_1 in place of q and p_2 in place of p_1;

Step 6: if none of the primes up to 23 is a divisor of n, or the test factor q, stop.

There are three points to make here before considering a numerical example.

The first is, because $23 \times 23 = 529$, if n is going to factor, it will have a prime divisor ≤ 23. So the number of prime divisors that have to be tested is limited.

The second is that if we find a least prime divisor, labelled p_1 above, then the other factor, labelled q above, will be much smaller than n and the next computation will be simpler because

$$q = n \div p_1 \leq n \div 2$$

since $2 \leq p_1$.

The third point is that because p_1 is the **least** prime divisor of n, none of its predecessors in the list of primes can be a divisor of the residual factor q. This is why in Step 4 we can repeat the procedure starting at p_1 instead of 2 without having to worry we will miss a divisor.

Let's see how this procedure works in practice by factoring 375. We start with 2, which does not divide 375, because the *ones* digit in 375 is **odd**, that is, not divisible by 2. The next prime is 3 and 3 divides 375, since $375 = 3 \times 125$. Thus,

3 and 125 are factors of 375. While we know 3 is prime, we do not know about 125, so we apply the procedure to 125.

To factor 125, we can start with 3 instead of 2 because we already know 2 is not a factor of $375 = 3 \times 125$. Test dividing 3 into 125 gives $q = 41$ and $r = 2$. So 3 is not a divisor of 125. Trying 5, we find $125 = 5 \times 25$, whence $375 = 3 \times 5 \times 25$. Thus, $375 = 3 \times 5 \times 25$, where 3 and 5 are primes.

What is left to do is test 25, which we recognize as 5×5 from the multiplication table. Thus,

$$375 = 3 \times 5 \times 5 \times 5$$

where all the factors are prime.

Next, consider 243. As in the last example, since 243 ends with an odd digit, it is not divisible by 2. Thus, try 3. Division shows $243 = 3 \times 81$. Moreover, knowledge of the multiplication table gives $81 = 9 \times 9$ and $9 = 3 \times 3$, so we quickly arrive at:

$$243 = 3 \times 3 \times 3 \times 3 \times 3.$$

Lastly, consider 164. Since 164 is **even** (its *ones* digit is divisible by 2), we know 2 is a divisor, and we have $164 = 2 \times 82$, where 82 is again even. So another division by 2 gives $164 = 2 \times 2 \times 41$. Again, knowledge of the multiplication table tells us 41 is not in the body of the table, whence it does not factor and it must be prime. Thus,

$$164 = 2 \times 2 \times 41.$$

15.6   Lowest Terms Procedure

We are now in a position to give a procedure for putting any fraction in lowest terms. We assume the fraction is positive, because given a positive fraction $\frac{m}{n}$, its additive inverse can be written as $(-1) \times \frac{m}{n}$ and the factor -1 has no effect on whether the fractional part is in lowest terms.

Thus, let m, $n \in \mathcal{N}$ be any positive integers. We consider the fraction $\frac{m}{n}$. The following procedure will put $\frac{m}{n}$ in lowest terms:

Step 1: express m as a product of prime factors;

Step 2: express n as a product of prime factors;

Step 3: if no prime factor of m is a prime factor of n, then m and n have no common factor and $\frac{m}{n}$ is in lowest terms so we stop;

Step 4: if some prime factor p of m is also a factor of n, then $m = m_1 \times p$ and $n = n_1 \times p$, so $\frac{m}{n} = \frac{m_1 \times p}{n_1 \times p}$ which is not in lowest terms;

Step 5: restart the procedure to determine whether $\frac{m_1}{n_1}$ is in lowest terms.

We consider some examples.

Let's start with $\frac{18}{21}$. Using the previous procedure for determining prime factors, we find:

$$18 = 2 \times 3 \times 3 \quad \text{and} \quad 21 = 3 \times 7.$$

This takes us to Step 3, where we note 3 occurs in both factorizations, so we go to Step 4. In Step 4 we write the numerator as 6×3 and the denominator as 7×3, whence the revised fraction is

$$\frac{18}{21} = \frac{3 \times 6}{3 \times 7} = \frac{\not{3} \times 6}{\not{3} \times 7} = \frac{6}{7}.$$

Next we go to Step 5 which tells us to apply the process to the revised fraction. Repeating Steps 1 and 2, we have $6 = 2 \times 3$, and 7 is already prime. Since there is no common factor, we conclude

$$\frac{18}{21} = \frac{6}{7}$$

is in lowest terms.

As our next example, consider $\frac{63}{72}$. The prime factorizations for the numerator and denominator are:

$$63 = 3 \times 3 \times 7 \quad \text{and} \quad 72 = 2 \times 2 \times 2 \times 3 \times 3.$$

There is a common factor of 3 which can be eliminated as in:

$$\frac{63}{72} = \frac{3 \times 21}{3 \times 24} = \frac{21}{24}.$$

Applying the procedure to $\frac{21}{24}$ produces a factored numerator of 3×7 and a factored denominator of $2 \times 2 \times 2 \times 3$, which again share a common factor of 3. Eliminating the common factor produces the fraction $\frac{7}{8}$. Applying the process to this fraction shows $\frac{7}{8}$ is in lowest terms because the numerator is the prime 7 and it is not a factor in the denominator. Because of its importance, we present this calculation in complete detail:

$$
\begin{aligned}
\frac{63}{72} &= \frac{3 \times 3 \times 7}{2 \times 2 \times 2 \times 3 \times 3} \\
&= \frac{7 \times 3 \times 3}{2 \times 2 \times 2 \times 3 \times 3} \\
&= \frac{7 \times \not{3} \times \not{3}}{2 \times 2 \times 2 \times \not{3} \times \not{3}} \\
&= \frac{7}{2 \times 2 \times 2} = \frac{7}{8}.
\end{aligned}
$$

The summary calculation effectively factors the numerator and denominator, and then cancels all common factors. Obviously, this is quicker than repeating the steps in the procedure. The key is that only common *factors* are cancelled. In other words, the arithmetic operation must be multiplication. As greater knowledge of the multiplication table is acquired, even shorter calculations are possible, as in:

$$\frac{63}{72} = \frac{9 \times 7}{8 \times 9} = \frac{\cancel{9} \times 7}{8 \times \cancel{9}} = \frac{7}{8}.$$

The requirement to convert arbitrary fractions into lowest terms materially increases knowledge and skill requirements. Specifically, to master the procedures given, a child needs to be able to factor numbers and correctly implement the cancellation process. But there is more, as our next topic shows.

15.6.1 Reducing Sums to Lowest Terms

We have given a procedure for adding two fractions. This procedure, which is derived from R16 in Appendix B.9.1, **works for all fractions**, including algebraic fractions, which is why it's mastery is so important. However, it is incomplete, because it does not, in general, produce a result that is in lowest terms, even when the two fractions being added are in lowest terms. The following example illustrates this failure:

$$\frac{1}{2} + \frac{1}{4} = \frac{4 + 2}{8} = \frac{6}{8},$$

since $\frac{6}{8}$ is not in lowest terms due to there being a common factor of 2 in the numerator and the denominator. A revised computation which includes reducing is:

$$\frac{1}{2} + \frac{1}{4} = \frac{6}{8} = \frac{3 \times \cancel{2}}{4 \times \cancel{2}} = \frac{3}{4}.$$

There is one pitfall in reducing fractions that the reader should be aware of because it arises regularly in respect to fractional forms having algebraic numerators and denominators. Consider the 2 sitting inside the box in the numerator in the following computation:

$$\frac{x + \boxed{2}}{6} = \frac{x + \boxed{2}}{3 \times 2}.$$

Since the denominator contains a factor of 2, many children learning algebra, and some college students, experience an irresistible urge to cancel a 2 from the denominator where it is a **factor** against the $\boxed{2}$ in the numerator where it is a **summand**, not a factor.

Think of this temptation as a **sin**. It is always incorrect. Let's see why. We know from the Notation Equation that $\frac{x}{y} = x \times \frac{1}{y}$. This means

$$\frac{x + \boxed{2}}{6} = (x + \boxed{2}) \times \frac{1}{6}.$$

The parentheses have to be there because the numerator is a sum! Thus the addition has to be carried out **before** the multiplication. This is why

only factors appearing in both the numerator and the denominator may be cancelled.

The 2 in the box in the numerator is not a factor and therefore cannot be cancelled.
A complete procedure for adding fractions

$$\frac{m}{n} + \frac{p}{k} = ?$$

would need to include a fourth step:

Step 1: Check whether the two denominators, n and k, are equal;

Step 2: If the two denominators are equal, that is, $n = k$, find $m + p$ and the sum is:

$$\frac{m}{n} + \frac{p}{n} = \frac{m + p}{n}$$

and go to Step 4;

Step 3: If the denominators are not equal, that is, $n \neq k$, find the products $n \times k$, $m \times k$ and $p \times n$, and the sum is:

$$\frac{m}{n} + \frac{p}{k} = \frac{m \times k + p \times n}{n \times k}$$

and go to Step 4;

Step 4: verify the sum found in Step 2 or 3 is in lowest terms; if not, reduce it to lowest terms.

Remember, when dealing with Steps 3 and 4 it is important to discuss how order of precedence affects the computations.
Consider the following:

$$\frac{13}{36} + \frac{5}{36} = \frac{18}{36} = \frac{1 \times 18}{2 \times 18} = \frac{1 \times \cancel{18}}{2 \times \cancel{18}} = \frac{1}{2}.$$

This example demonstrates that even in the simplest cases where both fractions have the same denominator and are in lowest terms, the sum will not automatically be in lowest terms. More importantly, it shows that when the numerator is a factor of the denominator, as in $36 = 2 \times 18$, we can represent the numerator as a product by writing it as 1×18. Seeing this spelled out completely, shows where the residual 1 in the numerator of $\frac{1}{2}$ comes from.

Even as revised, the procedure for adding fractions is not all that complicated. But there is one more twist, namely, the requirement that the addition process use the smallest denominator that will serve the purpose. Since this may well be part of you child's instruction, we turn to that now.

15.6.2 Least Common Denominator or Multiple

Suppose we want to add two fractions having the denominators 4 and 6. Using the procedure given above, the common denominator would be 4×6. The question is whether there is a smaller denominator that will do? It turns out that a number known as the **least common multiple** of the two denominators (also called the **least common denominator**) is the number we are looking for.

The procedure for finding the least common multiple of two counting numbers n and k is simply to list their multiples until the least common multiple is found. Since the least common multiple must be $\leq n \times k$, the listing process is sure to stop.

As a numerical example, consider 4 and 6. If we start listing multiples of 6 and 4, we have the following:

$$6,\ 12,\ 18,\ 24$$
$$4,\ 8,\ 12.$$

We list multiples of the larger number, 6, first, because there will always be fewer computations required to get to the product than if we start with the smaller number. To produce the entire list of multiples for 6 up to $6 \times 4 = 24$ has four numbers. The equivalent list for 4 contains six numbers, but our list for 4 stops at 12 because 12 is in both lists and is the least common multiple.

We use the least common multiple to add fractions in the following manner:

$$\frac{1}{4} + \frac{5}{6} = \frac{1 \times 3}{4 \times 3} + \frac{5 \times 2}{6 \times 2}$$
$$= \frac{3}{12} + \frac{10}{12} = \frac{13}{12}.$$

Let's compare the addition procedure using least common multiple with the computation given by the general rule set down as Step 3 in §14.2.3 above:

$$\frac{1}{4} + \frac{5}{6} = \frac{1 \times 6}{4 \times 6} + \frac{5 \times 4}{6 \times 4}$$

237

$$= \frac{6}{24} + \frac{20}{24} = \frac{26}{24}.$$

The answer has to be reduced since both the numerator and denominator have a common factor. We immediately have:

$$\frac{26}{24} = \frac{13 \times 2}{12 \times 2} = \frac{13 \times \cancel{2}}{12 \times \cancel{2}} = \frac{13}{12}.$$

In this case, the procedure using Step 3 appears to require the extra step of reducing to lowest terms which was not present in the first computation using the least common denominator, 12. However, reducing the final answer may still be required as the next example shows.

Find $\frac{1}{2} + \frac{1}{6}$. As before, to find the least common multiple, we begin by listing multiples of 6:

$$6, \ 12,$$

of which only two multiples are needed. Since 2 is a divisor of 6, we know that 6 is the least common multiple of 2 and 6. Thus,

$$\frac{1}{2} + \frac{1}{6} = \frac{1 \times 3}{2 \times 3} + \frac{1}{6} = \frac{3}{6} + \frac{1}{6} = \frac{4}{6}.$$

Notice that 4 and 6 are both even, so this fraction is not in lowest terms. It has to be reduced and this process gives:

$$\frac{4}{6} = \frac{2 \times 2}{3 \times 2} = \frac{2 \times \cancel{2}}{3 \times \cancel{2}} = \frac{2}{3}.$$

Again, we compare with Step 3 of the basic procedure in §14.2.3:

$$\frac{1}{2} + \frac{1}{6} = \frac{1 \times 6}{2 \times 6} + \frac{1 \times 2}{6 \times 2} = \frac{6}{12} + \frac{2}{12} = \frac{8}{12}.$$

Reducing this fraction involves two repetitions of cancelling a common factor of 2, or one step cancelling a common factor of 4, as in:

$$\frac{8}{12} = \frac{2 \times 4}{3 \times 4} = \frac{2 \times \cancel{4}}{3 \times \cancel{4}} = \frac{2}{3}.$$

The advantage of using the least common multiple of the denominators, n and k, instead of the product $n \times k$ resides totally in the fact that the least common multiple **may** be smaller than the product. But it may not be smaller, in which case the investment in making a list of multiples has been wasted.

15.7   Common Fractions: The Essentials

We have studied all the standard arithmetic computations with fractions. It is now possible to identify the essential facts that must be known to a child to succeed with fractions. The reader will note that we have dealt with the multiplication of fractions **before** their addition. In the author's view, this approach is simpler and more understandable to any child who knows what unit fractions are and understands the Fundamental and Notation Equations.

For purposes of this discussion, we assume $m, n, p, q \in \mathcal{I}$ and $n, q \neq 0$.

A child must know that the unit fraction $\frac{1}{n}$ is the result of dividing a whole into n equal parts. As such the Fundamental Equation results in:

$$n \times \frac{1}{n} = \frac{1}{n} \times n = 1.$$

A consequence of the Distributive Law, or equivalently that multiplication is repetitive addition, is the Notation Equation:

$$\frac{m}{n} = m \times \frac{1}{n}.$$

These first two items are absolutely essential to understanding the arithmetic of fractions.

To multiply fractions a child must know and be able to apply:

$$\frac{m}{n} \times \frac{p}{q} = \frac{m \times p}{n \times q}.$$

To divide fractions a child must understand that division is multiplication by the reciprocal. Using the standard notation which is introduced by Grade 6, the reciprocal of $\frac{m}{n}$ is given by:

$$\left(\frac{m}{n}\right)^{-1} \equiv \frac{1}{\frac{m}{n}} = \frac{n}{m}$$

and that to write this, we have the added requirement that $m \neq 0$. Using the reciprocal, we have

$$\frac{m}{n} \div \frac{p}{q} \equiv \frac{m}{n} \times \frac{1}{\frac{p}{q}} = \frac{m}{n} \times \frac{q}{p}.$$

where we have the added requirement that $p \neq 0$.

To succeed with addition, a child must know and be able to apply the following:

1. to find common denominators and reduce fractions:

$$\frac{k}{k} = 1 \quad \text{and} \quad \frac{m \times k}{n \times k} = \frac{m}{n};$$

239

2. to add fractions having the same denominator:

$$\frac{m}{n} + \frac{p}{n} = \frac{m+p}{n};$$

3. to add fractions having different denominators:

$$\frac{m}{n} + \frac{p}{q} = \frac{m \times q + n \times p}{n \times q}.$$

Finally, at the point that negative numbers are introduced, the child should understand that subtraction is a defined operation and that:

$$\frac{p}{q} - \frac{m}{n} \equiv \frac{p}{q} + \left(-\frac{m}{n}\right).$$

As we have seen, these equations can be converted to mechanical prescriptions for performing any required computation with fractions. The reasons why they are true are based on the Fundamental Equation and the Notation Equation which are explained in Chapter 13.

Most important is that these procedures apply generally. Inspection of the rules in Appendix B.9.12.6.1 shows that these same rules apply to numbers or algebraic expressions. So anyone who knows and can apply these equations can succeed at the manipulations of algebra which involve fractions, whether these fractions are symbolic or numerical. This includes your child!

15.8   Improper Fractions and Mixed Numbers

A fraction $\frac{m}{n}$, m, $n \in \mathcal{N}$, is called **improper** if $m > n$, that is if the numerator exceeds the denominator as in $\frac{15}{4}$. We can think of $\frac{m}{n}$ in terms of division, as in $m \div n$, and by applying the Division Algorithm we would obtain integers q and r such that

$$m = n \times q + r,$$

where $0 \le r < n$. Thinking this way allows us to write the numerator as a sum, which after applying the Division Algorithm to $\frac{15}{4}$ gives

$$15 = 4 \times 3 + 3.$$

Using this sum we can rewrite the fraction as:

$$\frac{m}{n} = \frac{n \times q + r}{n} = \frac{n \times q}{n} + \frac{r}{n} = q + \frac{r}{n}$$

which for $\frac{15}{4}$ produces

$$\frac{15}{4} = \frac{3 \times 4 + 3}{4} = \frac{3 \times 4}{4} + \frac{3}{4} = 3 + \frac{3}{4}.$$

Note that because $m > n$, q will be at least 1. The notation for $q + \frac{r}{n}$ is referred to as a **mixed number** because it consists of a whole number and a **residual fraction** that satisfies $0 \le \frac{r}{n} < 1$. The whole number will be the largest whole number that is less than or equal to our original fraction, $\frac{m}{n}$.

In splitting the sum into two fractions as in

$$\frac{n \times q + r}{n} = \frac{n \times q}{n} + \frac{r}{n}$$

the reader will observe that if we turn the equation around, we are simply adding two fractions having the same denominator using methods developed in §14.3.1:

$$\frac{n \times q}{n} + \frac{r}{n} = \frac{n \times q + r}{n}.$$

As such, it is simply another example of how every mathematical equation can be used in two directions.

The notation for mixed numbers can be ambiguous. For example, the last sum, $3 + \frac{3}{4}$, is often written suppressing the plus sign as:

$$3 + \frac{3}{4} = 3\frac{3}{4}$$

which makes clear why these are called mixed numbers. The notation with the plus suppressed is confusing because it could be misinterpreted as $3 \times \frac{3}{4}$ with the multiplication sign suppressed.

As another example, consider:

$$\frac{35}{8} = \frac{4 \times 8 + 3}{8} = \frac{4 \times 8}{8} + \frac{3}{8} = 4 + \frac{3}{8} = 4\frac{3}{8}.$$

15.8.1 Adding Mixed Numbers

Suppose we want to add these two mixed numbers:

$$4\frac{3}{8} + 3\frac{3}{4} = ?$$

No problems will arise provided the complete notation is used:

$$
\begin{aligned}
4\frac{3}{8} + 3\frac{3}{4} &= \left(4 + \frac{3}{8}\right) + \left(3 + \frac{3}{4}\right) = 7 + \left(\frac{3}{8} + \frac{3}{4}\right) \\
&= 7 + \left(\frac{3}{8} + \frac{6}{8}\right) = 7 + \frac{9}{8}.
\end{aligned}
$$

241

The reader will notice the fraction in the last mixed number is improper, because $9 > 8$. Since $\frac{9}{8} = 1 + \frac{1}{8}$, the final answer will be

$$7 + \frac{9}{8} = 8 + \frac{1}{8}.$$

In its most compact form, this would be written as the mixed number

$$8\frac{1}{8}.$$

Here is where the ambiguity in mixed numbers and the use of juxtaposition as a substitute for $\times$ can lead to error. The temptation to multiply can be overwhelming, particularly for one with knowledge of the Fundamental Equation. However, if the plus sign is not suppressed, there can be no question of what is meant, namely, $8 + \frac{1}{8}$. Alternatively, one can take as a rule that

juxtaposition never substitutes for times in a purely numerical expression.

Thus, $8\frac{1}{8}$, and the like, always means $8 + \frac{1}{8}$.

In summary, adding mixed numbers can be accomplished in a straight forward manner, as shown above:

add the whole numbers, add the fractions, reduce if necessary, record the result.

Multiplying mixed numbers is a different story.

15.8.2   Multiplying Mixed Numbers

There are two ways to multiply mixed numbers. Personally, I think the simplest is the one we give first.

The first procedure first converts the two mixed numbers to improper fractions by reversing the process of converting an improper fraction to a mixed number as described at the start of this section. Thus,

$$4\frac{3}{8} = \frac{4 \times 8}{8} + \frac{3}{8} = \frac{32}{8} + \frac{3}{8} = \frac{35}{8}$$

and

$$3\frac{3}{4} = \frac{3 \times 4}{4} + \frac{3}{4} = \frac{12}{4} + \frac{3}{4} = \frac{15}{4}.$$

We then find the product of the two improper fractions in the usual way:

$$
\begin{aligned}
4\frac{3}{8} \times 3\frac{3}{4} &= \frac{35}{8} \times \frac{15}{4} = \frac{35 \times 15}{8 \times 4} \\
&= \frac{525}{32} = \frac{16 \times 32 + 13}{32} \\
&= \frac{16 \times 32}{32} + \frac{13}{32} = 16 + \frac{13}{32} = 16\frac{13}{32}.
\end{aligned}
$$

The second calculation treats the mixed numbers as as sums, which is exactly what they are. Products of sums require the Distributive Law as we show by by finding the product of $a+b$ and $c+d$. Applying the Distributive Law to the sums gives:

$$
\begin{aligned}
(a+b) \times (c+d) &= a \times (c+d) + b \times (c+d) \\
&= (a \times c + a \times d) + (b \times c + b \times d).
\end{aligned}
$$

There are a total of four products to be found which then have to be summed. Let's see how this works out for the pair of mixed numbers above:

$$
\begin{aligned}
4\frac{3}{8} \times 3\frac{3}{4} &= \left(4 + \frac{3}{8}\right) \times \left(3 + \frac{3}{4}\right) \\
&= 4 \times \left(3 + \frac{3}{4}\right) + \frac{3}{8} \times \left(3 + \frac{3}{4}\right) \\
&= \left(4 \times 3 + 4 \times \frac{3}{4}\right) + \left(\frac{3}{8} \times 3 + \frac{3}{8} \times \frac{3}{4}\right) \\
&= \left(12 + \frac{4 \times 3}{4 \times 1}\right) + \left(\frac{3 \times 3}{8} + \frac{3 \times 3}{8 \times 4}\right) \\
&= (12 + 3) + \left(\frac{9}{8} + \frac{9}{32}\right) = 15 + \left(1 + \frac{1}{8} + \frac{9}{32}\right) \\
&= 16 + \left(\frac{4}{32} + \frac{9}{32}\right) = 16 + \frac{13}{32} = 16\frac{13}{32}.
\end{aligned}
$$

This computation is tedious, and there are many places in which errors can creep in. However, it is typical of how the Distributive Law is used and it is expected that children in Grade 6 will be able to perform similar computations.

15.9   What Your Child Needs to Know

The intention is that the arithmetic of common fractions be completed in Grades 4-5. For Grade 4 goals specific to multiplication of fractions, see §14.4.

15.9.1 Goals for Grade 4

The child should be able to

1. recognize that a whole number is a multiple of each of its factors. e.g., 21 is a multiple of 3 and 7;

2. determine whether any counting number ≤ 100 is prime or composite;

3. apply the Three Step Process for adding fractions to subtraction problems in which the subtrahend is smaller than the minuend.

15.9.2 Goals for Grade 5

A general objective for Grade 5 is that children become fully comfortable with the use and interpretation of algebraic formulae. These can be statements of principles, as in the Notation Equation, or rules for computation as in

$$\frac{m}{n} + \frac{p}{q} = \frac{m \times q + p \times n}{n \times q},$$

or formulae for computing area as in

$$A = \frac{1}{2} \times b \times h$$

for the area of a triangle.

The child should be able to

1. apply and extend the meaning of division to division of fractions by counting numbers;

   (a) interpret division of a unit fraction by a whole number computationally using the relationship between division and multiplication, e.g., $\frac{1}{4} \div 3 = \frac{1}{12}$, because $\frac{1}{12} \times 3 = \frac{1}{4}$;

   (b) solve real world problems involving division of unit fractions by whole numbers;

2. interpret and compute quotients of fractions, e.g., $\frac{3}{4} \div \frac{2}{3}$;

3. solve real world problems involving quotients of fractions, e.g., What is the width of a piece of land having length $\frac{3}{4}$ mi and area $\frac{2}{3}$ sq mi?, How many $\frac{3}{4}$ cup servings can be made from 3 and $\frac{1}{2}$ cups of yogurt?, etc.

Chapter 16

Ordering Common Fractions

In Chapter 13 we discussed the order relations between fractions having the same denominator and between unit fractions having different denominators. Our considerations were based on the positions of numbers on the line. We avoided the problem of comparing arbitrary common fractions because we lacked the computational tools. We now have the computational tools to complete the discussion.

16.1 Comparing Unit Fractions

In elementary school, studying the order properties of fractions begins with ordering **unit fractions**. These have been studied extensively in Chapter 13. Unit fractions can be thought of as being the result of dividing a unit whole into some number of equal parts. As part of their initial study of fractions, every child should arrive at the conclusion that the more equal parts a given whole is divided into, the smaller each part must be. This fact about unit fractions is summarized as:

$$1 > \frac{1}{2} > \frac{1}{3} > \frac{1}{4} > \frac{1}{5} > \frac{1}{6} > \frac{1}{7} > \ldots > 0.$$

As seen in Chapter 13, each unit fraction represents a number in the unit interval, hence all unit fractions are greater than 0. This reason is geometric, as opposed to being derived from a set of axioms for the number system. Although there are order axioms from which all these relations can be derived, they are not discussed until later in the curriculum (see Appendix C). We shall continue based on the type of reasoning used in Chapter 13. Thus, we have the following general relation: for any pair of counting numbers, n, m, if $n < m$, then

$$0 < \frac{1}{m} < \frac{1}{n}.$$

In summary, comparing unit fractions is straight forward and an easily remembered rule is:

Given two unit fractions, the one having the larger denominator is the smaller of the two.

This is the what children in K-5 should remember about comparing unit fractions.

16.2 Comparing Fractions With the Same Numerator

Consider next the case of comparing two fractions having the same numerator as in $\frac{q}{m}$ and $\frac{q}{n}$. The Notation Equation tells us that

$$\frac{q}{m} = q \times \frac{1}{m} \quad \text{and} \quad \frac{q}{n} = q \times \frac{1}{n}.$$

But we also know that:

$$\frac{q}{m} = q \div m \quad \text{and} \quad \frac{q}{n} = q \div n$$

so that the same counting number q is being divided by two different counting numbers, m and n.

So, as with unit fractions, we have a fixed quantity that is being divided into equal parts, and we know the more parts we divide into, the smaller each part must be. This case simply extends our previous rule to arbitrary numerators.

Once again, this fact should be consistent with a child's initial learning experience:

Given any fixed quantity whatsoever, the more equal parts it is divided into, the smaller each part must be.

16.3 Comparing Fractions With The Same Denominator

Let's fix a counting number n as the single denominator and note that $0 < \frac{1}{n}$. Given any other counting numbers p and q, we know from the discussion in §13.7.1

$$q < p \quad \text{if and only if} \quad q \times \frac{1}{n} < p \times \frac{1}{n}.$$

The RHS is $\frac{q}{n} < \frac{p}{n}$ which is what we want and gives us the following rule:

Given two fractions having the same denominator, the one having the larger numerator will be the larger.

16.4 Comparing Arbitrary Fractions

Comparing arbitrary fractions is difficult because, as the computations in §16.2 and 16.3 show, increasing the numerator of a fraction has the opposite effect on the size of a fraction as increasing the denominator. For example, consider a fixed fraction, $\frac{q}{m}$ and that we want to know whether it is more or less than another fraction of the form

$$\frac{q+k}{m+j}$$

where k, j are counting numbers. If we consider the effects on the denominator and numerator separately, we know

$$\frac{q}{m+j} < \frac{q}{m} \quad \text{but} \quad \frac{q}{m} < \frac{q+k}{m}.$$

The first effect makes the denominator larger and the fraction smaller. The second makes the numerator larger and the fraction larger. But which will dominate? There is no way to tell without further computation.

There is a simple computational procedure that solves the problem and always works. Consider that if the two fractions had the same denominator, we immediately know which is larger by comparing the numerators. But given any pair of common fractions, we have a procedure for finding a pair of equivalent fractions having the same denominator, namely the product of the two original denominators. So all we have to do is find this pair of equivalent fractions and compare the numerators. Thus given the two common fractions $\frac{q}{m}$ and $\frac{p}{n}$, we have

$$\frac{q \times n}{m \times n} = \frac{q}{m} \times \frac{n}{n} \quad \text{and} \quad \frac{p}{n} \times \frac{m}{m} = \frac{p \times m}{n \times m}$$

are equivalent and have the same denominator, $m \times n$. For this reason,

$$\frac{q}{m} < \frac{p}{n} \quad \text{if and only if} \quad q \times n < p \times m.$$

In words, cross multiply and compare the resulting products (numerator of first times denominator of second with numerator of second times denominator of first).

The equivalence expressed above is also valid for the other order relations as listed below:

$$\frac{q}{m} < \frac{p}{n} \quad \text{if and only if} \quad q \times n < p \times m$$

$$\frac{q}{m} = \frac{p}{n} \quad \text{if and only if} \quad q \times n = p \times m$$

$$\frac{q}{m} > \frac{p}{n} \quad \text{if and only if} \quad q \times n > p \times m.$$

247

All of these relations are obtained simply by putting the two initial fractions over the same denominator, which is the product of the two original denominators. We note this is another example of the same procedure being used in multiple ways, so that one stone kills many birds.

In the next section we give a variety of numerical examples. But let's examine one example to illustrate how these facts are used. Suppose we want to know the relationship between $\frac{25}{26}$ and $\frac{26}{27}$. What we know is

$$\frac{25}{26} \boxed{?} \frac{26}{27} \quad \text{if and only if} \quad 25 \times 27 \boxed{?} 26 \times 26$$

where what goes in the box is either $<$, $=$, or $>$. Since

$$25 \times 27 = 675 < 676 = 26 \times 26,$$

we know what goes in the box is $<$.

Summarizing for the general case,

$$\frac{q}{m} \boxed{?} \frac{p}{n} \quad \text{if and only if} \quad n \times q \boxed{?} m \times p$$

where again what goes in the box is either $<$, $=$, or $>$.

16.4.1   Some Numerical Examples

Given counting numbers m, n, p, q, the following three steps provide a procedure for determining the order of the two common fractions, $\frac{q}{m}$ and $\frac{p}{n}$.

Step 1 if $q = p$ (**Equal Numerators**), the fraction having the smaller denominator is larger;

Step 2 if $m = n$ (**Equal Denominators**), the fraction having the larger numerator is larger;

Step 3 if $q \neq p$ and $m \neq n$ (**Unequal Numerators and Unequal Denominators**), cross multiply to obtain $q \times n$ and $p \times m$ and apply the summary line to the cross products to conclude;

    (a) if the cross products are equal, the fractions are equal;

    (b) if $q \times n < p \times m$, then $\frac{p}{n}$ is the larger, that is, $\frac{q}{m} < \frac{p}{n}$;

    (c) if $q \times n > p \times m$, then $\frac{q}{m}$ is the larger, that is, $\frac{q}{m} > \frac{p}{n}$;

    (c') if $p \times m < q \times n$, then $\frac{q}{m}$ is the larger, that is, $\frac{p}{n} < \frac{q}{m}$.

(We have included (c') because it uses **less than**.)

We apply this procedure to the following list of fractions:

$$\frac{1}{2}, \ \frac{1}{3}, \ \frac{3}{5}, \ \frac{3}{7}, \ \frac{3}{8}, \ \frac{6}{13}, \ \frac{7}{14}, \ \frac{11}{23}.$$

For $\frac{1}{2}$ and $\frac{1}{3}$, we have $\frac{1}{3} < \frac{1}{2}$ by Step 1.

To compare $\frac{1}{2}$ with the remaining fractions which have the form $\frac{p}{n}$, we must determine the order relation that goes in the box as shown:

$$\frac{1}{2} \quad \boxed{?} \quad \frac{p}{n}$$

for each fraction in the list. We find the relation by using cross multiplication as described in Step 3. Thus, we compare, $1 \times n$ with $p \times 2$, in other words, the denominator of the second fraction with the numerator of the second fraction multiplied by 2. The relationship between n and $p \times 2$ will be the same as the relationship between $\frac{1}{2}$ and $\frac{p}{n}$. Thus,

$$3 \times 2 > 5, \quad \text{so} \quad \frac{3}{5} > \frac{1}{2},$$
$$3 \times 2 < 7, \quad \text{so} \quad \frac{3}{7} < \frac{1}{2},$$
$$3 \times 2 < 8, \quad \text{so} \quad \frac{3}{8} < \frac{1}{2},$$
$$6 \times 2 < 13, \quad \text{so} \quad \frac{6}{13} < \frac{1}{2},$$
$$11 \times 2 < 23, \quad \text{so} \quad \frac{11}{23} < \frac{1}{2}.$$

Ordering $\frac{1}{3}$ with any of the remaining fractions, $\frac{p}{n}$, again uses cross multiplication by comparing the order relationship between $n = 1 \times n$ and $p \times 3$. Thus,

$$5 < 3 \times 3, \quad \text{so} \quad \frac{1}{3} < \frac{3}{5},$$
$$7 < 3 \times 3, \quad \text{so} \quad \frac{1}{3} < \frac{3}{7},$$
$$8 < 3 \times 3, \quad \text{so} \quad \frac{1}{3} < \frac{3}{8},$$
$$13 < 6 \times 3, \quad \text{so} \quad \frac{1}{3} < \frac{6}{13},$$
$$14 < 7 \times 3, \quad \text{so} \quad \frac{1}{3} < \frac{7}{14},$$
$$23 < 11 \times 3, \quad \text{so} \quad \frac{1}{3} < \frac{11}{23}.$$

To compare $\frac{3}{5}$ with $\frac{3}{7}$ or $\frac{3}{8}$, we use Step 1 which reveals

$$\frac{3}{8} < \frac{3}{7} < \frac{3}{5} \quad \text{since} \quad 5 < 7 < 8.$$

To compare $\frac{3}{5}$ with the remaining fractions requires Step 3. In this case the cross multiples are $3 \times q$ and $5 \times p$. Thus, $\frac{6}{13} < \frac{3}{5}$ since

$$6 \times 5 < 13 \times 3, \quad \text{so} \quad \frac{6}{13} < \frac{3}{5}.$$

We can also use previous relationships as in

$$\frac{11}{23} < \frac{7}{14} = \frac{1}{2} < \frac{3}{5}.$$

We have $\frac{3}{7} < \frac{6}{13}$, since $3 \times 13 = 39 < 42 = 6 \times 7$ and $\frac{6}{13}$ with $\frac{11}{23}$ since

$$6 \times 23 = 138 < 143 = 13 \times 11,$$

so $\frac{6}{13} < \frac{11}{23}$.

Combining all these facts and applying the Transitive Law gives:

$$\frac{1}{3} < \frac{3}{8} < \frac{3}{7} < \frac{6}{13} < \frac{11}{23} < \frac{1}{2} = \frac{7}{14} < \frac{3}{5}.$$

The procedures are computational and **always work**. For this reason, use of Steps 1-3 in the author's view are the simplest way of determining the relative order of common fractions. Visual representations, which we turn to next, are useful, but the do not substitute for computations, which must be learned anyway for other reasons.

16.5   Positioning Fractions on the Real Line

Successfully being able to make measurments demands that children understand the relationship of numbers to the line. For this reason, we return to the line and focus on the placement of fractions. In §13.1 we developed a graphical description of the whole number line called the real line and described its construction. For purposes of discussion and illustration, we reproduce our previous diagram:

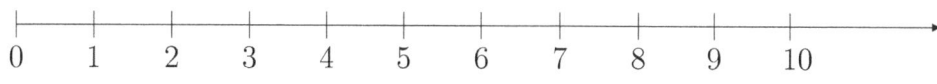

The positive half of the real line can be thought of as a measuring tape of unlimited length.

Construction of a measuring tape shows the critical step is the position of 1, since this determines the physical length of the **unit interval** in the diagram being constructed. The length of the unit interval then determines the position of all other counting numbers, since the distance between successive counting numbers must be the length of the unit interval.

Depending on the purpose of the diagram of the real line, we might choose the physical unit of length to be one inch, or one centimeter, or one foot. Either of the first two choices would be appropriate in making a diagram to fit in a book, but the last would not because it is too long. So it is clear exactly how the choice of unit affects the diagram, we redraw the diagram using a larger unit of length.

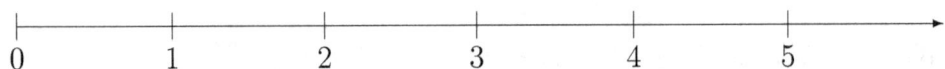

The positive half of the real line using a different unit of length.

16.5.1 Placing Fractions in the Unit Interval

A discussion of placing fractions on the line generally begins with proper fractions in which the numerator is less than the denominator, whence they satisfy

$$0 < \frac{q}{m} < 1.$$

Recalling the discussion in §13.2, we know the unit fraction $\frac{1}{m}$ (m a counting number) subdivides the unit interval into m equal parts (see diagrams in §13.2). This fact together with the Notation and Fundamental Equations determine the placement of each proper fraction with denominator m. For example, for $m = 4$, there are three proper fractions,

$$\frac{1}{4}, \quad \frac{2}{4} = \frac{1}{2}, \quad \text{and} \quad \frac{3}{4}$$

which are positioned in the unit interval as shown.

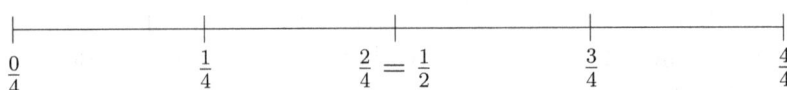

A graph showing the placement of proper fractions having denominator 4. The fractions divide the unit interval into four equal parts. Labelling starts at the left with $\frac{1}{4}$.

251

The graph below shows the position of the six proper fractions having denominator 7 in the unit interval.

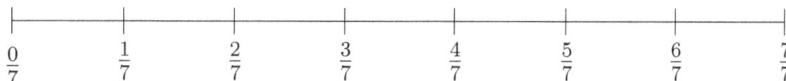

A graph showing the placement of fractions in the unit interval having denominator 7. These fractions divide the interval into seven equal parts. The naming starts from the left and uses the principles developed in §13.2.

Suppose we want to construct a single diagram showing both sets of fractions. The essential fact that governs everything is that fractions with denominator 4 divide the unit interval into four equal parts and those with denominator 7 divide the same interval into seven equal parts. Constructing a diagram for *sevenths* can be done by setting 7 unit lengths end-to-end to form a unit interval and marking each subdivision with the appropriate fraction. This will produce the last diagram. The required divisions for *fourths* can be added using the fact that $\frac{1}{2}$ must divide the whole interval into 2 equal parts, then $\frac{1}{4}$ and $\frac{3}{4}$ subdivide each of the intervals created by the placement of $\frac{1}{2}$ into 2 equal parts. This diagram is shown next.

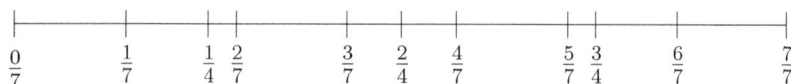

A graph showing the placement of fractions in the unit interval having denominator 7 followed by those with denominator 4.

To further illustrate the complex order relationships between fractions, we re-draw this diagram, this time including all the fractions with denominator 9. You might want to think about where fractions with denominator 8 would be placed, and whether knowing this would be helpful in respect to placing the fractions with denominator 9.

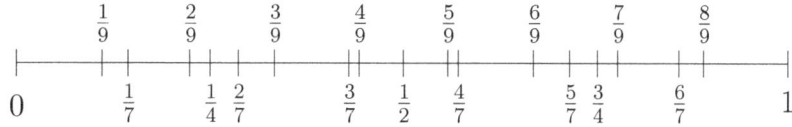

A graph showing the placement of fractions having denominator 4, 7 and 9 that are in the unit interval. The relative placement of fractions having different denominators requires calculation. Note the symmetry about $\frac{1}{2}$.

252

The fraction $\frac{1}{2}$ is referred to as a **benchmark**. This is because it divides the unit interval into two equal parts. If you consider the other fractions in the interval having the same denominator, for example fractions with denominator 7, they are distributed symmetrically about $\frac{1}{2}$. This means we can use the information from placing fractions less than $\frac{1}{2}$ on the line to tell us how to place fractions on the line that are greater than $\frac{1}{2}$.

While the idea of benchmarks is useful, it cannot substitute for knowledge of the principles and rules respecting the ordering of fractions, and the ability to correctly perform computations required to order common fractions. We see this when we try to place the list of fractions from our first example into a graph of the unit interval. While knowledge that most of the fractions have to be to the left of $\frac{1}{2}$ in the diagram, their exact placement requires calculations as discussed above.

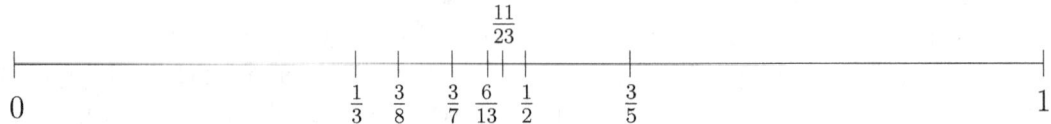

A graphical description of the fractions from the initial list of fractions to be ordered. As the reader can see, $\frac{1}{2}$ sits half way between 0 and 1. The remaining fractions are placed according to their fraction of the unit length, the determination of which requires calculation.

Review of all the diagrams in this section leads to the conclusion that deciding the order relationships among arbitrary fractions is difficult. That is why children need to have Step 3 to fall back on. It always works.

We will return to the problem of placing numbers on the line when we discuss decimals.

16.5.2 Placing Mixed Numbers on the Real Line

The reader will recall our discussion of positive mixed numbers in Section 15.6. There we asserted that the numerator of every improper fraction, $\frac{m}{n}$, can be written in the form: $q + \frac{r}{n}$, where q is an integer, and r is an integer satisfying, $0 \le r < n$. This fact is a result of applying the Division Algorithm to $m \div n$.

Suppose we consider a rational number, $\frac{m}{n}$, in the interval from 25 to 26 so that:

$$25 \le \frac{m}{n} < 26.$$

By applying the Division Algorithm to $m \div n$, we obtain $m = q \times n + r$ where $0 \leq r < n$. Substituting into $\frac{m}{n}$ gives

$$
\begin{aligned}
\frac{m}{n} &= \frac{q \times n + r}{n} \\
&= \frac{q \times \cancel{n}}{\cancel{n}} + \frac{r}{n} = q + \frac{r}{n}
\end{aligned}
$$

where q, r, $n \in \mathcal{N}$ and $0 < r < n$ so that $\frac{r}{n}$ is a proper fraction and hence is in the unit interval. Since $25 \leq q + \frac{r}{n} < 26$ and $0 \leq \frac{r}{n} < 1$, we have

$$
25 - \frac{r}{n} \leq q < 26 - \frac{r}{n}.
$$

Since there is only one integer in this interval, namely 25, we conclude that $q = 25$ and

$$
\frac{m}{n} = 25 + \frac{r}{n}
$$

and

$$
m = 25 \times n + r.
$$

Thus, to place the improper fraction, $\frac{m}{n}$, on the line, we would write the numerator as $m = q \times n + r$ using the Division Algorithm and understand that this means that $\frac{m}{n}$ is the sum of the quotient q and the residual $\frac{r}{n}$ which is a **proper fraction**.

Given any other fraction in the unit interval, that is, $\frac{p}{k}$ where p, $k \in \mathcal{N}$ and $p < k$, the Addition Law tells us that

$$
\frac{r}{n} < \frac{p}{k} \quad \text{if and only if} \quad 25 + \frac{r}{n} < 25 + \frac{p}{k}.
$$

Thus, the relative position on the line of the improper fraction $\frac{m}{n} = 25 + \frac{r}{n}$, in respect to the position of any **other** fraction $25 + \frac{p}{k}$ in the interval between 25 and 26, has to be the same as the relative position of the residual, $\frac{r}{n}$ in respect to the residual $\frac{p}{k}$ when these two fractions are considered as members of the unit interval between 0 and 1.

Of course there is nothing special about 25, so that every interval defined by consecutive integers merely repeats the unit interval in respect to the relative positions of its constituent numbers.

$25 + \frac{0}{4}$ $25 + \frac{1}{4}$ $25 + \frac{2}{4} = 25 + \frac{1}{2}$ $25 + \frac{3}{4}$ $25 + \frac{4}{4}$

A graph showing the placement of fractions having denominator 4 in the interval from 25 to 26. They are equally spaced and divide the interval into four equal parts. The last member coincides with 26. Written as improper fractions, the numbers are

$$\frac{100}{4}, \quad \frac{101}{4}, \quad \frac{102}{4}, \quad \frac{103}{4}, \quad \text{and} \quad \frac{104}{4}.$$

In summary, each interval between successive integers has the same order structure in respect to fractions as the unit interval. For this reason, the focus on the relative ordering of fractions can be confined to the unit interval. We will see further application of these ideas when we take up decimals, the topic of the next chapter.

16.6 What Your Child Needs to Know

16.6.1 Goals for Grade 3

It is expected that your child will be able to:

1. compare two fractions with the same numerator or the same denominator by using the methods and principles discussed;

2. recognize that numerical comparisons of real-world fractions are valid only when the two fractions refer to the same whole, i.e., we can not compare $\frac{1}{2}$ a cantaloupe with $\frac{1}{3}$ of a watermelon;

3. record the results of numerical comparisons in words, i.e., *less than*, *the same as* and *greater than*, and justify conclusions by making and/or using a visual fraction model.

16.6.2 Goals for Grade 4

It is expected that your child will be able to:

1. compare two fractions with different numerators and different denominators, e.g., by creating common denominators or numerators as in $\frac{2}{9} < \frac{3}{8}$ because $\frac{2}{9} < \frac{3}{9} < \frac{3}{8}$, or by comparing to a benchmark fraction such as $\frac{1}{2}$;

2. correctly record the results of comparisons with symbols $>$, $=$, or $<$;

3. recognize that numerical comparisons are valid only when the two fractions refer to the same whole, and numbers are in the same units, e.g. to compare $5°$ and $25°$ we must know both numbers refer to the same temperature scale.

Chapter 17

Decimal Numerals for Common Fractions

This chapter extends the Arabic System of numeration to common fractions. As we already know, the utility of the Arabic System is that it facilitates the computations of arithmetic. What we also know is that the computations of arithmetic when applied to common fractions can be difficult and look nothing like the methods developed in Chapters 8, 9, 11 and 12. By extending the Arabic System, that changes and it becomes possible for children to master the procedures of decimal arithmetic in a single year in Grade 5.

Understanding decimal numerals is an essential pre-requisite to mastering decimal arithmetic. The difficulty is that decimal numerals are best expressed using both positive and negative exponents and negative numbers are not treated until Grade 6 (see Chapter 18) and exponents not until Grade 7 in an A+ curriculum. Even though this is the case and exponents will be avoided when working with children, it is crucial that those teaching the procedures to children have a complete understanding of how decimals are used to represent numbers because these ideas are the key to why lining up or otherwise manipulating the decimal points in the numerals during computations produces correct answers.

17.1   Powers of 10 and Arabic Notation

Powers of 10 involve exponents the theory of which is discussed in Appendix C. For our purposes here, readers can simply treat the following equations as definitions:

$$10^0 = 1, \quad 10^1 = 10, \quad 10^2 = 10 \times 10, \quad 10^3 = 10 \times 10 \times 10, \quad 10^4 = 10 \times 10 \times 10 \times 10,$$

256

and

$$10^{-1} = \frac{1}{10}, \quad 10^{-2} = \frac{1}{10 \times 10}, \quad 10^{-3} = \frac{1}{10 \times 10 \times 10},$$

and so forth. The utility of the exponential notation will become evident when we discuss the operations of arithmetic in Chapter 18.

Let us recall the role of place in the Arabic Notation System for non-negative integers. Specifically, if we had a four digit number, the place of each digit had a different meaning, starting at the right and working left. Thus, in 8547, the 7 is in the *ones* place, the 4 in the *tens* place, the 5 in the *hundreds* place and the 8 is in the *thousands* place. The number represented by 8547 is the same as:

$$8547 = 8 \times 1000 + 5 \times 100 + 4 \times 10 + 7 \times 1,$$

as we discussed in Chapter 6. In that discussion, we used the names for each place, rather than powers of 10 as developed above.

There were two reasons for this. First, as taught to children, place is identified by its name, *ones, tens, thousands, ten thousands,* and so forth. Second, in order to discuss powers of 10 properly, we need a minimal knowledge of arithmetic which children in primary and elementary do not have, and which we did not have in Chapter 6. But now we have that knowledge, and so we can rewrite the equation above using exponents as:

$$8547 = 8 \times 10^3 + 5 \times 10^2 + 4 \times 10^1 + 7 \times 10^0.$$

Indeed, to express a much larger number, for example, we have:

$$\begin{aligned} 7653429 \;=\; & 7 \times 10^6 + 6 \times 10^5 + 5 \times 10^4 + 3 \times 10^3 \\ & + 4 \times 10^2 + 2 \times 10^1 + 9 \times 10^0. \end{aligned}$$

The ultimate simplicity of this system for expressing a positive integer is now apparent. First, each digit is multiplied by a power of 10 that is determined by its place. What is that power of ten? Starting with the right most digit, the power is 0, since $1 = 10^0$. With each step (place) to the left, we increment the power of 10 by 1. So, in our example, 7653429, the 5 is **four steps** to the left of the 9, so the power of 10 associated with that place is $0 + 4 = 4$, and the value of 5 in 7653429 is 5×10^4.

As we already know, the Arabic System provides a notation for each counting number and 0. This notation can and will be extended to include negative integers via the centered dash in Grade 6. But we know there are many numbers that are not integers (whole numbers), for example, $\frac{1}{2}$. The question is:

Can we extend the Arabic System to apply to all common fractions?

17.2   Representing Common Fractions

The Arabic System as discussed provides a notation for every counting number. More importantly, the Arabic System supports arithmetic computations. The problem we face is that the common fractions include numbers that are not counting numbers. Thus, we need to extend our system of notation to include common fractions, and to do so in a manner that supports numerical computations.

Recall that in §16.5 we used the Division Algorithm and the Fundamental Equation to show how every rational number could be expressed as the sum of an integer and a proper common fraction. Expressing this in mathematical form, we see that any common fraction, x, can be written as

$$x = m + \frac{q}{n}$$

where m is an integer and q, n are counting numbers with $q < n$. Since the RHS of this equation involves only integers, the reader may well ask:

> Isn't this a perfectly good notation for x?

The answer is yes, but as every reader who has ever dealt with fractions knows, computing with fractions is not easy in the sense that adding or multiplying integers is easy. Moreover, once we get to mixed numbers, things get really messy (see §15.8).

The problem of extending the Arabic System to the rationals can be solved provided we are willing to make a sacrifice. We can use the notation for fractions as ratios of integers and give up on *supports computations*, or we can support computations and give up on **having an exact notation for each rational number**. In other words to solve the problem, we have to make a trade-off.

The hope that we might achieve a system that supports computations for rational numbers comes from the fact that when two fractions have the same denominator, their sum is obtained simply by adding the numerators as in:

$$\frac{m}{n} + \frac{q}{n} = \frac{m + q}{n}.$$

Since the numerators are integers, and we have a good system for integers that supports computations, by sticking with the one denominator idea we can find a way to make things work.

17.3   Extending the Arabic System to Fractions

If we think about extending the Arabic System of notation to all common fractions, we see that an essential difficulty is how to extend the notion of place in respect to providing notations for positive mixed numbers. The difficulty arises because in any expression for an integer, the right-most place is the *ones* place and all places to the left are used by positive powers of 10.

To be clear, we need a numerical example, say 25. When we say all places to the left of the two are taken, what we mean is that if we add another digit, say 3, as in

$$325,$$

the 3 has a fixed, predetermined meaning, namely, 3×10^2 in this expression. Further, as we have seen, the predetermined meanings extend as many places to the left as we might try.

On the other hand, if we put the 3 on the right, as in

$$253,$$

the previous meanings attached to the 2 and 5 change, becoming 2×10^2 and 5×10^1, instead of 2×10^1 and 5×10^0, respectively.

More thought suggests that what is needed is a **marker**, that is, some notational device that marks the place of the *ones* digit.

The device that was chosen is the **decimal point** which is positioned immediately to the right of the digit intended to be the *ones* digit in a numeral.[1] The only function of the decimal point in Arabic numerals for numbers like 25.3 which are not integers is to locate the *ones* digit.

17.3.1 Interpreting Digits to the Right of the Decimal Point

Once we have figured out how to mark the place of the *ones* digit, we can add as many digits on the right as we choose. But we have to say how those digits will be interpreted.

The rule is simple:

> if a digit k in a numeral is n places to the right of the decimal point, its value is $k \times 10^{-n}$.

Thus, every digit to the right of the decimal point can be thought of as a fraction having a denominator that is a power of 10. Fractions having denominators that are powers of 10 are called **decimal fractions**. Numbers written in this form will be referred to as **decimal numbers**.

---

[1]The decimal point was introduced to the western world by the Persian mathematician al-Khwāizmhrī in the early 800s (see Wikipedia) based on Indian mathematics.

Let's see how this idea works out in practice. In the example just used, 25.3, we have

$$25.3 = 2 \times 10^1 + 5 \times 10^0 + 3 \times 10^{-1} = 25 + \frac{3}{10}$$

where the expression on the far right emphasizes that

> **every decimal number is the sum of an integer and a residual decimal fraction found in the unit interval**.

We will use this fact repetitively in what follows.

The integer part of a decimal number is determined by the digits to the left of the decimal point interpreted in their usual manner. Since we already know how to work with integers expressed in Arabic notation, we will concentrate on calculations with the residual, that is, the number represented by a decimal fraction that comes from the unit interval.

For example, consider:

$$
\begin{aligned}
0.205 &= 0 \times 10^0 + 2 \times 10^{-1} + 0 \times 10^{-2} + 5 \times 10^{-3} \\
&= \frac{2}{10} + \frac{0}{100} + \frac{5}{1000} \\
&= \frac{200}{1000} + \frac{0}{1000} + \frac{5}{1000} \\
&= \frac{205}{1000}.
\end{aligned}
$$

The last line of this expression illustrates how the residual ends up being a single decimal fraction even though each place to the right of the decimal point involves a different negative power of 10.

The names of the places in which digits to the right of the decimal point occur are simple; we use the name of the unit fraction multiplier. Thus, the first place to the right of the decimal point is the *tenths* place, the second digit to the right is the *hundredths* place, the third place to the right is the *thousandths* place, and so forth.

In the example 25.3, the 3 is in the *tenths* place. In 0.205, the 2 is in the *tenths* place, the 0 is in the *hundredths* place, and the 5 is in the *thousandths* place.

To summarize, every common fraction expressed by a decimal numeral is the sum of a whole number[2] and a residual that can be found in the unit interval, that is, between 0 and 1. The residual has an exact representation as a proper **decimal**

---

[2]For convenience, we take the whole numbers to consist of 0 and the counting numbers. Later we will include negative numbers.

fraction, that is, a fraction having a power of 10 in the denominator and a numerator that is less than the denominator. For clarity, three numerical examples are

$$25.47 = 25 + \frac{47}{100}, \quad 1.9863 = 1 + \frac{9863}{10000}, \quad \text{and} \quad 4.247 = 4 + \frac{247}{1000}.$$

17.3.2 The Problem of $\frac{1}{3}$

In extending the Arabic System of notation to common fractions, we indicated that a trade-off was being made: the simple fact is that there are some fractions that cannot be expressed as an exact decimal number. What do we mean by this? Any number for which we can find a decimal expression involving a finite number of digits is said to be an **exact decimal number**. Such a number can be expressed as a decimal fraction, i.e., $\frac{p}{q}$ where q is a power of 10. A very simple example is:

$$\frac{1}{5} = \frac{2}{10} = 0.2,$$

so we would say $\frac{1}{5}$ has an exact decimal representation.

There is a simple test for whether a given rational $\frac{m}{n}$ has an exact decimal representation. If it does, it means we can find a decimal number which we call d such that

$$d - \frac{m}{n} = 0.$$

Performing this calculation is straightforward. First convert d to a decimal fraction, then do the subtraction using the standard methods for subtracting fractions. Clearly, the decimal number above representing $\frac{1}{5}$ has this property.

But the rational number $\frac{1}{3}$ does not have such an exact representation. Suppose we consider 0.3 as a decimal number candidate to represent $\frac{1}{3}$. We apply the test to find

$$0.3 - \frac{1}{3} = \frac{3}{10} - \frac{1}{3} = \frac{9}{30} - \frac{10}{30} = -\frac{1}{30}.$$

The result is not zero. We might try $.33$. Performing the required calculation gives

$$0.33 - \frac{1}{3} = \frac{33}{100} - \frac{1}{3} = \frac{99}{300} - \frac{100}{300} = -\frac{1}{300}$$

which is closer to zero, but still not zero. Indeed, whatever decimal number we try, we will not get zero. The essential reason is that 3 is not a divisor of any power of 10, and hence no power of 10 will serve as a common denominator with 3 (see §16.4.2).

We will return to the problem of representing arbitrary rational numbers in Chapter 18 when we consider **repeating decimals**. There we will demonstrate that the exact form of the rational number can be recovered from the repeating decimal expansion.

17.4 Placing Decimals on the Real Line

In what follows, it will be convenient to have a notational scheme for discussing arbitrary decimal fractions. Similar to the notational scheme set up in §9.3.1, we shall denote digits to the right of the decimal place by n_t for *tenths*, n_h for *hundredths* and n_{th} for *thousandths*. Thus,

$$0.n_t n_h n_{th} = \frac{n_t}{10} + \frac{n_h}{100} + \frac{n_{th}}{1000},$$

where the numerator in each case is one of the digits 0 - 9. Now let us return to the problem of finding decimals on the real line.

We have already discussed (Section 17.7) the placement of fractions on the real line. So, in a sense, we already know how to do this. Nevertheless, a careful description of how decimal numbers are placed on the real line will be helpful from several perspectives.

Since every decimal number, e.g., 27.6, is the sum of an integer and a decimal fraction, we need only explore the position and relationship of decimal fractions found in the unit interval. So consider

$$0.736 = 0 + \frac{7}{10} + \frac{3}{100} + \frac{6}{1000} = \frac{736}{1000}.$$

In terms of our notational scheme, $n_t = 7$, $n_h = 3$ and $n_{th} = 6$.

To interpret the portion of decimal notation to the right of the decimal point, we work left to right, the opposite of the way we work with integers. So the first place to the right of the decimal point is the *tenths* place and the digit in this place is 7. The contribution to the sum from this place is:

$$.7 = \frac{7}{10}.$$

It is the largest individual contribution from any digit to the right of the decimal point and its position is shown in the illustration below showing the unit interval subdivided into 10 equal parts.

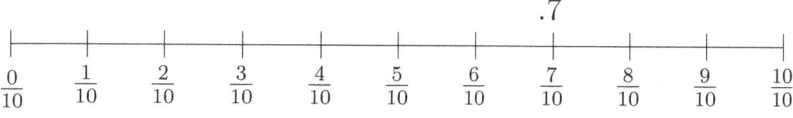

262

A graph showing the placement of fractions having denominator 10 in the unit interval. Each subinterval must have the same length, $\frac{1}{10}$, making the total length 1. The position of the decimal $.7$ is also shown.

Next consider the *hundredths* place. Because the digit n_h must be one of the digits from 0 to 9, the decimal $.7\,n_h$ satisfies

$$.7 \le .7\,n_h = \frac{7}{10} + \frac{n_h}{100} < \frac{7}{10} + \frac{10}{100} = \frac{8}{10} = .8.$$

For $n_h = 3$, we have

$$.7 \le .73 = \frac{7}{10} + \frac{3}{100} < \frac{8}{10} = .8.$$

To be completely clear as to why this is true, we put all fractions over 100:

$$.7 = \frac{70}{100} < \frac{73}{100} < \frac{80}{100} = .8$$

where we are applying the standard procedure for ordering fractions having the same denominator (§17.6.3).

Now the really critical point that needs to be recognized is that the fractions $\frac{71}{100}$, $\frac{72}{100}$, ..., $\frac{79}{100}$, subdivide the interval from 0.7 to 0.8 into ten equal parts in exactly the same way that the fractions $\frac{1}{10}$, $\frac{2}{10}$, ..., $\frac{9}{10}$, subdivide the interval from 0 to 1 into ten equal parts. We show this, together with the placement of $.73$, graphically as follows:

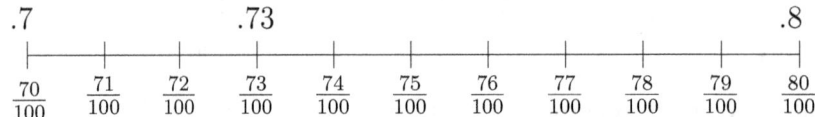

A graph showing the placement of fractions having denominator 100 in the interval between $.7$ and $.8$. The decimal $.73$ is also shown. The reader should also notice that the fractions with denominator 100 divide the interval into ten equal parts each having length $\frac{1}{100}$.

Lastly, consider the *thousandths* place in 0.736. Analogous to what we have already observed for *hundredths*, we have:

$$.73 \le .73\,n_{th} \le \frac{73}{100} + \frac{n_{th}}{1000} < \frac{73}{100} + \frac{10}{1000} = \frac{74}{100} = .74,$$

because n_{th} must be one of the digits from 0 to 9, so that in the example where $n_{th} = 6$:

$$.73 \le .736 = \frac{73}{100} + \frac{6}{1000} < \frac{74}{100} = .74.$$

263

For clarity, we put all fractions over 1000 and apply the rules for ordering fractions:

$$.73 = \frac{730}{1000} < \frac{736}{1000} < \frac{740}{1000} = .74.$$

Again, we make the critical point, namely, that the fractions $\frac{731}{1000}$, $\frac{732}{1000}$, ..., $\frac{739}{1000}$, subdivide the interval from 0.73 to 0.74 into ten equal parts in exactly the same way that the fractions $\frac{71}{100}$, $\frac{72}{100}$, ..., $\frac{79}{100}$, subdivide the interval from 0.7 to 0.8 into ten equal parts. We show this, together with the placement of .736, graphically as follows:

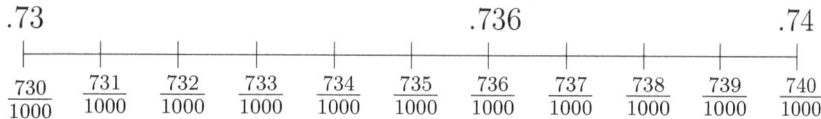

A graph showing the placement of fractions having denominator 1000 in the interval between .73 and .74. The placement of 0.736 is also shown.

The series of graphs above show ever finer divisions by concentrating on smaller intervals. Each interval being represented has a length that is one tenth the length of the preceding interval. Thus the interval from .73 to .74 has a length that is one tenth the length of the interval from .7 to .8, and so on. It is clear that this process of subdividing each interval into ten equal parts can continue for as long as we want. For children in Grade 5, we can stop at *thousanths*.

The following diagram shows the placement of 0.736 in the original unit interval:

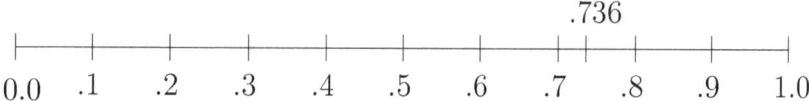

A graph showing the placement of 0.736 in the unit interval. Fractional forms have been replaced by their decimal equivalents.

To review, consider a decimal numeral $0.n_t n_h n_{th}$ where n_t, n_h and n_{th} are any of the digits from $0-9$. We know this numeral identifies a fraction in the unit interval. For clarity of exposition, we will take $n_t = 2$, $n_h = 9$ and $n_{th} = 5$, so we may think of our number as 0.295.

1. We know the unit interval is divided into ten equal parts by

$$.1, \ .2, \ .3, \ .4, \ .5, \ .6, \ .7, \ .8 \text{ and } .9;$$

the interval from .2 to .3 is divided into ten equal parts by

264

$$.21\,,\ .22\,,\ .23\,,\ .24\,,\ .25\,,\ .26\,,\ .27\,,\ .28\,,\ .29\,,$$

and the interval from .29 to .30 is divided into ten equal parts by

$$.291\,,\ .292\,,\ .293\,,\ .294\,,\ .295\,,\ .296\,,\ .297\,,\ .298\,,\ .299\,.$$

2. Given the *tenths* digit is 2, we know that 0.295 must lie in the interval between 0.2 and 0.3.

3. Given the *hundredths* digit is 9, we know 0.295 must lie in the interval between 0.29 and 0.30.

To summarize, we have

$$0.n_t n_h n_{th} = \frac{n_t n_h n_{th}}{1000},$$

so that each such decimal in the unit interval is equivalent to a decimal fraction having denominator 1000. More generally, each decimal in the unit interval is equivalent to a proper fraction having the denominator be a power of 10.

The discussion above tells us how to interpret any decimal number of the form: $0.n_t n_h n_{th}$. But what about numbers of the form $205.n_t n_h n_{th}$, or any other decimal form having non-zero digits to the left of the decimal point? Here, we simply use the fact that:

$$205.n_t n_h n_{th} = 205 + 0.n_t n_h n_{th} = 205 + \frac{n_t n_h n_{th}}{1000}.$$

In other words simply find 205 on the real line and treat the interval from 205 to 206 as though it were the unit interval to place 0.295. The position found in this interval will be the position of $205.n_t n_h n_{th}$. The essential fact which needs to be stressed here is that the interval between any two successive integers, k and $k + 1$, looks exactly like the unit interval. The only difference is its position on the real line.

The above tells us how to interpret decimal notation. It does not tell us how to represent particular numbers in decimal notation. For example, we know we have a number which is represented by the fraction, $\frac{1}{4}$. Does this number have a decimal notation? If so, is there a procedure for finding it? The answer to both questions are: yes. But to provide that answer, we have to discuss the arithmetic of decimals.

17.5   What Your Child Needs to Know

17.5.1   Goals for Grade 5

It is expected your child will be able to

1. incorporate decimal fractions into the place value system;

2. recognize that in a multi-digit decimal number, a digit in one place represents 10 times as much as it represents in the place to its right and $\frac{1}{10}$ of what it represents in the place to its left;

3. explain patterns in the number of zeros of the product when multiplying a number by powers of 10;

4. explain patterns in the placement of the decimal point when a decimal is multiplied or divided by a power of 10;

5. use whole-number exponents to denote powers of 10;

6. read, write, and compare decimals to thousandths;

   (a) read and write decimals to thousandths using base-ten numerals, number names, and expanded form, e.g.,

   $$347.392 = 3 \times 100 + 4 \times 10 + 7 \times 1 + 3 \times \frac{1}{10} + 9 \times \frac{1}{100} + 2 \times \frac{1}{1000};$$

   (b) compare two decimals to thousandths based on meanings of the digits in each place, using $>$, $=$, and $<$ symbols to record the results of comparisons.

There are no goals related to decimals in higher grades. Thus, it is expected that on completion of Grade 5 your child has a complete understanding of the decimal system of arithmetic and can fluently perform the operations of addition, subtraction, multiplication and division using the standard algorithms which we cover in the next chapter.

Chapter 18

Arithmetic Operations with Decimals

There are two ways to think about decimal numerals. In one, we think about the numeral as being a sum of a whole number and a decimal fraction and in the other we think about the number as being a product of a whole number and a power of 10, usually a negative power of 10. Obviously, the second approach to decimal numerals represents a higher level of abstraction and more substantial knowledge of the arithmetic of numbers.

Now there are two ways to teach decimal arithmetic to children. The first is simply to teach the procedures with little or no explanation. This is how children were taught when I went to school. The second provides an explanation as to why the procedures work based on an understanding of decimal numerals. The difficulty with the second approach is that an A+ curriculum expects children to master the procedures of decimal arithmetic in Grade 5, before children have seen negative whole numbers and exponents. The only reason that this is feasible is that the new procedures using decimal numerals are the old procedures for counting number numerals with an added requirement that a decimal point has to be placed in the final answer. The rules for placing the decimal point are straightforward for addition and subtraction and the theory can be explained completely using the sum representation. All of this is within the ken of Grade 5 students learning from an A+ curriculum.

Multiplication and division are another kettle of fish. Although the basic computational procedures are the same as for counting numbers, constructing explanations is much more difficult and, in the case of multiplication, only becomes clear in the context of exponents. That said, the actual procedures are accessible to children simply by narrowing the focus to the computational procedures that children need to master and leaving the explanation as to why these procedures are correct for a later time.

In what follows, we present both the theory underlying such explanations and examples of the computational procedures for children. Readers should always keep in mind that understanding in children follows computational skill, not the other way around.

18.1 Two Forms for Decimal Numerals

Recall that an arbitrary decimal numeral having three places on the right of the decimal point is written as:

$$n_{1000}n_{100}n_{10}n_1.n_t n_h n_{th}$$

where the entry in each place is one of the single digits from the list $0, 1, \ldots, 9$ multiplied by an integer power of 10. Thus in the numerical example 785.316,[1]

$$n_{100} = 7 \quad \text{and} \quad n_h = 1.$$

In this form, the integer power of 10 (see §17.1) is not explicitly shown but is determined by the position of each digit in relation to the others and to the decimal point. Making this information explicit, we have

$$n_{100} = 7 \times 100 = 7 \times 10^2, \quad \text{and} \quad n_h = \frac{1}{100} = 1 \times 10^{-2}.$$

Thus, if we write out our numerical example in complete detail, we have

$$
\begin{aligned}
785.316 \;&=\; 7 \times 100 + 8 \times 10 + 5 \times 1 + \frac{3}{10} + \frac{1}{100} + \frac{6}{1000} \\
&=\; 7 \times 10^2 + 8 \times 10^1 + 5 \times 10^0 + 3 \times 10^{-1} + 1 \times 10^{-2} + 6 \times 10^{-3}
\end{aligned}
$$

where the second line now makes full use of the exponential notation. Clearly, the decimal numeral on the LHS is more compact. That a decimal numeral, such as the one on the LHS, conveys the same information as the expression on the RHS is due entirely to the fact that there is a common agreement as to which integer power of 10 is associated with each place. Thus, anywhere in the world if one were to ask: What power of 10 multiplies the second digit to the right of the decimal point?, the answer will always be the same: 10^{-2}. This fact is crucial to performing addition.

---

[1]Restricting the discussion to *thousandths* is consistent with A+ requirements for Grade 5 but is not necessary in general.

There is an alternate way to think about decimal numerals which is also useful and was mentioned above. Consider the following manipulation of our arbitrary decimal numeral:

$$n_{1000}n_{100}n_{10}n_1.n_tn_hn_{th} = n_{1000}n_{100}n_{10}n_1 + \frac{n_tn_hn_{th}}{1000}.$$

The sum on the RHS puts the fractional parts over the common denominator which is 1000. Now, the Addition Law for Exponents ($a^m \times a^n = a^{m+n}$, see Appendix C 5.1) tells us to add the exponents 3 and -3 to obtain $3 + (-3) = 0$ (see §19.1) so that (see Chapter 17)

$$10^3 \times 10^{-3} = 10^0 = 1.$$

We use this to recast the RHS of the last decimal numeral equation as:

$$
\begin{aligned}
n_{1000}n_{100}n_{10}n_1 + \frac{n_tn_hn_{th}}{1000} &= n_{1000}n_{100}n_{10}n_1 + (n_tn_hn_{th}) \times 10^{-3} \\
&= n_{1000}n_{100}n_{10}n_1 \times (10^3 \times 10^{-3}) + (n_tn_hn_{th}) \times 10^{-3} \\
&= (n_{1000}n_{100}n_{10}n_1000 + n_tn_hn_{th}) \times 10^{-3} \\
&= (n_{1000}n_{100}n_{10}n_1n_tn_hn_{th}) \times 10^{-3}.
\end{aligned}
$$

Notice the use of the Distributive Law to factor out 10^{-3} in the fourth line. The last expression is the product of an integer and a negative power of 10. The point is that

every decimal number can be expressed as the product of an integer and a negative power of 10 that records the number of places to the right of the decimal point in the decimal numeral for that number

(in our case three), as in:

$$n_{1000}n_{100}n_{10}n_1.n_tn_hn_{th} = (n_{1000}n_{100}n_{10}n_1n_tn_hn_{th}) \times 10^{-3}.$$

A numerical example will be helpful, so consider 8136.207. Applying the sequence above gives

$$
\begin{aligned}
8136.207 &= 8136 + \frac{207}{1000} \\
&= 8136 \times (10^3 \times 10^{-3}) + 207 \times 10^{-3} \\
&= (8136 \times 10^3) \times 10^{-3} + 207 \times 10^{-3} \\
&= (8136000 + 207) \times 10^{-3} \\
&= 8136207 \times 10^{-3}.
\end{aligned}
$$

All we have done is to create a common factor, 10^{-3}, which we can then pull out using the Distributive Law to obtain a single integer times a negative power of 10:

$$8136.207 = 8136207 \times 10^{-3}.$$

A simple rule is being applied here, namely,

> count the places to the right of the decimal point, in this case 3, remove the decimal point to obtain an integer and then multiply that integer by 10 raised to -1 times the integer count (see §19.3.3), in this case, 10^{-3}.

The next three examples illustrate this rule:

$$31.4 = 314 \times 10^{-1}, \quad 1.57 = 157 \times 10^{-2}, \text{ and } 963.882 = 963882 \times 10^{-3}.$$

Finally, the reader will note that this process can be applied in both directions. Given an integer and a negative of a power of 10, we can immediately convert their product to a decimal number, for example:

$$2651 \times 10^{-2} = 26.51, \quad 87946 \times 10^{-4} = 8.7946,$$

and

$$3248 \times 10^{-5} = .03448.$$

The value of the exponent tells us how many places we need in every case. Then count the places starting at the right and put in the decimal point.

18.2 Theory of Decimal Arithmetic: Addition

Recall the first representation of a general decimal number as a sum given in the §17.3.1:

$$n_{1000}n_{100}n_{10}n_1.n_t n_h n_{th} = n_{100} \times 100 + n_{10} \times 10 + n_1 \times 1 + n_t \times \frac{1}{10} + n_h \times \frac{1}{100} + n_{th} \times \frac{1}{1000}.$$

Suppose we want to add two decimal numbers written in this form, for example:

$$785.316 = 7 \times 100 + 8 \times 10 + 5 \times 1 + 3 \times \frac{1}{10} + 1 \times \frac{1}{100} + 6 \times \frac{1}{1000}$$

and

$$147.375 = 1 \times 100 + 4 \times 10 + 7 \times 1 + 3 \times \frac{1}{10} + 7 \times \frac{1}{100} + 5 \times \frac{1}{1000}$$

The critical thing to notice is that in every decimal numeral digits in the same place in respect to the decimal point have the same power of 10 as a multiplier.

Thus, in the two numbers above starting at the right, the 6 and the 5 are both multiplied by $\frac{1}{1000} = 10^{-3}$. Moving one place to the left, the 1 and the 7 are both multiplied by $\frac{1}{100} = 10^{-2}$, and so forth. The fact that the multiplier is the same is the key that makes the addition algorithm work. The underlying reason is Rule 15 which tells us how to add fractions with the **same** denominator. Thus, by carefully aligning the digits in the same place starting at the right, we are always adding **apples to apples**. So for example,

$$6 \times \frac{1}{1000} + 5 \times \frac{1}{1000} = (6 + 5) \times \frac{1}{1000} = 11 \times \frac{1}{1000}.$$

Of course the sum of $6 + 5 = 11$ is not a single digit, but the Distributive Law and Rule 14 tell us that

$$11 \times \frac{1}{1000} = (10 + 1) \times \frac{1}{1000} = 10 \times \frac{1}{1000} + 1 \times \frac{1}{1000}$$

$$= 1 \times \frac{1}{100} + 1 \times \frac{1}{1000}$$

which simply means we have an additional 1 to be added in the *hundredths* place. Thus, the revised sum of the digits in the *hundredths* place is

$$1 \times \frac{1}{100} + 1 \times \frac{1}{100} + 7 \times \frac{1}{100} = (1 + 1 + 7) \times \frac{1}{100} = 9 \times \frac{1}{100}.$$

For *tenths* we have

$$3 \times \frac{1}{10} + 3 \times \frac{1}{10} = (3 + 3) \times \frac{1}{10} = 6 \times \frac{1}{10},$$

which completes the addition to the right of the decimal place. Now for *ones* we have

$$5 \times 1 + 7 \times 1 = (5 + 7) \times 1 = 12 \times 1 = 1 \times 10 + 2 \times 1,$$

which leaves one unit of 10 to carry to the *tens* place. The *tens* total is

$$1 \times 10 + 8 \times 10 + 4 \times 10 = (1 + 8 + 4) \times 10 = 13 \times 1 = 1 \times 100 + 3 \times 10$$

and finally which includes one unit of 100 carried from the *tens* computation

$$1 \times 100 + 7 \times 100 + 1 \times 100 = (1 + 7 + 1) \times 100 = 9 \times 100.$$

Expressing this sum in the usual setup would be:

$$785.316$$
$$+ \quad 147.375$$
$$\overline{932.691}$$

As the reader can see, simply by lining up the two numbers, one under the other so that **the decimal points are in the same column**, and then adding the columns starting at the right ensures we are repeating the processes described above. The utility of aligning the decimal points will be evident as we work some numerical examples because aligning the decimal points causes all the other places to be aligned, one above the other, for example *hundredths* above *hundredths*, and so forth.

18.2.1 Addition of Decimals: Numerical Examples

Example 1. Add 132.165 and 25.204.

In this first example we will carefully explain each step so you understand how the theory is applied. Note that this explanation uses the representation of a decimal numeral as the sum of a whole number and decimal fractions. For this reason, it can be used to show children in Grade 5 why the simple fact of aligning the decimal points will produce a correct computation.

Grade 5 Explanation of Addition

Recalling the meaning of the decimal notation, the two numbers can be written as:

$$132.165 = 1 \times 100 + 3 \times 10 + 2 \times 1 + 1 \times \frac{1}{10} + 6 \times \frac{1}{100} + 5 \times \frac{1}{1000}$$

and

$$25.204 = 2 \times 10 + 5 \times 1 + 2 \times \frac{1}{10} + 0 \times \frac{1}{100} + 4 \times \frac{1}{1000}.$$

Performing the addition by adding terms having the same place multiplier gives:

$$
\begin{aligned}
132.165 + 25.204 &= (1+0) \times 100 + (3+2) \times 10 + (2+5) \times 1 + \\
&\quad (1+2) \times \frac{1}{10} + (6+0) \times \frac{1}{100} + (5+4) \times \frac{1}{1000} \\
&= 1 \times 100 + 5 \times 10 + 7 \times 1 + 3 \times \frac{1}{10} + 6 \times \frac{1}{100} + 9 \times \frac{1}{1000} \\
&= 157.369.
\end{aligned}
$$

The last sequence illustrates again that the algorithm works because the digits in each place have the same power of 10 as a multiplier.

Grade 5 Procedure for Addition

Now let's repeat the process in the fluid manner we expect children to master in Grade 5. The first step is to position the two numerals one above the other so the decimal points are aligned in a column as shown:

$$
\begin{array}{r}
132.165 \\
+25.204 \\
\hline
\end{array}
$$

Notice that by aligning the decimal points, *tenths* are above *tenths*, *hundredths* above *hundredths*, and *thousandths* above *thousandths*. Thus, aligning the decimal points assures that the digits from the two numerals are in the **correct place**. The addition begins by applying the standard procedure to the right-most column and continuing from there. For the three columns on the right, this produces:

$$
\begin{array}{r}
132.165 \\
+25.204 \\
\hline
369
\end{array}
$$

where the 3 is in the *tenths* column, so it must be marked on the left with a decimal point. This is accomplished by bringing down the decimal point as shown below.

$$
\begin{array}{r}
132.165 \\
+25.204 \\
\hline
.369
\end{array}
$$

The remainder of the process continues using the standard procedure to give:

$$
\begin{array}{r}
132.165 \\
+25.204 \\
\hline
157.369
\end{array}
$$

The central issue in using the revised procedure is making sure **the decimal points are aligned in a single column at the point of setup**.

The next problem requires carrying and illustrates that the same procedure applies.

Example 2. Perform the following addition:

$$
\begin{array}{r}
2.8 \\
+\quad 5.6 \\
\hline
\end{array}
$$

Starting on the right, we sum the *tenths* column. The result is 14, so we record the 4 below the line and carry the 1. This 1 is in fact ten *tenths*, and clearly belongs in the *ones* column and this is where we put it. In doing this, we carry the 1 across the decimal point to the next column consisting of digits exactly as if the decimal point were not there. But, **the decimal point must be recorded below the line in the same column**, as shown in the intermediate result.

$$
\begin{array}{r}
1 \\
2.8 \\
+\ \ 5.6 \\
\hline
.4
\end{array}
$$

The last step is to sum the *ones* column with the final result shown below.

$$
\begin{array}{r}
1 \\
2.8 \\
+\ \ 5.6 \\
\hline
8.4
\end{array}
$$

The examples above involve adding numbers having the same number of places to the right of the decimal point. We address the added complexity of a different number of places in our last example.

Example 3. Add 374.9 and 8.234. The setup is

$$
\begin{array}{r}
374.9 \\
+\ 8.234 \\
\hline
\end{array}
$$

The important thing children need to remember here is an empty place on the right of the decimal point can be filled with a 0 without changing the value of the number. We illustrate this:

$$374.9 = 374.90 = 374.900.$$

Rewriting the above using 374.900 and finding the sums in the first two columns working right to left gives:

$$
\begin{array}{r}
374.900 \\
+\ 8.234 \\
\hline
34
\end{array}
$$

The sum of the *tenths* column is 11, so put a 1 below the line, bring down the decimal point, and carry a 1 to the top of the *ones* column, as shown below:

$$\begin{array}{r} 1 \\ 374.900 \\ +\ 8.234 \\ \hline .134 \end{array}$$

The sum of the *ones* column is 13, so we write the 3 below the line in the *ones* place and carry a 1 to the top of the *tens* column, as shown:

$$\begin{array}{r} 11 \\ 374.900 \\ +\ 8.234 \\ \hline 3.134 \end{array}$$

Summing the remaining columns gives:

$$\begin{array}{r} 11 \\ 374.900 \\ +\ 8.234 \\ \hline 383.134 \end{array}$$

18.3 Theory of Decimal Arithmetic: Subtraction

In Chapter 19 we develop the full arithmetic of the integers and show that subtraction is really addition of additive inverses (see §19.2.5). Thus, the theory developed for addition of decimals applies to subtraction of decimals and the reasons why things work for addition must also apply to subtraction. However, in Grade 5 negative numbers are not available and subtraction is still taught as take-away. Thus, the only problems we need consider take the form $p - n$ where p and n are decimal numbers satisfying

$$0 < n < p.$$

The procedures are the same as for addition, namely, line up the decimal points and do the subtraction in the usual way. We do two examples of the Grade 5 procedure illustrating how the theory is applied.

18.3.1 Subtraction of Decimals: Numerical Examples

Example 4. Suppose we want to compute

$$\begin{array}{r} .854 \\ -.623 \\ \hline \end{array}$$

Because $.623 < .854$, the subtraction procedure is take away and hence we use exactly the process discussed in Chapter 9 with the added requirement that we must keep track of the decimal point. Thus,

$$
\begin{array}{r}
.854 \\
- .623 \\
\hline
.231
\end{array}
$$

The key to the computation is that by lining the decimal points up in a single column, we ensure that all the various places, *tenths*, *hundredths*, etc., are aligned.

Example 5. Suppose we want to find:

$$
\begin{array}{r}
7 \\
- \quad 6.23 \\
\hline
\end{array}
$$

Since $7 > 6.23$, the calculation can be done as take-away, but the fact that 7 is an integer appears to be a problem. However, as shown in §18.2.1 we can always put a 0 in an empty place on the right of the decimal point as in, $7 = 7.00$, so the computation is rewritten as:

$$
\begin{array}{r}
7.00 \\
- 6.23 \\
\hline
\end{array}
$$

Since the decimal points are aligned, we perform the subtraction using the standard procedure, borrowing across two columns and the decimal point as shown:

$$
\begin{array}{ccccc}
 & 6 & & 9 & \\
7\!\!\!/. & & \cancel{0} & & 10 \\
- \ 6. & & 2 & & 3 \\
\hline
\end{array}
$$

Notice that the first borrow of 1 put an extra 10 *tenths* in the *tenths* column, and the second borrow of $\frac{1}{10}$ put an extra 10 *hundredths* in the *hundredths* column by virtue of:

$$
\frac{1}{10} = \frac{10}{100}.
$$

At this point it is possible to complete the subtraction:

$$
\begin{array}{ccccc}
 & 6 & & 9 & \\
7\!\!\!/. & & \cancel{0} & & 10 \\
- 6. & & 2 & & 3 \\
\hline
0. & & 7 & & 7 \\
\end{array}
$$

Again, we remind the reader that the decimal point is written below the line in the same column.

18.4 Theory of Decimal Arithmetic: Multiplication

The theory underlying multiplication of decimal numbers is easy to explain and understand if we use the alternate representation of a decimal number as a product of an integer and a negative power of 10. The key theoretical reason is that products in general are subject to the Associative and Commutative Laws and products of powers of 10 use the Addition Law for Exponents (§C.5.1). Let's see how.

Consider multiplying two decimal numbers, p and n.

Step 1: Express the two numbers in standard decimal notation;

$$p_{1000}p_{100}p_{10}p_1.p_tp_hp_{th} \quad \text{and} \quad n_{1000}n_{100}n_{10}n_1.n_tn_hn_{th};$$

Step 2: rewrite each numeral as an integer times a negative power of 10 using the methods of the last section as in

$$p_{1000}p_{100}p_{10}p_1p_tp_hp_{th} \times 10^{-3} \quad \text{and} \quad n_{1000}n_{100}n_{10}n_1n_tn_hn_{th} \times 10^{-3};$$

Step 3: compute the product, call it m, of the two integers using the standard algorithm;

Step 4: compute the sum, call it k, of the two negative exponents in the powers of 10;

Step 5: introduce a decimal point into m by counting k places to the left of the rightmost digit in m.

Step 4 is an application of the Addition Law for Exponents. The introduction of the decimal point in Step 5 uses the procedure in §19.1.

Thus multiplication of decimals comes down to counting places to the right of the decimal point and the process is easily mastered by any child who knows how to multiply integers using the standard algorithm. All that has to be recognized is that the position of the decimal point in the answer is obtained by counting the total number of places to the right of the decimal points in the two factors comprising the product. The reason is that the total will be the sum of the negative powers of 10 in the original decimal numbers. It is not expected that children in Grade 5 will understand why the procedure works, but they can understand why as soon as the learn about exponents and a teacher who knows why could provide an explanation to a really curious student.

18.4.1 Multiplication of Decimals: Numerical Examples

Example 6. In this example, we apply the procedure in detail to illustrate the underlying theory. The explanation will involve exponents and as such will not be accessible to children in Grade 5. Nevertheless, those teaching need to know why things work at a deeper level than: it's what's in the book.

Typically, a decimal multiplication problem would appear as:

$$\begin{array}{r} 5.76 \\ \times\ 4.8 \\ \hline \end{array}$$

Following Step 1, we write the two numbers as 5.76 and 4.8. Step 2 converts these to

$$576 \times 10^{-2} \quad \text{and} \quad 48 \times 10^{-1}.$$

Finding the product of the two integers amounts to performing:

$$\begin{array}{r} 576 \\ \times\ 48 \\ \hline \end{array}$$

which we already know how to do using exactly the procedure developed in Chapter 9.

Applying the Addition Law for Exponents (see §C.5.1) we have

$$10^{-2} \times 10^{-1} = 10^{-3}$$

since $(-2) + (-1) = -3$ (see §19.4) and we put the decimal point three places to the left. Applying these procedures gives

$$27648 \times 10^{-3} = 27.648$$

which completes Step 5.

Example 7. Let's do another example, this time in a manner which you would teach to children fluent with the standard algorithm for multiplication.

$$\begin{array}{r} 2.6 \\ \times\ .5 \\ \hline \end{array}$$

Instead of going through all the steps of writing out the numbers as integers and powers of 10, let's just do the multiplication ignoring the decimal points as we would have in Chapter 9. Then we would have:

$$\begin{array}{r} 2.6 \\ \times\ .5 \\ \hline 130 \end{array}$$

The original problem had a total of two decimal places. What we know is that each digit to the right of the decimal point counts for one power of 10^{-1}. There are a total of two digits to the right of the decimal points in the two numbers, so there will be two places to the right of the decimal point in the answer as determined by Step 5. Thus the complete solution to the problem requires us to insert the decimal point in the answer as shown:

$$
\begin{array}{r}
2.6 \\
\times\ .5 \\
\hline
1.30
\end{array}
$$

Example 8. Use the simplified procedure to find:

$$
\begin{array}{r}
.46 \\
\times\ .13 \\
\hline
\end{array}
$$

Now, simply ignore the decimal points and perform the multiplication as integers. After multiplying by the digit 3, we have the intermediate result

$$
\begin{array}{r}
.46 \\
\times\ .13 \\
\hline
138
\end{array}
$$

The next step calls for multiplying by 1 and carefully placing the result in the *tens* column to obtain

$$
\begin{array}{r}
.46 \\
\times\ .13 \\
\hline
138 \\
+\ 46 \\
\hline
\end{array}
$$

Performing the indicated addition gives

$$
\begin{array}{r}
.46 \\
\times\ .13 \\
\hline
138 \\
+\ 46 \\
\hline
598
\end{array}
$$

At this point, the integer multiplication is complete. To place the decimal point, we count the total number of places to the right of the decimal points in the two factors. There are 4 places, so the complete solution is:

$$\begin{array}{r} .46 \\ \times\ .13 \\ \hline 138 \\ +\ 46 \\ \hline .0598 \end{array}$$

Notice that the *tenths* place contains a 0. We had to put a 0 in the *tenths* place to obtain 4 places to the right of the decimal point.

In summary, for children we can revise the 5 steps to a simpler two-step procedure for computing the product of two decimal numbers, m and n:

Step 1: Compute $m \times n$ using the methods in Chapter 9 and ignoring the decimal points.

Step 2: Count the number of places to the right of the decimal point in m, and the number of places to the right of the decimal point in n. The total number of places in both factors is the number of places to the right of the decimal point in $m \times n$.

These two steps are easily learned by children fluent with the standard multiplication algorithm.

18.5   Theory of Decimal Arithmetic: Division

Readers may remember the process learned in school for division involving decimal numbers. This process begins by moving the decimal point in the divisor to the right so that it becomes an integer. Simultaneously, the decimal point in the dividend must also be moved the same number of places to the right. For example, in $1.1 \div .02$, the initial setup is:

$$.\underline{02}\ |\ \overline{1.1}$$

and becomes

$$\underline{02}.\ |\ \overline{1\underline{1}0.}$$

after moving the decimal point in the divisor. Since $55 \times 2 = 110$, the quotient is 55. Why this process produces correct results is explained in this section. Unfortunately, like the theory underlying decimal multiplication, the explanation uses exponents. For this reason it is not accessible to children in Grade 5, but must be delayed until at least Grade 7. However, the procedure is accessible to children in Grade 5 who know the standard procedure (see §12.4) for finding quotients and remainders for counting numbers. Examples for children in Grade 5 explaining the setup and using the procedure are given in the next two sections.

In Chapter 12 we studied division for counting numbers. Thus, given a **dividend**, n, and a **divisor**, d, the procedure developed in §12.4.1 found a **quotient**, q, and a **remainder**, r, such that

$$n = d \times q + r,$$

where $0 \leq r < n$. In the case where $r = 0$, we wrote

$$n \div d = q.$$

The procedure used to find q and r was called the Division Algorithm (see §12.4).

In §15.2, the process of division was extended to common fractions x and $y \neq 0$, by defining the operation of division in terms of multiplication via the relations:

$$x \div y \equiv x \times y^{-1} = \frac{x}{y}$$

where y^{-1} is the multiplicative inverse of y (see §A.6). In making this definition, it was carefully explained why we should think of $x \times y^{-1}$ as a quotient (see §13.6.5). Our task here is to extend the Division Algorithm from integers to arbitrary decimal numbers.

Consider then the two decimal numbers $x = 33.74$ and $y = 2.1$. As shown in §18.1 each decimal number has a representation as a product of an integer and a negative power of 10:

$$x = 3374 \times 10^{-2} \quad \text{and} \quad y = 21 \times 10^{-1}.$$

Combining the representations with the definition of division and applying the rules governing exponents and multiplication of fractions, we convert $33.74 \div 2.1$ to:

$$
\begin{aligned}
33.74 \div 2.1 \ &\equiv \ (3374 \times 10^{-2}) \times \left(21 \times 10^{-1}\right)^{-1} \\
&= \ \frac{3374}{10^2} \times \left(\frac{21}{10^1}\right)^{-1} \\
&= \ \frac{3374}{10^2} \times \frac{10^1}{21} \\
&= \ \frac{3374}{21} \times \frac{10^1}{10^2} \\
&= \ (3374 \div 21) \times 10^{-1}.
\end{aligned}
$$

281

The key fact that results from this sequence is that $3374 \div 21$ can be computed using the Division Algorithm as developed in §12.4 and this enables us to find q and r such that

$$n = d \times q + r \quad \text{where} \quad 0 \le r < d.$$

For $3374 \div 21$ the procedure in §12.4 yields:

$$3374 = 21 \times 160 + 14.$$

But notice that we still have a power of 10 to deal with because

$$33.74 \div 2.1 = (3374 \div 21) \times 10^{-1} = \frac{3374}{21} \times 10^{-1}.$$

The factor 10^{-1} is applied to the numerator: $3374 \times 10^{-1} = 337.4$. This produces a revised setup for the original division problem, namely:

$$21 \,|\, \overline{337.4}$$

The reason this is a correct setup is that

$$\frac{3374}{21} \times 10^{-1} = \frac{337.4}{21} = \frac{33.74}{2.1}.$$

Further, for $q = 160$ and $r = 14$, as found above, we have

$$
\begin{aligned}
337.4 &= 3374 \times 10^{-1} = (21 \times 160 + 14) \times 10^{-1} \\
&= (21 \times 160) \times \frac{1}{10} + 14 \times \frac{1}{10} \\
&= 21 \times 16 + 1.4
\end{aligned}
$$

where the second line uses the Distributive Law and the third uses the Associative Law. So when we include the factor of 10^{-1} in the numerator, the quotient is 16 instead of 160 and the remainder is 1.4 instead of 14. Since 1.4 is not an integer, we split up the remainder into an integer and a residual as follows:

$$r = 1.4 = 1 + 0.4.$$

The Division Algorithm applied to the integers $337 \div 21$ gives a quotient of 16 and a remainder of 1 as shown below:

$$337 = 21 \times 16 + 1.$$

Putting all the pieces together in respect to the original division problem, $337.4 \div 21$, we have

$$337.4 = (21 \times 16 + 1) + .4$$

and we know where each piece comes from.

18.5.1 Setup of Division of Decimals: Examples

The following examples deal with setting up division problems in the context of decimals using the theory developed above.

Example 9. Suppose we want to find $5.1 \div 3.02$. Following the above scheme, we know

$$5.1 \div 3.\underline{02} \equiv (51 \times 10^{-1}) \times (302 \times 10^{-2})^{-1}$$
$$= \frac{51}{302} \times 10^{2-1} = \frac{510}{302}.$$

The procedure results in a ratio of integers and a setup that looks like:

$$3\underline{02} \mid \overline{\ 51\underline{0}\ }$$

It is important to realize that the power of 10 which we will multiply into the numerator may be positive or negative. We have underlined the two places where we moved the decimal point to the right in the divisor and the corresponding two places we must move the decimal point in the numerator. In this case, we are forced to add a zero creating a new *ones* place.

Example 10. As another example, consider $4.3 \div .05$. Again, following the scheme we have

$$4.3 \div 0.\underline{05} \equiv (43 \times 10^{-1}) \times (5 \times 10^{-2})^{-1}$$
$$= \frac{43}{5} \times 10^{2-1} = \frac{430}{5}.$$

Thus, the computational setup is:

$$\underline{05} \mid \overline{\ 43\underline{0}\ }$$

Example 11. Find $0.04 \div 0.3$. Applying the scheme gives:

$$.04 \div 0.\underline{3} \equiv (4 \times 10^{-2}) \times (3 \times 10^{-1})^{-1}$$
$$= \frac{4}{3} \times 10^{-2+1} = \frac{.4}{3}.$$

Thus, the computational setup is:

$$\underline{3} \mid \overline{\ 0.4\ }$$

where the dividend is a decimal number in the unit interval.

These examples follow a simple rule for the setup. Do the setup as in §12.4.1. Then move the decimal point in the divisor as many places to the right as needed to obtain an integer; move the decimal point in the dividend the same number of places to the right, adding zeros if and when required.

The underlines illustrate where the decimal points have been moved in each of the examples above.

18.5.2 The Division Algorithm Revised: Numerical Examples

Example 11 shows that we need to revise the Division Algorithm to accommodate dividends that are decimal numerals, not merely integers. As we will see, a revised computational version of the Division Algorithm that allows us to divide positive integers into decimal numbers is nothing more than the old algorithm with a properly placed decimal point. Because there is no substantial addition to the underlying theory, we will proceed by looking at some sample computations.

Example 12. We begin with the simplest possible example: $1 \div 2$. We put this in the usual format to apply the Division Algorithm, except we write 1 as 1.0 and we place a decimal point on the quotient line **directly above the decimal point in the dividend** as a marker. Note that the decimal point in the quotient is **exactly aligned** with the decimal point in the dividend: 1.0.

$$
2 \mid \overline{\,1\ \ .0}
$$

Let us ask ourselves how we would proceed if the problem we were required to solve was:

$$
2 \mid \overline{\,1\ \ \ 0}
$$

In such a case we would say 2 does not divide 1, but it does divide 10, the quotient being 5. So we place the five directly above the 0 in 10, as shown:

$$
2 \mid \overline{\,1\ \ \ 0}^{\ \ \ \ \ 5}
$$

Given the presence of the decimal point, we simply do the same thing, except that the reasoning is marginally different. In this case we say 2 does not divide 1, but it does divide 1.0, since

$$
2 \times .5 = .5 + .5 = 1.0.
$$

The point is the quotient is no longer an integer but a decimal number. We record this by writing the 5 above the line in the place to the right of the decimal:

$$\begin{array}{r} .5 \\ \hline 2\,|\ \ 1\ \ .0 \end{array}$$

The next step in the procedure is to multiply $2 \times .5 = 1.0$ and record the result. The 0 must be in the same column as the 0 in 1.0. This is the same as simply requiring the decimal points to be aligned, so we have:

$$\begin{array}{r} .5 \\ \hline 2\,|\ \ 1\ \ .0 \\ 1\ \ .0 \end{array}$$

The last step in the procedure is to subtract, as shown and this produces 0, so we stop., as shown.

$$\begin{array}{r} .5 \\ \hline 2\,|\ \ 1\ \ .0 \\ -\ \ 1\ \ .0 \\ \hline 0\ \ .0 \end{array}$$

So the Division Algorithm tells us that $1.0 \div 2 = .5$, which is a well known result. More precisely, we see that $\frac{1}{2}$ has an exact representation as a decimal number.

The reader should notice that the procedure for computing $10 \div 2$ to obtain 5 and the procedure for computing $1.0 \div 2$ to obtain $.5$ are the **same, except for the decimal point**. So in essence we can proceed as though the decimal point were not there, as long as we position it properly during the setup.

Example 13. Find $.4 \div 3$. Since the divisor is already an integer, the setup is shown below:

$$\begin{array}{r} . \\ \hline 3\,|\ \ .4\ 0 \end{array}$$

Now 3 divides 4 once, so we put a 1 above the 4 on the quotient line. Since the 4 was in the *tenths* place, the 1 is also in the *tenths* place. Now compute $.1 \times 3$ and put the product below, as shown with the 3 under the 4:

$$\begin{array}{r} .1 \\ \hline 3\,|\ \ .4\ 0 \\ -\ \ .3\ 0 \end{array}$$

This process automatically lines up the decimals points in the same column. After performing the indicated subtraction, we have:

$$\begin{array}{r} .1 \\ \hline 3 \mid .4\ 0 \\ -.3\ 0 \\ \hline .1\ 0 \end{array}$$

Notice that whether the difference is recorded as .1 or .10 is of no consequence because these are numerals for the same number. At this point, we know that

$$\frac{.4}{3} = .1 + \frac{.1}{3}.$$

The numerator of the fraction, 0.1 is a remainder term, exactly as we had when dividing counting numbers. And exactly analogous to that situation, we have

$$.4 = .1 \times 3 + .1.$$

To continue the computational process, we divide 3 into .10 to obtain .03. We find .03 by the same process of test multiplications analogous to what was described in §12.4.1. The difference is that in this case we test multiply by .01, .02, .03, until we find the largest multiplier whose product with 3 is less than .1 , as the following shows:

$$.03 \times 3 = .09 \le .1 < .04 \times 3 = .12.$$

The product $3 \times .03 = .09$ is recorded as shown:

$$\begin{array}{r} .1\ 3 \\ \hline 3 \mid .4\ 0 \\ -.3\ 0 \\ \hline .1\ 0 \\ -.0\ 9 \end{array}$$

Alternatively, we can obtain the same result by ignoring the decimal point and observing that

$$3 \times 3 = 9 \le 10 < 4 \times 3 = 12.$$

We record the 3 above the 0 to give exactly the result obtained from the previous, more complicated calculation. Simply keeping the decimal points aligned in a column takes care of everything!

Performing the indicated subtraction gives:

$$\begin{array}{r} .1\ 3 \\ \hline 3 \mid .4\ 0 \\ -.3\ 0 \\ \hline .1\ 0 \\ -.0\ 9 \\ \hline .0\ 1 \end{array}$$

The continued computation revises our previous expression for $.4 \div 3$ to:

$$\frac{.4}{3} = .13 + \frac{.01}{3}.$$

So our new expression for $.4$ is

$$.4 = .13 \times 3 + .01 = .39 + .01$$

and the remainder term, $.01$, is now smaller by a factor of $\frac{1}{10}$.

We may make the remainder term even smaller still by continuing the division process. To do this, we merely add another 0 to the dividend, and continue as shown:

$$
\begin{array}{r}
.1\,3 \\
\hline
3\,\vert\ \ .4\,0\,0 \\
-\ \ .3\,0 \\
\hline
.1\,0 \\
-\ \ .0\,9 \\
\hline
.0\,1\,0
\end{array}
$$

This time we want to divide 3 into $.010$. We find that

$$.003 \times 3 = .009 \le .01 < .004 \times 3 = .012,$$

so we record the $.003$ in the quotient and the $.009$, below as shown:

$$
\begin{array}{r}
.1\,3\,3 \\
\hline
3\,\vert\ \ .4\,0\,0 \\
-\ \ .3\,0 \\
\hline
.1\,0 \\
-\ \ .0\,9 \\
\hline
.0\,1\,0 \\
-\ \ .0\,0\,9 \\
\hline
\end{array}
$$

Performing the last subtraction gives:

$$
\begin{array}{r}
.1\,3\,3 \\
\hline
3\,\vert\ \ .4\,0\,0 \\
-\ \ .3\,0 \\
\hline
.1\,0 \\
-\ \ .0\,9 \\
\hline
.0\,1\,0 \\
-\ \ .0\,0\,9 \\
\hline
.0\,0\,1
\end{array}
$$

where we now have:

$$\frac{.4}{3} = .133 + \frac{.001}{3}$$

and our revised expression for .4 is

$$.4 = .133 \times 3 + .001 = .399 + .001.$$

At this stage, the remainder, .001, has been reduced in size by another factor of $\frac{1}{10}$. But the fraction, $\frac{.001}{3}$, is unchanged; that is, it continues to be $\frac{1}{3}$ times a negative power of 10. Only the negative power of 10 multiplier changes with each additional division. And we can continue this process indefinitely by adding zeros to the dividend. However, since we are always, in essence, dividing 3 into 10 and getting a remainder of 1, the result will simply add another 3 to the quotient and leave a remainder of 1 times an increased power of $\frac{1}{10}$.

The situation above, in which the computation repeats itself forever, produces what are referred to as **repeating decimals**. The notation for repeating decimals is to place an over-line above the digits that repeat. We give some examples:

$$\frac{.4}{3} = .1\overline{3}$$

$$\frac{1.4}{9} = .1\overline{5}$$

$$\frac{1}{12} = .08\overline{3}$$

$$\frac{5}{7} = .\overline{714285}$$

In each case, the pattern of digits under the over-line repeats endlessly. The number of places in the repeating pattern can be of any length. It is a fact that every common fraction will generate a decimal expression that either terminates, as in, $\frac{1}{2} = .5$, or repeats, as in the cases above. To illustrate how a multi-place pattern arises, we show the computation for $\frac{5}{7}$.

Example 14. Find the repeating decimal expression for $5 \div 7$. We apply the procedure discussed above, showing only the final step.

$$
\begin{array}{r}
.714285 \\
7\,|\;\overline{5\;.0000000} \\
-\;\;4\;.90 \\
\hline
.10 \\
-\;\;.07 \\
\hline
.030 \\
-\;\;.028 \\
\hline
.0020 \\
-\;\;.0014 \\
\hline
.00060 \\
-\;\;.00056 \\
\hline
.000040 \\
-\;\;.000035 \\
\hline
.0000050
\end{array}
$$

$$
\frac{5}{7} = .71428 + \frac{.000005}{7}
$$

so that

$$
5 = .71428 \times 7 + .000005.
$$

To see why the calculation repeats, consider what happens when the computation is extended. Continuing the computation means we need to find $.000005 \div 7$ which we know is equivalent $5 \div 7$ and multiplying the result by 10^{-6}. We have already computed $\frac{5}{7}$, and the result is shown. As we can see, the same result will be achieved if we repeat the cycle, only the place of the digits will change. Alternatively, we see that to continue the computations shown, when we bring down the next 0, we will be dividing 7 into $.0000050$, which means we will put a 7 in the quotient one place to the right of the 5, and end up subtracting $.0000049$ from $.0000050$, which is essentially where we started.

18.6 Recovering a Common Fraction from a Repeating Decimal

The procedure for recovering fractions involves solving a linear equation and requires a complete understanding of decimal representations. For these reasons the procedure is not introduced until Grade 8 in an A+ curriculum and many readers may wish to skip this section.

As we have discussed in §17.7, every unit interval between two consecutive integers looks like every other such interval in respect to the kind and position of fractions.

289

Thus, when it comes to recovering a common fraction from a repeating decimal, we need only consider proper fractions. What that means in terms of the decimal representation for that fraction is that all the digits to the left of the decimal point are zeros.

Now an arbitrary repeating decimal that represents a proper fraction consists of two finite strings of single digits, which we label r and s. We label the number of digits in the strings n and m, respectively. The repeating decimal then looks like:

$$0.s\overline{r}$$

where the m digits in the s string occur once and the n digits in the r string repeat forever. An example would be

$$0.12456\overline{714}$$

where $n = 3$ for $r = 714$ and $m = 5$ for $s = 12456$.

Recovering the rational requires solving a linear equation and complete knowledge of decimal representations. Moreover, the arithmetic that results can be quite complex as you will quickly see if you try to recover the rational in the above example.

For r, s, n and m as above, let x denote the unknown rational number we want to recover so that $x = 0.s\overline{r}$. Now we multiply x by different powers of 10 so that we get two different numbers both having $.\overline{r}$ to the right of the decimal place:

$$x \times 10^m = s.\overline{r} \quad \text{and} \quad x \times 10^{m+n} = sr.\overline{r}.$$

These two equations have **different** numbers on the RHS, but these two numbers have the same repeating sequence of digits to the right of the decimal place, $0.\overline{r}$. Subtracting the first equation from the second yields a whole number, not a decimal:

$$x \times 10^{m+n} - x \times 10^m = sr.\overline{r} - s.\overline{r} = sr - s.$$

Solving for x yields

$$x = \frac{sr - s}{10^{m+n} - 10^m}$$

which is a rational number. The key thing to notice is that the denominator, $10^{m+n} - 10^m$, is not a power of 10 as the simple example $100 - 10 = 90$ shows.

Example 15. Find the rational number associated with $0.\overline{3}$. We simply follow the instructions in the setup: $r = 3$, $n = 1$, s is the empty string and $m = 0$, so that

$$x \times 10^{1+0} - x \times 10^0 = 3.\overline{3} - 0.\overline{3} = 3$$

Simplifying the LHS gives

$$x \times 10 - x = 9x = 3$$

so that

$$x = \frac{3}{9} = \frac{1}{3}$$

which is what we knew it was.

Example 16. Find the rational number associated with $0.1\overline{3}$. We simply follow the instructions in the setup: $r = 3$, $n = 1$, $s = 1$ and $m = 1$, so that

$$x \times 10^{1+1} - x \times 10^1 = 13.\overline{3} - 1.\overline{3} = 12$$

Simplifying the LHS gives

$$x \times 100 - x \times 10 = 90x = 12$$

so that

$$x = \frac{12}{90} = \frac{2}{15}.$$

Although these two examples have been carefully chosen to simplify the arithmetic, they illustrate the method completely.

18.7   What Your Child Needs to Know

18.7.1   Goals for Grade 5

It is expected your child will be able to

1. fluently add and subtract multi-digit decimal numbers using the standard algorithm and explain the position of the decimal point;

2. fluently multiply multi-digit decimal numbers using the standard algorithm, including properly placing the decimal point;

3. fluently find decimal number quotients of decimal numbers with up to three decimal place dividends and two decimal place divisors using the standard algorithm;

4. be able to convert a common fraction into a decimal by dividing and recognizing the repeating decimal.

Chapter 19

The Negative Whole Numbers

Negative whole numbers and the set of integers, $\mathcal{I}$, are introduced in Grade 6 in two thirds of A+ countries; they are studied in 100% of A+ countries in Grade 7. The CCSS-M standards treat negative whole numbers extensively in Grade 6.

Negative numbers add much utility to the the number system and many difficulties simply disappear with their introduction. In addition, by Grade 6 students will be familiar with the use of equations in the form of the Associative, Commutative and Distributive Laws so that discussing ideas in the context of equations is meaningful. For this reason we adhere to the introduction of negative numbers in Grade 6 as required in the CCSS-M.

There are two ways to introduce negative numbers:

- computationally;

- geometrically.

In the end, they are equivalent. However, in the long run the issue for children is whether they are able to successfully perform computations. For this reason we stress the computational aspects because these are what contribute to algebraic skill that will be critical in the future.

19.1   Completing the Whole Numbers

Recall, in our discussion of subtraction of counting numbers in §9.2 we found that

$$n - m = p \text{ if and only if } n = p + m.$$

Thus, the operation of subtraction of $n - m$ is defined by an addition equation involving the unknown p. When these algebraic ideas are introduced to children in

Grade 3, the addition equation is written as

$$\boxed{?} + m = n.$$

In the case that $n < m$, it is a simple fact that there is no **counting number**, p, with the property that $p + m = n$ as the following numerical example shows:

$$p + 20 = 4.$$

Equations like $p + 20 = 4$ were considered to be nonsensical by European mathematicians well into the 17th century because there is no counting number solution.[1] Moreover, so long as we only have counting numbers available, it is clear why $m \leq n$ is required.

As we have remarked, all numbers, including counting numbers, are abstract ideas. So creating new numbers, beyond the counting numbers and zero, is not a problem. All we have to do is think of them!

19.1.1   The Notion of Additive Inverse

To follow the computational path, let's start with what we and all children know, namely, given any counting number n:

$$n - n = 0.$$

Here we are thinking of the process as being take away. But we want to think of the process in a different way, namely as addition, but with the same result. To do this, we rewrite the equation as follows:

$$n + (-n) = 0$$

and think of $(-n)$ as a new type of number; one that has the property set out in the equation, namely,

its sum with n must be 0.

When we think this way, it is clear that the quantity symbolized by $(-n)$ has to be something new, as the following numerical example shows

$$6 + (-6) = 0.$$

The quantity symbolized by (-6) is neither a counting number nor 0, so this quantity has to be something new.

---

[1]See Wikipedia on Negative numbers.

What we are now demanding is that given **any** counting number n, the **addition** equation

$$\boxed{?} + n = 0$$

should have a solution.

Again, what we all know is that this equation **does not have a solution** in the counting numbers, $\mathcal{N}$. Thus, the requirement, that for every counting number, n, the equation $\boxed{?} + n = 0$ has a solution, forces us to create new numbers, one for each counting number n. We call the solution to this equation the **additive inverse** of n and use the notation $(-n)$ for the solution. We require the Commutative Law apply so that

$$n + (-n) = (-n) + n = 0.$$

This equation is **fundamental** and so we will speak of this equation as the **Additive Inverse Equation** to emphasize its importance. We would stress that the notion of additive inverse is an algebraic (computational) property defined by the Additive Inverse Equation. The Additive Inverse equation specifies a behavioral relationship between two numbers and every child should understand the additive inverse concept in terms of this equation.

19.1.2 Defining the Whole Numbers Computationally

We can now say precisely what we mean by the set of **whole numbers**, or **integers**, which we denote by $\mathcal{I}$. Something is a whole number exactly if it satisfies one of the following three conditions:

1. it is a counting number, i.e., a member of $\mathcal{N}$ (see §8.1);

2. it is 0 (see §7.1);

3. it is the additive inverse of a counting number.

Recall, each counting number corresponds to the cardinal number of a collection and the counting numbers are naturally ordered by the process of adding one more element to a collection, a process we referred to as forming the **successor**.

For a number $k \in \mathcal{I}$ to satisfy the last condition means that we can produce a counting number, n, such that the sum

$$n + k = 0.$$

We stress that since n is a counting number and the operation is addition, the number k can not be either a counting number or 0.

It is important to understand what is happening here. We are identifying a number, namely the additive inverse of a counting number, by its behavior. Thus, when we speak of a particular number as being the additive inverse of some other number, in respect to behavior, we know with certainty that when we add these two numbers together, we must get 0 as the result. Another example of defining a number by its behavior occurred in respect to 0 which was required to satisfy the equation

$$0 + n = n + 0 = n.$$

This equation can, in fact, be taken as the defining property of 0.

We will use $\mathcal{I}$ to denote the set of whole numbers, integers. We note that the additive inverses of the counting numbers are generally referred to as **negative integers**, and the counting numbers are called **positive integers**.

We will use the descriptors *integers* and *whole numbers* interchangeably as names for $\mathcal{I}$.

19.1.3 Ordering the Whole Numbers

The counting numbers are ordered. To decide which counting number is the larger, m or n, the first principles procedure requires construction of a collection having m members, a collection having n members, and checking whether there is an exact pairing between the collections. If not, the collection having unpaired members has more and tells us which counting number is the larger (see §6.3).

We can use this as the basis for defining an order relation on the entirety of the whole numbers. We use the symbol $<$ as short hand for **less than** and for m and n any whole numbers what so ever, we write

$$m < n \quad \text{if and only if} \quad \text{for some } p \; m + p = n.$$

This definition of **less than** extends the first principles meaning of **more** to all integers and will be the basis for all future discussion of order on numbers.

19.1.4 Historical Note

According to Wikipedia, Chinese mathematicians knew about negative numbers in 200 BCE, but not about zero. Negative numbers were recognized as a measure of debt in commerce, and it was known how to calculate correctly with negative numbers. These ideas reached India by 400 CE and thence to the Middle East where they were being used by Islamic mathematicians in respect to debts by the 10th century. As we have already noted, European mathematicians resisted their use into the 17th century.

Fractions, as ratios of counting numbers, were known to the Pythagoreans, and predate negative numbers in the historical development of ideas.

19.1.5 Notation for Additive Inverse

Recall, that in order to make counting numbers useful, we had to come up with a system of notation. At this point we have notations for numbers that are counting numbers, and a notation for zero, but no notation for the new numbers that arise as additive inverses of counting numbers. We can deal with this problem by agreeing on a single notation for **all** additive inverses.

Let n be any integer. We will use n preceded by a centered dash to denote the additive inverse of n. Thus, $-n$, is the additive inverse of n, and the proof of this fact is that

$$n + (-n) = 0$$

because satisfaction of the Additive Inverse Equation is what defines *additive inverse*. We will refer to $-n$ as **minus** n. Thus, minus n will be the name of the additive inverse of n where we stress that n can be any integer whatsoever, positive, negative, or zero.

Concrete examples of this notation are:

$$-10, \quad -501, \quad -1, \quad -0, \quad -(-25),$$

and so forth. The respective behavioral equations are:

$$10 + (-10) = 0, \quad 501 + (-501) = 0, \quad 1 + (-1) = 0, \quad 0 + (-0) = 0,$$

and

$$(-25) + (-(-25)) = 0.$$

Other notations, for example, $^-1$ denoting -1, have been tried over the years for denoting the additive inverses of counting numbers. However, this notation, which uses parentheses and a centered dash, seems to have the fewest problems when it comes to rules of precedence which specify the order in which operations are to be performed.

19.2 Properties of Addition on the Integers

Now that we have a notation for additive inverse, we can give a precise definition of $\mathcal{I}$ in standard set theoretic notation:

$$\mathcal{I} = \{k : k \in \mathcal{N} \text{ or } k = 0 \text{ or } (k = (-n) \text{ for some } n \in \mathcal{N})\}$$

where $(-n)$ denotes the additive inverse of the counting number n. This definition asserts that the integers consist of exactly the three kinds of numbers identified in §19.1.2.

The next issue that must be addressed is extending the operation of addition to all integers. As remarked above, $\mathcal{N}$ comes complete with operations: **successor**, **addition** and **multiplication**. Recall that successor is the basis for counting and addition (see Chapters 6-8). This means that $+$ is already defined on $\mathcal{N}$ and that $+$ is commutative and associative on $\mathcal{N}$. Since $\times$ is defined as repetitive addition, $\times$ is also already defined on $\mathcal{N}$. Our objective is to extend these definitions to all of $\mathcal{I}$. Thus, given any two whole numbers, n and m, we have to be able to say what the sum of n and m is.

Our sole purpose here is that our arithmetic must model the real world. Thus, if our extended system of arithmetic, which now includes the negative integers, is to satisfy this condition, it must preserve everything that has been developed to date. To be clear, if n and m are counting numbers, then their sum as integers, $n + m$, must be the same as their sum as counting numbers, and we must be able to find this sum using existing procedures.[2]

In order to satisfy this requirement, the operation of addition as extended to **all** the integers must satisfy the following laws or axioms. What this means is that for n, m and k in $\mathcal{I}$, that is, arbitrary integers (whole numbers), then the following laws must hold:

1. the **Commutative Law**:
$$n + m = m + n;$$

2. the **Associative Law**:
$$(n + m) + k = n + (m + k);$$

3. the **Identity Law**:
$$n + 0 = 0 + n = n.$$

The definition of addition derived from counting (successor) ensures these laws hold for counting numbers and zero. In §19.2.4 below we will show how addition is extended to all integers in a manner that traces the operation back to counting and makes sure these laws hold. The extension process must preserve all existing sums and makes maximal use of what children know from their experience with counting. However, it is not a truly deductive approach such as found in FoA.

---

[2]For those comfortable with the jargon of software, as we extend addition from the counting numbers to the integers, we should have **forward compatibility**.

There is one other general property that holds in $\mathcal{I}$ together with $+$ which we state in the form of a theorem:

Theorem 19.1. Let n be any member of $\mathcal{I}$. Then n has an additive inverse.

Proof. Let n be an arbitrary member of $\mathcal{I}$. From the definition of $\mathcal{I}$, there are three possibilities concerning n.

Case 1: $n \in \mathcal{N}$. Such an n is a counting number, and for every such counting number $(-n) \in \mathcal{I}$ where by definition, $(-n)$ satisfies

$$(-n) + n = 0.$$

This means $(-n)$ is the additive inverse of n, so if $n \in \mathcal{N}$, n has an additive inverse in $\mathcal{I}$.

Case 2: $n = 0$. Since $0 + 0 = 0$, we know previous facts about addition continue to be true. Thus, 0 satisfies the additive inverse equation and is its own additive inverse.

Case 3: $n = (-p)$ where $p \in \mathcal{N}$. In this case, substituting $n = (-p)$ into the following, we have

$$n + p = (-p) + p = 0.$$

So p qualifies as an additive inverse for n. $\boxed{\text{qed}}$ [3]

It is not expected that children in Grade 6 would be presented with this proof. But it is expected that a child in Grade 6 could understand why particular numbers from each category must have an additive inverse.

For example, -9 is an integer because it is the additive inverse of the counting number 9, whence

$$9 + (-9) = 0$$

by definition of additive inverse, whence

$$(-9) + 9 = 0$$

by the Commutative Law so that 9 is the additive inverse of (-9). Going through particular examples like this should reinforce the basic laws so they become second nature. The proof above uses nothing more than what is contained in the definition of $\mathcal{I}$ and what is meant by additive inverse. The key fact that is required about additive inverses is that with the exception of 0, additive inverses come in pairs, each member of the pair being the additive inverse of the other member of the pair. This is true because the pair, m and n, satisfy the Additive Inverse Equation which has the form:

$$n + m = m + n = 0.$$

---

[3]The $\boxed{\text{qed}}$ symbol will be used to mark the end of a proof.

The Commutative Law tells us that if n is the additive inverse of m, then m is also the additive inverse of n. Moreover, to check this assertion about any pair of numbers, we simply compute the sum and see if that sum is 0.

Consider the following numerical example

$$(-7) + 7 = 0,$$

so 7 is the additive inverse of (-7) and (-7) is the additive inverse of 7. We emphasize that once we have created an additive inverse for each positive integer, all integers will have an additive inverse because additive inverses come in pairs and each member of the pair is the additive inverse of the other member of the pair. Every child should be aware of this fact and that each member of the pair is referred to as the additive inverse of the other member of the pair.

To summarize this section we note there are four key properties that will drive most of the future development of arithmetic. These are the Commutative Law, the Associative Law, the Identity Law and the Additive Inverse Equation. Every child needs to know these facts about addition and have a good sense of where they come from.

19.2.1   Additive Inverses Are Unique

Moving from a number to its additive inverse is essentially a unary operation. For this reason, we need to know that each integer has only one additive inverse. We provide a complete argument, including the reasons why we can make each step.

Theorem 19.2. Let $n \in \mathcal{I}$. Then n has exactly one additive inverse.

Proof. Let n be any integer. Suppose, n has more than one additive inverse. Let's call one of these $-n$ and the other k. Now since $-n$ is an additive inverse of n, we know

$$n + (-n) = 0,$$

and since k is an additive inverse of n, it must also be true that

$$n + k = 0.$$

Since *things equal to the same thing are equal to each other* (see E3, Transitive §8.4), we have

$$n + (-n) = n + k.$$

The Commutative Law tells us that:

$$(-n) + n = k + n.$$

Further, since $-n = -n$ and *equals added to equals are equal* (see E4, Addition §8.4), we can add $-n$ to both sides of the equation to obtain

$$((-n) + n) + (-n) = (k + n) + (-n).$$

The Associative Law tells us

$$(-n) + (n + (-n)) = k + (n + (-n)).$$

Since $n + (-n) = 0$, carrying out the indicated computation inside the parenthesis gives:

$$(-n) + 0 = k + 0.$$

Finally, since $(-n) + 0 = -n$, and $k + 0 = k$,

$$(-n) = k,$$

so that $(-n)$ and k are names for the same number. Thus, each integer n has only one additive inverse, for which the standard notation is $-n$. $\boxed{\text{qed}}$

Since there is only one additive inverse for any number, we say

additive inverses are unique.

On first reading, this argument may seem complicated; but its structure is worth reviewing because it is typical of methods used to demonstrate uniqueness. In essence, one takes two of something with the required property and demonstrates the two things must be equal to each other, as in $(-n) = k$ above.

The key thing the reader should take from this is that

for each integer, there is only one other integer with which it will sum to 0.

19.2.2 Using Uniqueness to Demonstrate Equality

Uniqueness of additive inverses will be used repetitively in what follows as a means of demonstrating equality. It is essential that the reader understand the demonstration process.

The first step in the process will be the identification of two, or more, quantities that we think might be equal. These quantities can be anything at all. There is only one requirement: they have to have names that make them identifiable. For example, they could be $a + b$, $q \times r$, or simply m.

The second step in the process is that we can find a single other number with which both, or all three, of the previous quantities sum to 0. This number also has to have a name. For purposes of this discussion, let's call this other number n. But remember, this number could have any name at all, just so long as we have a means of identification.

Now the computation $n + \boxed{?}$ is performed, as in:

$$n + (a + b), \quad n + (q \times r), \quad \text{and} \quad n + m.$$

If the result of any one of these computations is 0, we know we have found an additive inverse of n. Thus, for example, if

$$n + (a + b) = 0,$$

then we know that $a + b$ is an additive inverse of n. But we now know more than this. Because of uniqueness, we know there is in fact only one integer having the property that its sum with n is 0. The standard notation for this number is $-n$. So, once we know $n + (a + b) = 0$, this enables us to write:

$$a + b = -n.$$

Further, if we also find that $n + (q \times r) = 0$, then we also know that

$$q \times r = -n,$$

and therefore that

$$a + b = q \times r,$$

since both quantities are equal to the same quantity, $-n$. Finally, if it is also the case that $n + m = 0$, then we can write

$$a + b = q \times r = m.$$

For illustration purposes, we use this process to find the sum of (-5) and (-3). The numerical computation of $(-5) + (-3)$ is performed in detail.

Example 1. Find $(-5) + (-3)$. We know that

$$8 = 3 + 5.$$

Adding $(-5) + (-3)$ to both sides of the last equation produces

$$8 + ((-5) + (-3)) = (3 + 5) + ((-5) + (-3)).$$

Starting with the RHS and applying the Associative Law twice gives

$$
\begin{aligned}
(3 + 5) + ((-5) + (-3)) &= ([3 + 5] + (-5)) + (-3) \\
&= (3 + [5 + (-5)]) + (-3).
\end{aligned}
$$

301

Since (-5) is the additive inverse of 5, the RHS becomes

$$(3 + [5 + (-5)]) + (-3) = (3 + 0) + (-3).$$

Applying the Identity Law to $3 + 0$ followed by the fact that (-3) is the additive inverse of 3 gives

$$(3 + 0) + (-3) = 3 + (-3) = 0.$$

Recalling the original starting point, we conclude:

$$8 + ((-5) + (-3)) = 0,$$

so that $(-5) + (-3)$ is a name for the additive inverse of 8. Applying Theorem 19.2 and using the standard name for the additive inverse of 8 gives

$$(-5) + (-3) = -8.$$

Applying the Commutative Law to the LHS above tells us that

$$(-3) + (-5) = -8$$

as well.

19.2.3 Consequences of Uniqueness

Using the standard centered dash notation for the additive inverse of n, the Additive Inverse Equation states that

$$n + (-n) = (-n) + n = 0.$$

Theorem 19.3. $0 = -0$, hence 0 is its own additive inverse.

 Proof. From the Identity Law, we know that for every integer n, $n + 0 = 0 + n = n$. Taking $n = 0$, we have $0 + 0 = 0$ which is just the Additive Inverse Equation for 0. By uniqueness of additive inverses, $0 = -0$. $\boxed{\text{qed}}$

Theorem 19.4. For every integer n, $n = -(-n)$.

 Proof. Let n be an arbitrary integer. Observe, $-(-n)$ is the standard name for the additive inverse of the integer $-n$. We also know $-n$ is the additive inverse of n. Thus, the Additive Inverse Equation applied to n and $-n$ is

$$(-n) + n = n + (-n) = 0.$$

This equation tells us that n is an additive inverse for $-n$. Thus, n and $-(-n)$ are both names for the additive inverse of $-n$. Hence, by the uniqueness of additive inverses, Theorem 19.2, asserts

$$n = -(-n).$$

The symmetry property of equality, E2 in §8.4, lets us write

$$-(-n) = n,$$

as well. $\boxed{\text{qed}}$

A simple numerical example of this fact is $-(-40) = 40$. In words, **minus, minus 40 is equal to** 40.

19.2.4   Finding the Sum of Two Arbitrary Integers

In this section we establish the formula:

$$(-m) + (-n) = -(m+n),$$

which in words says

the sum of additive inverses is the additive inverse of the sum

and generalizes the fact expressed in Example 1. Knowing how to apply this result in concrete situations is an essential expectation of Grade 6.

This formula makes computations with negative integers trivial for any one who knows how to add counting numbers. For example

$$(-23) + (-61) = -(23 + 61) = -84.$$

As the reader can see, the computation comes down to adding a pair of counting numbers, namely, $23 + 61$.

Let's consider why

$$(-m) + (-n) = -(m+n)$$

might be true. On the RHS we have the additive inverse of $m + n$. If the LHS is also an additive inverse for $m + n$, then uniqueness of additive inverses will witness the truth of the equality. Let's turn this into a proof.

Theorem 19.5. For all integers n and m, $(-m) + (-n) = -(m+n)$.

Proof. We begin by observing that $m+n = n+m$, whence $-(m+n) = -(n+m)$, a fact we use below.

To show that $(-m) + (-n)$ is also an additive inverse for $n + m$, we start by applying the Associative Law twice as follows:

$$\begin{aligned}
((-m) + (-n)) + (n+m) &= ([(-m) + (-n)] + n) + m \\
&= ((-m) + [(-n) + n]) + m.
\end{aligned}$$

Since n and $-n$ are additive inverses, as are m and $-m$, we have

$$
\begin{aligned}
((-m) + [(-n) + n]) + m &= ((-m) + 0)) + m \\
&= (-m) + m \\
&= 0.
\end{aligned}
$$

Thus, $(-n) + (-m)$ is an additive inverse for $n + m = m + n$ and by uniqueness of additive inverses,

$$(-m) + (-n) = -(m + n)$$

is established. $\boxed{\text{qed}}$

Theorem 19.5 tells us how to compute sums of negative integers. For example,

$$(-55) + (-25) = -(55 + 25) = -80.$$

This is an example of where theory really becomes helpful in performing practical computations.

The same theorem will also tell us how to compute the sum of an arbitrary positive integer with an arbitrary negative integer.

Theorem 19.6. Let m, $n \in \mathcal{I}$ such that n is positive and $m = -k$ is negative, where k is a positive integer. If $k \leq n$ then $n + m = n - k$, where this computation is ordinary subtraction. If $n < k$, then $n + m = -(k - n)$, where again $k - n$ is found by subtraction.

Proof. That m is a negative integer means that $m = -k$ where k is a positive integer. After replacing m by $-k$, we see that we need to compute

$$n + (-k) = n + m.$$

There are two possibilities regarding k, namely, that $k \leq n$, or that $n < k$.

Case 1: $0 < k \leq n$. In this case, we know

$$n + m = n + (-k) = n - k,$$

where, $n - k$ is computed using the method for subtraction in §9.3.[4]

Case 2: $0 < n < k$. We want to find $n + m = n + (-k)$ where $0 < n < k$. To perform this computation, we will twice use the fact that for any integer p, $-(-p) = p$ (see lines 2 and 4 of the computation). Also we will use the fact that the additive inverse of a sum is the sum of the additive inverses to obtain line 3.

---

[4]Given this fact, the centered dash is used in two ways, to indicate subtraction, and to indicate finding the additive inverse. Some argue this is confusing. However, proper use of parentheses avoids any confusion.

Let's begin:

$$\begin{aligned} n + m &= n + (-k) \\ &= -(-[n + (-k)]) \\ &= -([-n] + [-(-k)]) \\ &= -([-n] + k) = -(k + [-n]). \end{aligned}$$

Thus,

$$n + m = -(k + [-n]).$$

The expression inside parentheses on the RHS, $k + (-n)$, satisfies the hypothesis of this case, $0 < n < k$, so that we now know how to perform the computation: simply use subtraction as in §9.3. Therefore, when $0 < n < k$, simply compute $k - n$ and the additive inverse of this number (see RHS of last equation) will be the required sum, $n + m$. $\boxed{\text{qed}}$

A couple of numerical examples would be helpful.

Example 2. Find

$$18 + (-16) \quad \text{and} \quad 18 + (-25).$$

For $18 + (-16)$, since $16 < 18$, we proceed under Case 1 above:

$$18 + (-16) = 18 - 16 = 2.$$

To find $18 + (-25)$, since $18 < 25$, we proceed under Case 2 as follows:

$$18 + (-25) = -(25 - 18) = -7$$

To summarize, we now know that the sum of any two integers is another integer, and we have a procedure for finding that sum using methods already developed for addition and subtraction of counting numbers. As a result, we say the integers are **closed** under addition. Finally, we note that because the operation of addition on integers reduces to addition of counting numbers, every addition computation with integers can be replicated with jars of buttons!

19.2.5   Subtraction as a Defined Operation

The rationale for negative numbers was founded on the idea of take away. As a result, the reader may wonder whether it is possible to subtract arbitrary integers from one another. The answer to this question is: Yes. The reason is because we replace the old operation of take away by a new operation which is defined in terms of addition for any pair of integers.

305

Thus, let m and n be any two integers. Then $m - n$ is defined by

$$m - n \equiv m + (-n).$$

As shown in previous sections, the quantity on the RHS can always be computed and it is always another integer. Further, if m and n are counting numbers with $n \leq m$, then $m - n = m + (-n)$ is exactly the result which would be obtained by the take away process discussed in Chapter 9.

The use of the centered dash, $-$, as both the subtraction symbol and the additive inverse symbol has a built in level of ambiguity. However, this ambiguity is tremendously reduced by the intention that subtraction, as an operation, should be eliminated and replaced by addition of the additive inverse. Indeed, this intention is captured by the rule that was taught to children when I was going to school, namely, subtraction means: **change the sign and add**. Understanding this intention makes it clear why the notion that the additive inverse can be both positive and negative must be made clear.

19.3   Multiplication of Integers

In Chapter 11, we studied multiplication of counting numbers which was formulated as **repetitive addition**. We found that, as applied to counting numbers, multiplication satisfied the following essential properties:

1. the **Commutative Law**:

$$n \times m = m \times n;$$

2. the **Associative Law**:

$$(n \times m) \times k = n \times (m \times k);$$

3. the **Identity Law**:

$$1 \times n = n \times 1 = n;$$

4. the **Two-sided Distributive Law**:

$$n \times (m + k) = n \times m + n \times k;$$

and,

$$(m + k) \times n = m \times n + k \times n.$$

All of the above properties can be established for the structure consisting of $\mathcal{N}$ and successor. Similarly, these properties were observed to hold in respect to addition as defined using counting numbers and real-world collections. The reason is straight forward: arithmetic is intended to model the real world, and multiplication of integers is a form of addition. Thus, these properties must also hold in respect to multiplication of integers.

In previous sections, we found that addition on the integers extended to all integers the definition of addition on the counting numbers derived from successor. By requiring these properties to hold in respect to multiplication, we ensure that multiplication on the integers extends the definition of multiplication on the counting numbers to all integers. In what follows, we will treat these properties as **axioms**. However, we note that they can all be proved as theorems from Peano's Axioms (see FoA).

19.3.1 Multiplication is Still Repetitive Addition

Much of the remainder of this book will be concerned with what happens to arithmetic as we add new kinds of numbers. In the present case we have added negative integers. One of the features of our arithmetic as developed for **counting numbers** was that the operation of multiplication was repetitive addition. For example, in §11.1 the product 4×5 was defined by the equation

$$4 \times 5 = 4 + 4 + 4 + 4 + 4.$$

It is important to realize that this interpretation of multiplication remains true for integers, and will remain true in the future so long as **one factor in a product is an integer**. There is a simple reason why this must be so. It is a consequence of the Identity Law for multiplication and the Distributive Law. To see why, consider that

$$
\begin{aligned}
4 + 4 &= 4 \times 1 + 4 \times 1 \\
&= 4 \times (1 + 1) \\
&= 4 \times 2
\end{aligned}
$$

where the last step simply uses the definition that $2 = 1 + 1$. This is how these two laws ensure that 2×4 must be $4 + 4$.

The identical sequence of steps applies to any integer, n, thus,

$$
\begin{aligned}
n + n &= n \times 1 + n \times 1 \\
&= n \times (1 + 1) \\
&= n \times 2
\end{aligned}
$$

Moreover, in any system of arithmetic that has a multiplicative identity and a Distributive Law, we can use the same sequence to obtain

$$
\begin{aligned}
x + x &= x \times 1 + x \times 1 \\
&= x \times (1 + 1) \\
&= x \times 2.
\end{aligned}
$$

where x represents an arbitrary number in the system being discussed. Again, the only requirement is that one factor in the product can be obtained by the repetitive addition of the identity element to itself, as in $(1 + 1 + 1 + 1) \times x$ to give another example. Thus, in any such system, there will be products where multiplication operation is repetitive addition.

There is one other feature of this discussion worth noting. In each of the last three computations, the reasoning was the same. The only difference between the first and last example was the change from 4 to x. The rest of the reasoning is identical. Once you realize this you will see that the same reasoning is used over and over again, not only here, but in many other situations as well. You should be on the look out for repeated use of the same reasoning because it makes coming to terms with these ideas much easier.

19.3.2 Multiplication by 0

A key fact about multiplication found in Chapter 9 was

$$
0 \times n = n \times 0 = 0
$$

where n was any non-negative integer. Since multiplication is still repetitive addition, this equation should continue to be true and apply to all integers. We give a proof from what we have taken to be axioms.

Theorem 19.7. Let n be any integer. Then $0 \times n = n \times 0 = 0$.

Proof. Let n be an arbitrary integer. Since 0 is the additive identity, we have $0 = 0 + 0$. We multiply through this equation by n to obtain

$$
0 \times n = (0 + 0) \times n.
$$

Applying the Distributive Law to the RHS gives

$$
0 \times n = 0 \times n + 0 \times n
$$

where the standard rule of precedence applies on the RHS (see §11.4.2). Now every integer has an additive inverse. Using the standard name for an additive inverse and E4 (see §8.4), we add $-(0 \times n)$ to both sides of the last equation to obtain

$$
0 \times n + [-(0 \times n)] = (0 \times n + 0 \times n) + [-(0 \times n)].
$$

308

The Additive Inverse Equation asserts that the sum on the LHS is 0 and applying the Associative Law to the RHS gives:

$$0 = 0 \times n + (0 \times n + (-[0 \times n)]).$$

Applying the Additive Inverse Equation to the RHS reduces the RHS to the center, to which we apply the fact that 0 is the additive identity to obtain the RHS below:

$$0 = 0 \times n + 0 = 0 \times n.$$

The Transitive property for equality (see §8.4) and the Commutative Law tell us that

$$0 = 0 \times n = n \times 0.$$

qed

    Let's review the facts required to obtain this theorem. First we have to have 0 as the additive identity. Next we need additive inverses. All the rest is due to the Distributive, Associative and Commutative Laws. Thus you should expect to see a theorem like this whenever you have a structure that has these kinds of features.

19.3.3   Multiplication by -1

Products of the form $(-1) \times n$ are special and have great utility as the next theorem shows.

Theorem 19.8. Let n be any integer. Then $(-1) \times n$ is the additive inverse of n.

    Proof. Fix an arbitrary integer n. The Additive Inverse Equation tells us that $1 + (-1) = 0$. Applying E5 (see §8.4) to multiply through this equation by n yields

$$(1 + (-1)) \times n = 0 \times n.$$

Applying the Distributive Law to LHS gives

$$(1 + (-1)) \times n = 1 \times n + (-1) \times n = n + (-1) \times n.$$

Since Theorem 19.7 tells us that the RHS, $0 \times n = 0$, the Transitive property for equality (E3, §8.4) yields

$$n + (-1) \times n = 0.$$

This means $(-1) \times n$ is an additive inverse for n. Since additive inverses are unique (Theorem 19.2), we have

$$(-1) \times n = -n.$$

qed

    We state this in words

the product of any number and minus 1 is the additive inverse of the number.

An immediate consequence of this is the fact that

$$(-n) \times m = n \times (-m) = -(n \times m) = (-1) \times (n \times m)$$

because using the Associative and Commutative Laws the (-1) can be moved anywhere in the product without affecting the answer as the next line shows:

$$((-1) \times n) \times m = n \times ((-1) \times m) = (-1) \times (n \times m).$$

Thus, all the expressions above are notations for the same number, namely, $-(n \times m)$, the additive inverse of $n \times m$.

Another useful result involving (-1) is:

Theorem 19.9. $(-1) \times (-1) = 1$.

   Proof. Theorem 19.8 asserts

$$(-1) \times (-1) = -(-1),$$

that is that the quantity on the LHS names the additive inverse of (-1). Theorem 19.4 asserts that

$$-(-1) = 1,$$

so that $(-1) \times (-1) = 1$ follows from the transitivity of equality. $\boxed{\text{qed}}$

19.3.4   Computing Arbitrary Products of Integers

A non-zero integer is either a counting number or the additive inverse of a counting number and we know how to find the product of any pair of counting numbers. Thus, let n and m be arbitrary counting numbers. If we have methods that enable us to find the products,

$$n \times (-m), \quad (-n) \times m, \quad \text{and} \quad (-n) \times (-m),$$

then we can find the product of any pair of integers whatsoever.

   Using the Theorem 19.8-9 and the Associative Law, we have the following:

$$n \times (-m) = (-m) \times n = ((-1) \times m) \times n = (-1) \times (m \times n),$$

$$(-n) \times m = ((-1) \times n) \times m = (-1) \times (n \times m),$$

and

$$(-n) \times (-m) = [(-1) \times n] \times [(-1) \times m] = (-1) \times (n \times [(-1) \times m])$$
$$= (-1) \times ([n \times (-1)] \times m) = (-1) \times ([(-1) \times n] \times m)$$
$$= [(-1) \times (-1)] \times (n \times m) = 1 \times (n \times m) = n \times m.$$

Each of the products comes down to finding the product of two positive integers, n and m. In two cases, we have to go another step and multiply by -1, which simply amounts to finding the additive inverse of the product. The point is that the computation simply uses previously learned skills for multiplying counting numbers as the following examples show.

Applying these results to particular computations gives:

$$7 \times (-8) = (-7) \times 8 = (-1) \times (7 \times 8) = -56$$

and

$$(-9) \times (-6) = (-1) \times (-1) \times (9 \times 6) = 54.$$

We are now able to find the product of any pair of integers because every non-zero integer is either a counting number or the additive inverse of a counting number. Moreover, we also know the product of any pair of integers is again an integer, whence the integers are **closed** under multiplication. Of course, since the integers are closed under addition, and multiplication is repetitive addition, the integers must also be closed under multiplication.

19.3.5 Summary of Arithmetic Properties of Integers

The **integers** (whole numbers), $\mathcal{I}$, consist of exactly three types of numbers:

1. counting numbers, also referred to as **positive** integers;

2. the number 0, also known as the **additive identity**;

3. numbers n for which there is a specific counting number, m, with the property that
$$n + m = 0.$$

   Numbers of this type are called **negative** integers and are additive inverses of counting numbers.

In set theoretic notation $\mathcal{I}$ is defined by

$$\mathcal{I} = \{k : k \in \mathcal{N} \text{ or } k = 0 \text{ or } (k = (-n) \text{ for some } n \in \mathcal{N})\}.$$

We note that this definition of $\mathcal{I}$ forces $0 \neq 1$.

The operations of addition, $+$, and multiplication, $\times$ are defined on $\mathcal{I}$ as extensions of the respective operations on $\mathcal{N}$ and satisfy the following for n, m, and k arbitrary members of $\mathcal{I}$:

1. **Closure**: $m + n$ is an integer, and $m \times n$ is an integer;

2. the **Commutative Laws**:

$$m + n = n + m \text{ and } m \times n = n \times m;$$

3. the **Associative Laws**:

$$(m + n) + k = n + (m + k) \text{ and } (m \times n) \times k = n \times (m \times k);$$

4. the **Two-sided Distributive Law**:

$$(m + n) \times k = n \times k + m \times k \text{ and } k \times (n + m) = k \times n + k \times m;$$

5. the **Additive Identity Law**:

$$0 + n = n + 0 = n;$$

6. the **Additive Inverse Law**: for each n we can find m with the property,

$$m + n = n + m = 0;$$

7. the **Multiplicative Identity Law**:

$$1 \times n = n \times 1 = n.$$

Structures consisting of a set containing distinct members 0 and 1 and on which there are two operations $+$ and $\times$ which satisfy the above axioms are called **rings**.

The following arithmetic rules hold for all integers n and m. We note $-n$ is the standard name for the additive inverse of n. Each rule is either a theorem or a direct consequence of a theorem.

I1 $-n$ denotes the unique additive inverse of n;

I2 $-0 = 0$;

I3 $n = -(-n)$;

I4 $\quad -(n + m) = (-n) + (-m)\,;$

I5 $\quad n \times 0 = 0 \times n = 0\,;$

I6 $\quad -n = (-1) \times n\,;$

I7 $\quad 1 = (-1) \times (-1)\,;$

I8 $\quad -(n \times m) = ((-1) \times n) \times m = n \times ((-1) \times m)\,;$

I9 $\quad n \times m = (-n) \times (-m)\,;$

I10 $\quad -(n \times m) = (-n) \times m = n \times (-m)\,.$

We stress that these laws and rules are universal. As we will see, they apply to all the mathematical quantities[5] that turn up in the modern world. They form the basis of how we think about, and manipulate, these quantities. It is their universal application that makes them so powerful and enabling.

19.4 The Whole Number Line

In Chapter 10 we discussed length as a numerical attribute and how this attribute could be used to model the addition process. In §13.1 we developed a graphical description of the positive half of the line as a measuring tape of unlimited length:

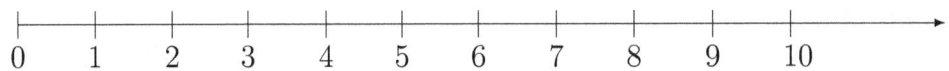

The positive half of the real line can be thought of as a measuring tape of unlimited length.

This representation of the counting numbers provides a visual interpretation of counting and the addition of counting numbers by virtue of juxtaposition.

What seems clear is that if it is reasonable to think about a measuring tape of unlimited extent going off to the right, as shown above, it should be equally possible to think about such a tape extending off to the left, or indeed, in both directions as in:

An unmarked line of unlimited extent in two directions.

---

[5]By mathematical quantities we mean anything which is measured by a real or complex number.

Our purpose is to create a geometric representation of the integers using the attribute of length. The process amounts to a construction according to the following procedure.

First, at an arbitrary place on the line we mark the position of 0. We then mark the place of 1 subject only to the requirement that this place be to the right of 0. That the place of 1 is to the right of 0 is purely a matter of convention and it is quite possible to construct a whole number line so that the place of 1 is to the left of 0. However, the convention is to put 1 on the right side of 0, and that is what we do:

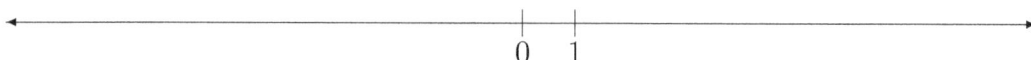

The first two steps in the construction process.

The distance between 0 and 1 now becomes the standard unit of distance so that the distance between n and $n+1$ is exactly the same as the distance between 0 and 1. All the other positive integers are positioned consecutively on the line using this fact so that the right half of the line looks like:

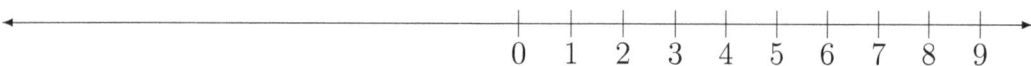

The right half of the whole number line.

Turn now to finding positions of negative numbers. Consider that n is a counting number and as such has a position on the right side of the line. It is paired with its additive inverse which is denoted $(-n)$. The property that defines the relationship between these two numbers is:

$$n + (-n) = (-n) + n = 0.$$

For this to be true in the sense of our additive model, $-n$ has to be exactly the same distance to the left of 0 as n is to the right of 0. If you think about it, you will see that this requirement must be in place if n and $n+1$ are to be the same distance apart as 0 and 1 for every pair of integers n and $n+1$. This means that the whole number line looks like:

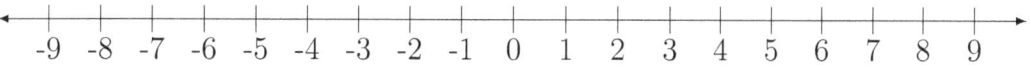

The whole number line.

All addition and subtraction computations with whole numbers can be reproduced using the line as a visual aid. In this respect, the whole number line plays the same role for $\mathcal{I}$ and $+$, as collections do for $\mathcal{N}$ and $+$. They serve as a concrete model. Collections cannot fill this role because the sum of two counting numbers is always another counting number, whereas the sum of two whole numbers may be positive, negative or zero.

While the line can be an excellent source of intuition, is cannot serve as a substitute for computational skill and knowledge of the Additive Inverse Equation:

$$n + (-n) = (-n) + n = 0$$

and what it means for two numbers to be additive inverses. This equation and the additive inverse concept are the bedrock on which understanding of negative numbers in relation to positive numbers is built and are what children need to know to succeed at algebra.

19.5   What Your Child Needs to Know

This chapter appears heavily theoretical. The reader might well wonder what it is doing in a book that is intended to help parents support children in primary and elementary? The answer is that the concepts discussed are introduced early in a manner appropriate to children in a given Grade. For example in Grade 1 ideas are expressed using whole numbers in the range $0 \le n \le 20$; in Grade 2 this restriction is relaxed to $0 \le n \le 100$. By the time a student completes the arithmetic portion of the math curriculum, it is expected the student knows and can apply the rules given above in all situations and is comfortable with their formulation using standard mathematical notation such as found in this book.

19.5.1   Goals for Grade 5

Graphing.

In Grade 5 children should be introduced to the first quadrant of the Cartesian coordinate system so that they are able to:

1. use a pair of perpendicular number lines, called axes, to define a coordinate system, with the intersection of the lines (the origin) arranged to coincide with the 0 on each line and a given point in the plane located by using an ordered pair of numbers, called its coordinates;

2. understand that the first number indicates how far to travel to the right of the origin on the horizontal axis, and the second number indicates how far to travel in the vertical direction on the second axis, with the convention that the names of the two axes and the coordinates correspond (e.g., x-axis and x-coordinate, y-axis and y-coordinate);

3. represent real world and mathematical problems by graphing points in the first quadrant of the coordinate plane, and interpret coordinate values of points in the context of the situation.

19.5.2   Goals for Grade 6

By the end of Grade 6 it is expected your child will:

1. understand that positive and negative quantities are used together to describe real situations, for example, monetary assets and debt, temperature scales, altitudes in relation to sea level;

2. understand the role of 0 in relation to positive and negative quantities in real situations, e.g., debt, temperature, sea level, etc.;

3. know that the equation $-(-n) = n$ is valid for all integers and understand why it must be true in terms of the Additive Inverse Equation;

4. write and evaluate numerical expressions involving whole-number exponents;

5. write, read, and evaluate expressions in which letters stand for numbers;

   (a) write expressions that record operations with numbers and with letters standing for numbers, e.g, express the calculation "Subtract y from 5" as $5 - y$;

   (b) identify parts of an expression using mathematical terms (sum, term, product, factor, quotient, coefficient);

   (c) view one or more parts of an expression as a single entity, e.g., describe the expression $2(8+7)$ as a product of two factors and recognize the quantity $(8 + 7)$ as both a single entity and a sum of two terms;

   (d) evaluate expressions at specific values of their variables;

   (e) evaluate expressions that arise from formulas used in real-world problems such as volume formulae;

(f) perform arithmetic operations, including those involving whole-number exponents, in the conventional order when there are no parentheses to specify a particular order (Order of Operations/Precedence);

(g) for example, use the formulas $V = s^3$ and $A = 6s^2$ to find the volume and surface area of a cube with sides of length $s = 1/2$ (see Chapter 12);

6. apply the properties of operations to generate equivalent expressions, e.g, apply the Distributive Law to the expression $3(2 + x)$ to produce the equivalent expression $6 + 3x$, and to the expression $24x + 18y$ to produce the equivalent expression $6(4x + 3y)$; apply the defining property of multiplication to $y + y + y$ to produce the equivalent expression $3y$.

7. identify when two expressions are equivalent (i.e., when the two expressions name the same number regardless of which value is substituted into them), e.g., the expressions $y + y + y$ and $3y$ are equivalent because they name the same number regardless of which number y stands for.

8. understand solving an equation or inequality as a process of answering a question: which values from a specified set, if any, make the equation or inequality true?

9. use substitution to determine whether a given number in a specified set makes an equation or inequality true;

10. use variables to represent numbers and write expressions when solving a real-world or mathematical problem;

11. understand that a variable can represent an unknown number, or, depending on the purpose at hand, any number in a specified set;

12. solve real-world and mathematical problems by writing and solving equations of the form $x + p = q$ and $px = q$ for cases in which p, q and x are all nonnegative rational numbers.

13. use variables to represent two quantities in a real-world problem that change in relationship to one another;

(a) write an equation to express one quantity, thought of as the dependent variable, in terms of the other quantity, thought of as the independent variable;

(b) analyze the relationship between the dependent and independent variables using graphs and tables, and relate these to the equation;

(c) for example, in a problem involving motion at constant speed, list and graph ordered pairs of distances and times, and write an equation such as $d = 65t$ to represent the relationship between distance and time.

The reader should think about these goals in terms of the development of ideas. Consider the notion of equality which is introduced in Grade 1 using the phrase:

is the same as.

There, the expectation is that children will deal with purely numerical expressions which might be supported by pictures of the type presented in the early chapters of this work. In Grade 3, the symbol for equality is introduced as mathematical shorthand. The expectation is that when children see an equation like $5 + 2 = 7$, they will think $5 + 2$ is the same as 7. By Grade 6, there is an expectation that the child can deal with equations like

$$y + y + y = 3y,$$

which reflect a far higher level of abstraction and state of knowledge. Indeed, equations like this one are the foundation of algebraic computations that will be presented in later grades. It is critical that your child becomes comfortable with this level of abstraction by the end of Grade 6.

Appendix A

Arithmetic of Real Numbers

All the important rules for algebraic computations are developed in these appendices and collected in §A.9.1. It is a short list, but all are facts that children should know.

In respect to an A+ curriculum, this material is in Grades 7-8. The A+ catergory is identified as *properties of rational/real numbers* and within this body of knowledge, justification for all of the rules and procedures of arithmetic can be found. A description of that portion of the curriculum is in ACC and links to the paper can be found at my website[1] or the EPC website.[2]

A.1 Key Arithmetic Properties of Integers

As a first step in the development of the real numbers, $\mathcal{R}$, we review the essential properties of the integers. We remind readers that as discussed in §10.4, the whole number line can used to reproduce addition computations with counting numbers.

The **integers** (whole numbers), $\mathcal{I}$, consist of exactly three types of numbers:

1. counting numbers, also referred to as **positive** integers;

2. the number 0, also known as the **additive identity**;

3. numbers n for which there is a specific counting number, m, with the property that
$$n + m = 0.$$

Numbers of this type are called **negative** integers and are additive inverses of counting numbers.

---

[1]http://www.math.mun.ca/ hsgaskill/
[2]http://education.msu.edu/epc/

The operations of addition, $+$, and multiplication, $\times$ are defined on the integers and satisfy the following for n, m, and k, arbitrary members of $\mathcal{I}$:

1. **Closure**: $m + n$ is an integer, and $m \times n$ is an integer;

2. the **Commutative Laws**:

$$m + n = n + m \text{ and } m \times n = n \times m;$$

3. the **Associative Laws**:

$$(m + n) + k = n + (m + k) \text{ and } (m \times n) \times k = n \times (m \times k);$$

4. the **Two-sided Distributive Law**:

$$(m + n) \times k = n \times k + m \times k \text{ and } k \times (n + m) = k \times n + k \times m;$$

5. the **Additive Identity Law**:

$$0 + n = n + 0 = n;$$

6. the **Additive Inverse Law**: for each n we can find m with the property,

$$m + n = n + m = 0;$$

7. the **Multiplicative Identity Law**:

$$1 \times n = n \times 1 = n.$$

Key facts to remember are that the whole numbers arise through counting and that all computations with counting numbers can be validated by counting collections of buttons of appropriate sizes.

In addition, the integers are ordered by $<$ which is defined for m, $n \in \mathcal{I}$ by

$m < n$ **if and only if for some counting number** (positive integer) k,
$$m + k = n.$$

This ordering of the integers is described graphically by:

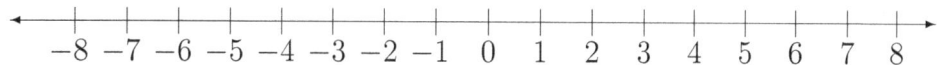

A graphical description of the integers constructed in §19.4 and referred to as the **number line**. The key feature is that the distance between every pair of successive integers is the same fixed length.

In what follows, we will treat the arithmetic properties above as axioms and use these properties to drive the continued development of our ideas.

A.2 What Are Real Numbers?

Real numbers are the numbers that are used to describe numerical quantities in science, engineering, and commerce. In short, if there is a physical thing in the modern world that has a number attached to it, that number is almost certainly a real number. This is the reason that real numbers have an essential place in the curriculum.

In the same way that counting numbers were a subset of the integers, we will show that the integers are a subset of real numbers; this is why we began this appendix by recalling the key properties of integers. However, unlike the integers which were derived from our notions about collections, the real numbers arise from our notions about geometry and length.

A.2.1 The Geometric Motivation for the Real Numbers

All of us have had to make measurements of physical things in our lives. These might include measuring the size of a room before going to buy carpet, measuring the width of a space in the kitchen to figure out what size of new fridge would fit, or simply recording the height of a child as he/she grows up. What we know from making such geometric measurements is that there are lots of numbers wandering around that do not correspond to whole numbers.

In school curricula, children are introduced to the idea that there are other kinds of numbers through the same process: making measurements. Such activities begin early in the Kindergarten curricula by having children identify numerical attributes of objects such as height and weight, and continue in later grades by having children make measurements using various standard units, for example, feet and meters for height, and pounds and kilograms for weight. One thing that children discover by making such measurements is that in most instances the result of a measurement is not a counting number. So the notion that there are numbers that are not whole numbers becomes clear very early in a child's school experience.

A.2.2 The Algebraic Necessity for Real Numbers

The arithmetic necessity for numbers that are not whole arises as soon as children are introduced to fractions in Grade 3. The physical necessity for fractional parts arises very early in a child's experience through the process of splitting a candy bar with a friend, or other similar experiences. Even if the initial experiences do not involve numbers, numbers are very quickly introduced.

In terms of arithmetic, finding solutions to equations like

$$2 \times x = 1$$

show why the whole numbers are insufficient. Other equations, for example,

$$x^2 - 2 = 0,$$

demonstrate further insufficiencies.

So the fact of the matter is that for both geometric and arithmetic reasons the whole numbers are insufficient to numerically describe the real world. In this appendix we develop the rules for working with real numbers. The rules incorporate all of what we have learned about counting numbers and whole numbers. Remarkably, there is only one additional key fact (axiom) that is required to generate the full power of arithmetic. Once we have all the axioms, we generate all the rules required to enable anyone to succeed at Algebra I, which is the keystone course prerequisite to all of the higher level courses. The ideas developed here are the reason why the methods for manipulating fractions, and all the other methods of algebra, work.

A.3 The Real Numbers from Geometry

We start by reiterating the point that all numbers are abstractions and as such do not exist in the world. This is why axiomatic treatments of the real numbers do not try to say what real numbers are: they focus on their behavior. Our intention is to develop the real numbers concretely using geometry as a guide.

In geometry, every line is defined by two points. Lines have unlimited extent, or can be extended as far as we want in either direction. The construction process for the whole number line (see §19.4) consisted of marking points on the line in such a way that all pairs of adjacent marks were the same distance apart. We may assume this length is determined by the initial two points defining the line.

A line showing equally spaced hash marks.

Unlike the whole number line, the line above has no labels. The hash marks simply indicate extent. Introducing numerals as labels did not occur until the 1600's (see Wikipedia entry for René Descartes).

The complete the construction process for the whole number line is presented in §19.4, and we do not repeat it here. We assume the reader knows how to arrive at the following graphic:

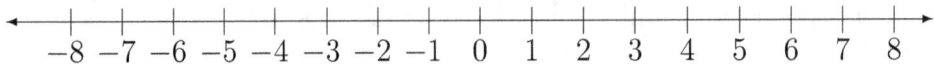

The whole number line depicting the integers. The labels identify numbers with places on the line.

Next, consider the line redrawn as follows:

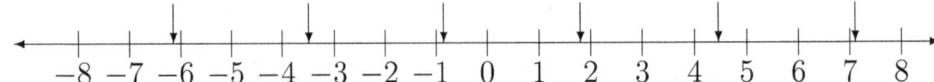

The number line again. Vertical arrows indicate identifiable places on the line that are **not** associated with integers. Each such place identifies a real number. This line is referred to as the **real line**.

We have added vertical arrows that point to places on the line. A fundamental question is whether each such arrow must also identify a number? The conclusion reached by thinkers in previous generations was that

> **wherever an arrow might point to on the line, there should be a unique number associated with that place.**[3]

The set of numbers that can be found by identifying all possible places on the real line is called the **real numbers** and is denoted by $\mathcal{R}$. This set includes not only the integers which are labelled on our diagram above, but also fractions and numbers like the square root of 2. These numbers are the principal objects that serve as the basis for science and technology in the modern world. This is why knowledge of their arithmetic is critically important.

As shown in §10.4, addition computations with counting numbers can be replicated on the line. Alternatively, as discussed in §19.4, the integers together with the operation of addition can be considered as a model for the line and length. The relation between the integers under addition and the number line as a physical reality does not depend on the distance between 0 and 1. For this reason we can use the physical model as a guide for much of our thinking. It is also why we began this appendix by recalling the properties of $\mathcal{I}$.

Our task in the remainder of this appendix is to develop the laws and rules of arithmetic as they apply to real numbers. In performing this task there are three ideas that will drive the development.

1. **The sum of any two real numbers must again be a real number.**

---

[3]As it is, this is not a statement of mathematics. However, its intent is clear and it can be turned into precise mathematics as shown in either FOA (§5) or ERA (§0.4).

2. **The product of any two real numbers must also be a real number.**

3. **The operations of addition and multiplication should satisfy the same laws that apply to integers.**

These are obvious facts about arithmetic of integers and their truth will guide our thinking about the arithmetic of $\mathcal{R}$.

A.4 Real Numbers not in $\mathcal{I}$

Based on the construction and labelling above, we know that every integer should correspond to a real number. Thus,

$$\mathcal{I} \subseteq \mathcal{R}.$$

However, as indicated by the vertical arrows in our last diagram, we know there must be at least some real numbers that fall in between the labelled hash marks if we are going to make measurements of length that are useful. So what kinds of numbers can we identify that lie in the spaces between integers?

A.4.1 The Common Fractions

The arithmetic of common fractions was presented in Chapters 13-15. By the end of Grade 5, children should understand common fractions as arising from the unit fraction concept as indicated by the following numerical example:

$$\frac{4}{7} \equiv \frac{1}{7} + \frac{1}{7} + \frac{1}{7} + \frac{1}{7}.$$

In respect to unit fractions children should be comfortable with the Fundamental Equation:

$$n \times \frac{1}{n} = \frac{1}{n} \times n = 1$$

and know that this equation reflects the physical fact that when a whole is divided into n equal parts and these parts are brought back together, the whole is restored (see §13.4).

Children must also recognize that the general common fraction notation reflects the fact that so long as one factor in a product is a whole number, multiplication arises from repetitive addition so that the $\frac{m}{n}$ results from adding $\frac{1}{n}$ to itself m times, we have:

$$\frac{m}{n} = m \times \frac{1}{n}.$$

324

This equation was given the name Notation Equation and is discussed at length in §13.6.

These two equations, Fundamental and Notation, are the keys to successfully understanding the arithmetic of fractions and children must be comfortable with their use.

Lastly, children must also know that the common fraction $\frac{m}{n}$ can equally well be thought of as arising through the process of division so that:

$$m \div n \equiv m \times \frac{1}{n} = \frac{m}{n}.$$

In Chapter 16 we discussed ordering the common fractions. As part of that discussion, it was shown that each common fraction had a unique position on the positive half of the line. A consequence of this fact is that each common fraction corresponds to a real number.

Now as every child who completes Grade 4 should know, the sum and product of two common fractions are again common fractions. Further, these operations are commutative, associative and distributive, so they satisfy the axioms listed in §A.1.

A.4.2   Other Real Numbers

There are still other numbers that turn up in the real world that are neither counting numbers nor fractions. Some such numbers arise as lengths. Others arise from geometric considerations having to do with circles. And still others arise from considerations having to do with the accumulation of interest on debts. Thus, the identification of these other numbers is based on the practical considerations of human activity, as the following example shows.

Long before the development of the Arabic system of numeration that has been critical to our development, it was shown that there are numbers arising as lengths that cannot be expressed as ratios of whole numbers (fractions). For example, consider the square shown in the figure.

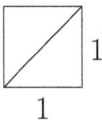

Given the sides of the square have unit length, what is the length of the diagonal, d?

It was known to the followers of Pythagoras in 500 BC that the length of the diagonal of a square was incommensurate with the length of its side (which, in effect, means the length cannot be expressed as a fraction). From the diagram it is clear that the length of the diagonal is longer than the length of one side, but shorter than the length of two sides. This idea is expressed by

$$1 < d < 1 + 1,$$

which is the next whole number greater than 1. So the number associated with the length of the diagonal is neither a fraction, nor a whole number.

Even so, the length of the diagonal can be used to identify a place on the real line, and hence a real number. So we take as a given that there is a number corresponding to this length. Using d to denote the length of the diagonal of the unit square and what readers may recall as the Pythagorean Theorem,[4] we have

$$d \times d = 1 \times 1 + 1 \times 1 = 1^2 + 1^2 = 2.$$

In other words, d is a solution to an arithmetical equation of the form

$$x^2 = x \times x = 2.$$

The number d that solves this equation is called the square root of 2.[5] The equation $x^2 = n$, where n is a counting number always has a solution in $\mathcal{R}$. However, unless the counting number is a perfect square, like 1, 4, 9, or 16, the solution will not be a fraction or a whole number. Numbers of this type are called **irrational numbers**. It is a remarkable fact that most real numbers are neither fractions nor whole numbers. But such considerations are beyond the scope of this book and we confine ourselves to simply giving an example of a familiar real number ($\sqrt{2}$) that is neither a fraction nor a whole number.

A.4.3 Combining Counting and Geometry

In the beginning we started with the counting numbers and collections. Now we have introduced geometry and measurement as a source for numbers. The immediate question is:

What is the relationship between numbers from these two different sources?

It is a fact that it is possible to develop the real numbers constructively from the Peano Axioms as in Landau's excellent book (see FoA). Constructing the real numbers as

---

[4]The readers may recall $a^2 + b^2 = c^2$ as the relation holding between the sides of a right triangle.

[5]Approximations to $\sqrt{2}$ date to 1000 years earlier. Wikipedia has an incredible picture of a Babylonian clay tablet showing how to approximate the square root of 2 to six decimal places.

an extension to $\mathcal{N}$ would automatically make $\mathcal{N}$ a subset of the reals. But, that is not the approach followed here. Instead, we will consider the real numbers as a model for the real line and lengths, that is, as the numerical measure of a directed line segment from 0 to an identifiable point on the line. Because we are dealing with **directed line segments** the numerical measure carries an algebraic sign, $+$ if the segment points to the right and $-$ if the segment points to the left.

To develop the real numbers in this manner, we need a set, which we denote by $\mathcal{R}$, operations $+$ and $\times$, and some assumptions (axioms) about how the operations behave on the set. (Recall the analogy with the set $\mathcal{N}$, successor to model the behavior of counting numbers and collections.) Proceeding this way, we will have to make sure that we also capture all the properties of the counting numbers. The way we will do this is by identifying a subset of $\mathcal{R}$ and an operation on that subset that we can think of as successor, and then showing that the important properties of the arithmetic of counting numbers are satisfied. Since we already know that the whole number line can replicate all addition computations with counting numbers, what we have to do is ensure our axioms actually capture **all** the important **arithmetic properties** of the whole number line.

A.5 Arithmetic (Algebraic) Properties of $\mathcal{R}$ Derived from $\mathcal{I}$

In the remainder of this appendix, we will develop arithmetic rules obeyed by all real numbers. What is amazing is how few of these rules there are and how much power they provide those who master their use. It is fact that essentially all children in A+ countries can learn and master the use of these rules. To succeed in teaching them to children, teachers/mentors must have a thorough knowledge of how and why these rules work. That knowledge is provided in what follows.

A.5.1 Axioms and Structures: a Digression

We have been thinking about arithmetic in two ways.

First, there are structures: the counting numbers, the integers, and the subject of this appendix, the real numbers. These structures include not only the numbers, but also the operations on the numbers. Thus, we think of the counting numbers and successor as a single unit which we refer to as a structure. Although $\mathcal{N}$ under successor is totally abstract, we have argued that it is sourced in the real world in the behavior of collections and the numerical attribute of cardinality.

Second, there are the rules these structures obey. When we use the term *obey*, we are referring to the entirety of the structure, that is, its elements and the operations on those elements. For example, we say the integers obey the Commutative Law of Addition. The real importance of the rules is they tell us how to do computations. The reader will recall that counting numbers without numerals were not very useful. Similarly, structures without rules for computation are not useful either.

There is a further point. Recall that we can define addition on $\mathcal{N}$, but that the addition on $\mathcal{N}$ does not satisfy all the properties satisfied by addition on $\mathcal{I}$. This observation produces a classification system for structures. Namely, identify a list of properties that are interesting and give that list a name. Then look for all the structures that satisfy that particular list. The five axioms (properties) listed below are called the **ring** axioms. Thus any structure that satisfies these axioms is referred to as a ring.

If we are really lucky when we select a group of axioms, there will only be one structure that satisfies them. As we will see, the real numbers are such a structure, but it takes some work to get there.

A.5.2 $\mathcal{R}$ Satisfies the Ring Axioms

The properties (axioms) identified below have all been studied previously in the context of the integers and are listed in §19.3.5. The properties were identified from the behavior of counting numbers associated with collections. In this sense, they can be considered experimental facts.

The following five ring axioms are true about $\mathcal{R}$ and the binary operations of $+$ and $\times$:

1. Closure: let $x,\ y \in \mathcal{R}$, then $x + y \in \mathcal{R}$, and $x \times y \in \mathcal{R}$;

2. let $x,\ y,\ z \in \mathcal{R}$, then the Commutative, Associative and Distributive properties hold as shown:

$$x + y = y + x \quad \text{and} \quad x \times y = y \times x$$
$$(x + y) + z = x + (y + z) \quad \text{and} \quad (x \times y) \times z = x \times (y \times z)$$
$$(x + y) \times z = x \times z + y \times z \quad \text{and} \quad z \times x + z \times y = z \times (x + y);$$

3. $\mathcal{R}$ has an **additive identity**, denoted by 0, with the property that for any $x \in \mathcal{R}$
$$0 + x = x + 0 = x;$$

4. every element, x in $\mathcal{R}$ has an **additive inverse**, y, that satisfies
$$x + y = y + x = 0;$$

5. $\mathcal{R}$ has an **multiplicative identity**, denoted by $1 \neq 0$, with the property that for any $x \in \mathcal{R}$

$$1 \times x = x \times 1 = x.$$

With two exceptions, 0 and 1, the axioms do not specify what things are in $\mathcal{R}$ or what real numbers are. To be specific, we cannot even say that $\mathcal{N}$ is a subset of $\mathcal{R}$. The axioms only specify how things behave in respect to the operations of addition and multiplication defined on $\mathcal{R}$.

Our task now is to identify the important theorems which are consequences of these axioms.

A.5.3   Additive Inverses and the Real Line

In Chapter 19 we proved the following theorem:

Theorem 19.1. Let n be any member of $\mathcal{I}$. Then n has an additive inverse.

Let's review how this theorem arose. First we had the counting numbers and 0 and the operations of addition and multiplication defined on them. We then created a collection of new numbers, $-n$, one for each counting number n, and required each new number to satisfy an equation of the form:

$$n + (-n) = 0.$$

It was after creating these new numbers that we proved Theorem 19.1 which asserted that now **every** number had an additive inverse.

In the present case, our approach is different. Axiom 4 (§a.1) asserts that every real number has an additive inverse. In other words, Theorem 19.1 is taken as an axiom, which means it is considered to be **unquestionably true**. But our purpose here is to understand, and that means we need to understand the connection between concrete geometry and abstract computations.

A.5.4   Representing Addition on the Line

In §A.3 we identified the real numbers with places on the real line. So imagine that we have identified a place on the real line to the right of 0 associated with a number which we will call w. We know that the length of an arrow that starts at 0 and ends at w is said to be w because, as discussed in §19.4, all numbers to the right of 0 are positive by choice. So when we give this place the name w, we are saying that the number named by w is the positive distance w units from 0 to the place on the line associated with w, where the unit of measure is determined by the length of the interval from 0 to 1. Now as discussed in §10.3, we know that if we measure the length from w to 0, the result will still be w because it cannot matter which end of the line segment we start at when it comes to determining length. Thus, an arrow that initiates at 0 and ends at w will be associated with a positive distance of w units.

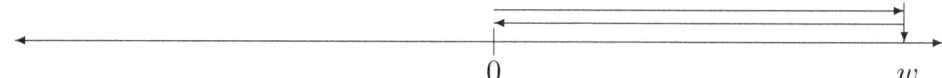

Two arrows, same length, namely w.

Suppose now we consider a second real number located on the left side of 0. Call this number z. As discussed in §19.4, z is a negative number, that is, not positive. Thus, if we measure the length of the line segment from 0 to z, it cannot be z because z is not positive and length is always positive. So what is the length of an arrow from 0 to z? We know that this arrow must have the same length as an arrow that initiates at z and ends at 0. Further, if we take this arrow and slide it to the right so that its tail is at 0, it will still have the same length and will identify that length as a point on the line that we are calling z^* in the following diagram:

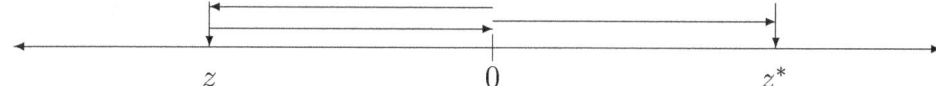

The horizontal arrows all have the same length, namely z^*.

This diagram establishes that there is an exact relation between z and z^* because they are both exactly the same distance from 0, just in opposite directions. A simple solution to this labelling problem is to refer to z as the **directed distance** from 0 to z. Thus, given any real number, z, we will refer to the number identified by an arrow that **initiates at 0 and terminates at z as the directed distance from 0 to z**. Unlike length, directed distance can be either positive or negative.

Recall that in §10.4 we discussed how to represent the addition of counting numbers. The procedure called for constructing two line segments of the required lengths, placing these line segments end-to-end and then finding the length of the resulting segment. We apply these ideas here.

Consider finding the sum of the real numbers z and w. First we find w and z on the line. Next we the construct arrows that initiate at 0 and end at w and z, respectively, as shown:

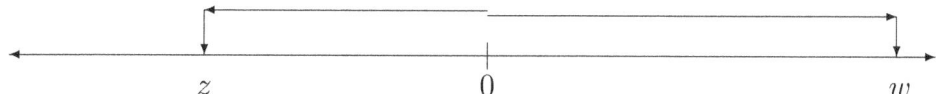

Directed line segments identifying w and z.

330

Then keeping one of the arrows fixed, align the tail of the other arrow with the head of the fixed arrow. In the case where z is fixed and w is moved the head of w now identifies $z + w$, as shown next:

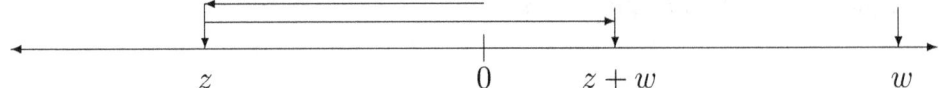

Finding the sum of $z + w$ using directed line segments. The directed line segment identifying w has been moved.

In a similar manner we can compute $w + z$ by keeping the arrow pointing to w fixed and sliding the arrow that points to z so it initiates at w:

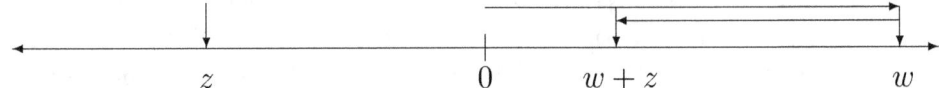

Finding the sum of $w + z$ using directed line segments. Note that the sum is the same in either case so addition must be commutative.

These diagrams show how all addition computations can be replicated geometrically using directed line segments.

Geometry and the Additive Inverse

Suppose we start with an arbitrary positive real number which we know has a position on the right of 0 on the real line. We can identify another real number, y, on the left of 0 that is the same distance to the left of 0 as x is to the right of 0 as shown in the next figure.

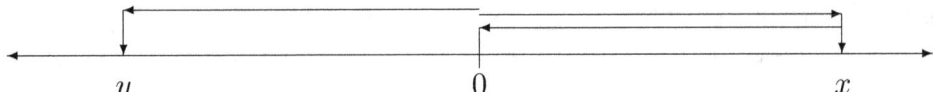

If a number identified by x is found at a place to the right of 0 on the line, then there must also be a number y that is the same distance to the left of 0 as x is to the right.

Recalling the process of obtaining z^* described above, we see that the number x that we started with is really y^*. Applying the ideas on addition to this diagram, we conclude that

$$x + y = y + x = 0,$$

whence using the language of §19.1, x and y are additive inverses of one another because their sum is 0. This is why we should believe that it is obviously the case that every real number has an additive inverse.

Given any real number x, the standard notation for the additive inverse of x employs the **centered dash** and is $-x$. Thus, given $x \in \mathcal{R}$, we know with certainty that

$$x + (-x) = (-x) + x = 0.$$

The second part of this equation, $(-x) + x$, captures the idea that starting at $-x$ and proceeding x units to the right puts us again at 0 which corresponds to keeping the left-pointing arrow fixed and moving the right-pointing arrow.

As the following numerical example shows, since x can be any real number, we immediately know that

$$\frac{4}{5} + \left(-\frac{4}{5}\right) = 0.$$

Theorem A.1. Let $x \in \mathcal{R}$. Then x has exactly one additive inverse.

Proof. Let x be any real number. Suppose, x has more than one additive inverse. Let's call one of these $-x$ and the other z. Now since $-x$ is an additive inverse of x, we know

$$x + (-x) = 0,$$

and since z is an additive inverse of x, it must also be true that

$$x + z = 0.$$

Since equality is transitive, (see E3, §8.2), we have

$$x + (-x) = x + z.$$

Further, since $-x = -x$ and *equals added to equals are equal* (see E4 §8.2), we can add $-x$ to both sides of the equation on the left to obtain

$$((-x) + x) + (-x) = ((-x) + x) + z.$$

Since $(-x) + x = 0$, carrying out the indicated computation inside the parenthesis gives:

$$0 + (-x) = 0 + z.$$

Finally, since $0 + (-x) = -x$ and $0 + z = z$,

$$-x = z,$$

so that $-x$ and z are names for the same number. $\boxed{\text{qed}}$

This theorem was proved for integers and extensively discussed in §19.2.1-2.2. It has great utility for showing two things are equal as we see below.

As noted in §19.4, geometry is an excellent source of intuition, but is cannot serve as a substitute for computational knowledge. For this reason we would avoid geometric terminology like **opposite** in reference to **additive inverse**. It is essential that children come to terms with the existence of additive inverses as being a key fact about arithmetic. Every child understands $x - x = 0$ as an equation about subtraction. The important idea to emphasize here is that $(-x)$ **names a new number** that is connected to x by addition through the equation

$$x + (-x) = 0.$$

A.5.5   Extending the Arithmetic Rules to $\mathcal{R}$

In Chapter 10 we developed a series of rules for doing arithmetic that applied to integers. These rules were stated as theorems. We repeat them here generalized to $\mathcal{R}$. Where it seems useful, we will provide a proof for real numbers. As in the case of integers, the importance of these theorems lies in the fact that they tell us how to perform computations, as for example finding $\frac{4}{5} + (-\frac{4}{5}) = 0$.

Theorem A.2. $0 = -0$, hence 0 is its own additive inverse.
   Proof. As in §19.2.3. $\boxed{\text{qed}}$

Theorem A.3. For every $x \in \mathcal{R}$, $x = -(-x)$.
   Proof. Let $x \in \mathcal{R}$ be arbitrary. Observe, $-(-x)$ is the standard name for the additive inverse of the real number $-x$. We also know $-x$ is the additive inverse of x. Thus, the Additive Inverse Equation applied to x and $-x$ is

$$(-x) + x = x + (-x) = 0.$$

This equation tells us that x is an additive inverse for $-x$. Thus, x and $-(-x)$ are both names for the additive inverse of $-x$. Hence, uniqueness of additive inverses, Theorem A.1, asserts

$$x = -(-x)$$

and we are done. $\boxed{\text{qed}}$

Theorem A.4. For all x, $y \in \mathcal{R}$, $(-x) + (-y) = -(x + y)$.
   Proof. Simply replace m and n in the proof of Theorem 19.5 by x and y and repeat the given argument. $\boxed{\text{qed}}$

Theorem A.5. Let $x \in \mathcal{R}$. Then $0 \times x = x \times 0 = 0$.
   Proof. Let $x \in \mathcal{R}$ be arbitrary. Since 0 is the additive identity, we have $0 = 0 + 0$. We multiply through this equation by x to obtain

$$0 \times x = (0 + 0) \times x.$$

Applying the Distributive Law to the RHS gives

$$0 \times x = 0 \times x + 0 \times x$$

where the standard rule of precedence applies on the RHS (see §9.2.2). Every real number, including $0 \times x$, has an additive inverse. Using the standard name, we add $-(0 \times x)$ to both sides of the last equation to obtain

$$0 \times x + [-(0 \times x)] = (0 \times x + 0 \times x) + [-(0 \times x)].$$

The Additive Inverse Equation asserts that the sum on the LHS is 0 and applying the Associative Law to the RHS gives:

$$0 = 0 \times x + (0 \times x + (-[0 \times x])).$$

Applying the Additive Inverse Equation to the RHS again reduces the RHS to the $0 \times x + 0$. Since 0 is the additive identity, we obtain the RHS below:

$$0 = 0 \times x + 0 = 0 \times x.$$

The transitive property for equality (see §8.2) and the Commutative Law tell us that

$$0 = 0 \times x = x \times 0.$$

qed

Theorem A.6. Let $x \in \mathcal{R}$. Then $(-1) \times x$ is the additive inverse of x.

Proof. Fix $x \in \mathcal{R}$. The Additive Inverse Equation tells us that $1 + (-1) = 0$. Multiplying through this equation by x and applying Theorem A.5 yield

$$(1 + (-1)) \times x = 0 \times x = 0.$$

Applying the Distributive Law followed by $1 \times x = x$ to LHS above gives the RHS below:

$$(1 + (-1)) \times x = 1 \times x + [(-1) \times x] = x + [(-1) \times x].$$

The transitive property for equality applied to the last two equations yields

$$x + [(-1) \times x] = 0$$

so that $(-1) \times x$ is an additive inverse for x. Since additive inverses are unique,

$$(-1) \times x = -x.$$

qed

Stated in words, Theorem A.6 is

> the product of any number and minus 1 is the additive inverse
> of the number,

and should be in every child's arithmetic toolbox because of its many applications, particularly when dealing with fractional forms. One application is:

Theorem A.7. Let $x, y \in \mathcal{R}$. Then

$$-(x \times y) = (-x) \times y = x \times (-y).$$

Proof. Using the Associative and Commutative Laws we have:

$$(-1) \times (x \times y) = ((-1) \times x) \times y = x \times ((-1) \times y).$$

Applying Theorem A.6 to each product gives:

$$-(x \times y) = (-x) \times y = x \times (-y).$$

qed

Theorem A.8. $(-1) \times (-1) = 1$.
   Proof. See §19.3.3. qed

A.6   Multiplicative Inverses: a New Arithmetic Property of $\mathcal{R}$

In §A.2 we pointed out that every common fraction was a real number because every common fraction had an identifiable place on the number line. This fact should have computational consequences, and it does.

Consider an equation like

$$\frac{4}{5} \times x = 7.$$

Here we are trying to find a real number that when substituted for x makes the equation true. We know this equation has no solution in the whole numbers, much less the counting numbers. But consider that we have full command of computations with fractions. Then we know that the result of dividing $\frac{4}{5}$ by itself will be 1 and since we also know that **equals divided by equals are equal**, we can divide both sides of the equation by $\frac{4}{5}$. Then the coefficient on x will be 1, so that:

$$\left(\frac{4}{5} \times x\right) \div \frac{4}{5} = 7 \div \frac{4}{5}.$$

Recall the rule for dividing by factions: invert and multiply, so that the equation becomes

$$\left(\frac{4}{5} \times x\right) \times \frac{5}{4} = 7 \times \frac{5}{4}.$$

Since multiplication is commutative and associative, we can rewrite the LHS as:

$$\left(\frac{4}{5} \times \frac{5}{4}\right) \times x = \frac{4 \times 5}{5 \times 4} \times x = 1 \times x = x$$

whence

$$x = \frac{7 \times 5}{4} = \frac{35}{4}.$$

A.6.1   The Notion of Multiplicative Inverse

One of the arithmetic properties of $\mathcal{R}$ derived from integers is that every real number has an additive inverse. Using standard notation, we have for every real number x, there is a number $-x$ with the property that

$$x + (-x) = (-x) + x = 0$$

where, 0 is the **additive identity**.

We also have a **multiplicative identity**, namely, 1. In addition, every unit fraction is a real number since every unit fraction has a place in the unit interval (see §13.7). Thus, given any counting number n, we know the Fundamental Equation is satisfied in $\mathcal{R}$:

$$n \times \frac{1}{n} = \frac{1}{n} \times n = 1.$$

This equation looks exactly like the additive inverse equation above with the operation of addition replaced by multiplication and 0 replaced by 1. For this reason, it must be the case that at least some real numbers satisfy an equation that could reasonably be referred to as the **multiplicative inverse** equation.

If every real number satisfied a multiplicative inverse property in the same way as they satisfy the additive inverse property, the following would be true.

Given $x \in \mathcal{R}$, we can find $y \in \mathcal{R}$ with the property that

$$x \times y = y \times x = 1.$$

This equation is not universally satisfied because if we take $x = 0$, then no matter how y is chosen, Theorem A.5 says:

$$0 \times y = y \times 0 = 0.$$

336

So we know 0 cannot have a multiplicative inverse. But as noted, all unit fractions and all counting numbers have multiplicative inverses. Therefore we might think that all real numbers **except** 0 ought to have multiplicative inverses and they do! We take this as our last axiom about arithmetic.

For every $x \in \mathcal{R}$, $x \neq 0$, there is a $y \in \mathcal{R}$ with the property that

$$x \times y = y \times x = 1.$$

The number y is referred to as the **multiplicative inverse** of x.

A.6.2 Summary of Arithmetic (Algebraic) Axioms for $\mathcal{R}$

The following are a list of the arithmetic (algebraic) axioms for $\mathcal{R}$. Structures satisfying this particular collection of axioms are called **fields**. In fact, there are many different fields and the real numbers are only one example so we will need more axioms to completely specify the structure $\mathcal{R}$ and eliminate the other fields. That said, the six field axioms given below are the ones that tell us how to do the kinds of calculations used in doing the ordinary work of the world such as solving equations, computing profit and loss statements, and so forth.[6] **The rules generated below apply in any field.**

Let x, y, $z \in \mathcal{R}$. Then the following six statements are axioms governing the two binary operations, $+$ and $\times$, on $\mathcal{R}$:

1. Closure: $x + y \in \mathcal{R}$, and $x \times y \in \mathcal{R}$;

2. the Commutative, Associative and Distributive properties hold as shown:

$$x + y = y + x \quad \text{and} \quad x \times y = y \times x$$
$$(x + y) + z = x + (y + z) \quad \text{and} \quad (x \times y) \times z = x \times (y \times z)$$
$$(x + y) \times z = x \times z + y \times z \quad \text{and} \quad z \times x + z \times y = z \times (x + y);$$

3. $\mathcal{R}$ has an **additive identity**, denoted by 0, with the property that for every $x \in \mathcal{R}$
$$0 + x = x + 0 = x;$$

4. every element $x \in \mathcal{R}$ has an **additive inverse**, y, that satisfies
$$x + y = y + x = 0;$$

---

[6]For a completely axiomatic approach to the development of $\mathcal{R}$, see ERA or FOA.

5. $\mathcal{R}$ has an **multiplicative identity**, denoted by $1 \neq 0$, with the property that for every $x \in \mathcal{R}$

$$1 \times x = x \times 1 = x;$$

6. for every $x \neq 0$, we can find a **multiplicative inverse** $y \in \mathcal{R}$ such that

$$x \times y = y \times x = 1.$$

As discussed in the next section, we will use x^{-1} as the standard notation for the multiplicative inverse of x. However, before proceeding with this discussion, we need to prove the uniqueness of multiplicative inverses.

Theorem A.9. Each non-zero real number has a unique multiplicative inverse.

Proof. We begin by fixing $x \neq 0$. Suppose that this x has two multiplicative inverses, y and x^{-1}, the latter being the one guaranteed to exist since $x \neq 0$. The following computation witnesses that y and x^{-1} are both multiplicative inverses for x:

$$x \times y = 1 = x \times x^{-1}.$$

Transitivity of equality tells us that

$$x \times y = x \times x^{-1}.$$

Multiplying both sides of this equality on the left by x^{-1} followed by applying the Associative Law, yields:

$$(x^{-1} \times x) \times y = (x^{-1} \times x) \times x^{-1}.$$

Since $x^{-1} \times x = 1$, substitution on the LHS and RHS gives

$$1 \times y = 1 \times x^{-1}.$$

Since 1 is the multiplicative identity, we have $y = x^{-1}$, so there was really only one multiplicative inverse after all. qed

A.6.3 Nomenclature and Notation

Knowledge of the nomenclature used in arithmetic is essential for understanding computations. Thus, **the content of this section is essential to what follows and a critical part of children's learning.**

We shall refer to the additive inverse of x as **minus** x, and use the symbols $-x$ to denote this number.

As in the case of integers (see §19.2.5), the operation of **subtraction** is then defined by:

$$x - y \equiv x + (-y).$$

Thus, there is no primary operation of subtraction, only a defined operation based on addition as shown. (Again, we use $\equiv$ which is read as: *is defined to be.*) The fact that subtraction is a form of addition gives rise to the computational rule for performing subtraction: *change the sign and add.*

Given $x \neq 0$, we shall refer to the **unique multiplicative inverse** of x as the **reciprocal** of x and denote it by the notational unit: x^{-1} where -1 is called an **exponent**. We stress that in its entirety, x^{-1}, is treated as the name of the unique number that satisfies

$$x \times x^{-1} = x^{-1} \times x = 1$$

where $x \neq 0$. The equation, $x \times x^{-1} = 1$, defines what is meant by x^{-1}. To give a numerical example,

$$5 \times 5^{-1} = 1.$$

Again, 5^{-1} is the notational identifier of the multiplicative inverse of 5. The reader will recall the Fundamental Equation which for counting number n states that

$$n \times \frac{1}{n} = \frac{1}{n} \times n = 1.$$

Thus, the unit fraction $\frac{1}{n}$ is also a notation for multiplicative inverse for n. Applying Theorem A.9 gives

$$\frac{1}{n} = n^{-1}.$$

This fact gives rise to the following definition of fractional notation. Let x denote any non-zero real number, then

$$\frac{1}{x} \equiv x^{-1}.$$

Every child should learn that $\frac{1}{x}$ is just one of the names for the multiplicative inverse of x.

In the same manner that we defined subtraction using: $x - y \equiv x + (-y)$, we can define division using multiplication. Given $y \neq 0$, we define **the operation of division** by:

$$x \div y \equiv x \times y^{-1}.$$

In this context, x is the **dividend**, y is the **divisor** and the product $x \times y^{-1}$ is the **quotient**. To see why we should regard the product of x and the multiplicative inverse of y, $x \times y^{-1}$, as a quotient, recall that for an integer quotient $q = n \div d$, dividend n, and one of its divisors integer d, we have:

$$q \times d = n.$$

In words, **the dividend, n, equals the quotient, q, times the divisor, d.** Now, for the reals x and $y \neq 0$, think of x as the dividend, y as the divisor and $x \times y^{-1}$ as quotient. Forming the product of the quotient times the divisor and applying the Associative Law to the LHS of the following equation gives

$$(x \times y^{-1}) \times y = x \times (y^{-1} \times y).$$

Since $y^{-1} \times y = 1$, we conclude the RHS above is just x, whence:

$$(x \times y^{-1}) \times y = x.$$

So the quantity $x \times y^{-1}$ behaves like a quotient in respect to y acting as a divisor of the dividend x.

Recalling the notation for the reciprocal, for counting numbers m and n, we have:

$$m \div n \equiv m \times n^{-1} = m \times \frac{1}{n}.$$

Recalling the Notation Equation for common fractions:

$$m \times \frac{1}{n} = \frac{m}{n}$$

which explains why $m \div n$ identifies the same place on the line as $\frac{m}{n}$.

It also suggests making the following general definition of fractional forms. Let $x, y \in \mathcal{R}$ with $y \neq 0$. Then

$$\frac{x}{y} = x \times y^{-1}.$$

This definition has great utility because it replaces fractional forms by products. In performing computations with products, the full power of the Associative, Commutative and Distributive Laws apply and we will make great use of this definition below.

The fractional notation merely extends to all real numbers, and all the expressions of algebra that represent real numbers, the notation developed for common fractions in §13.6. In respect to expressions of the form $\frac{x}{y}$, the standard nomenclature is retained. Thus, x is the **numerator** and y is the **denominator** of the fractional form: $\frac{x}{y}$.

There is an immediate ambiguity that arises out of the usage of the centered dash and the exponential notation, namely, what do we mean by:

$$-x^{-1} \ ?$$

Two different activities are specified. The centered dash tells us to find the additive inverse of x, while the exponent -1 tells us to find the multiplicative inverse of x. Which activity gets performed first? There is an order of precedence rule that removes the ambiguity. The rule says

all exponential operations are performed before addition and/or multiplication.

Because of these conventions, the expression on the LHS is not ambiguous and is equal to the RHS,

$$-x^{-1} = -(x^{-1}) = -\frac{1}{x},$$

which in words means the quantity being computed is the additive inverse of the multiplicative inverse of x. If the minus sign is moved inside the parentheses as in $(-x)$, the computation becomes

$$(-x)^{-1} = \frac{1}{-x}.$$

It turns out that for the case of $-x^{-1}$, the two expressions are equal by Theorem A.13 (see §13.9). However, if the exponent were divisible by 2, as in the expressions $-x^{-2}$ and $(-x)^{-2}$, the computation would result in a pair of additive inverses. (Exponents will be discussed in Chapter 18.)

Useful rules governing the behavior of these forms will be developed below.

A.7   Notation for the Integers as a Subset of $\mathcal{R}$

The axioms for $\mathcal{R}$ and the operations of $+$ and $\times$ do not mention successor. However, they do mention both 0 and 1 and give them both very specific properties which we take advantage of to provide notations for all the numbers we have been discussing in previous chapters. If we want to find $\mathcal{N}$ and $\mathcal{I}$ inside $\mathcal{R}$, we need names for all these things as specific members of $\mathcal{R}$. In other words, we need a system of numeration.

We define the numeration system on $\mathcal{R}$ by the following:

$$2 \equiv 1+1 \quad \text{and} \quad 3 \equiv 2+1 \quad \text{and} \quad 4 \equiv 3+1$$
$$5 \equiv 4+1 \quad \text{and} \quad 6 \equiv 5+1 \quad \text{and} \quad 7 \equiv 6+1$$
$$8 \equiv 7+1 \quad \text{and} \quad 9 \equiv 8+1 \quad \text{and} \quad 10 \equiv 9+1$$

and so forth. This process supplies the usual names from the Arabic System of notation. In particular, there is a notation for each integer in $\mathcal{I}$, where $\mathcal{I}$ is generated from $\mathcal{N}$ as set out in Chapter 19. The reason this is so is that the naming process preserves the **successor** idea in the form of adding the multiplicative identity 1 as in the successor of 4 is $4 + 1 \equiv 5$.

The naming process assigns each numeral to a unique member of $\mathcal{R}$, that is, each numeral is paired with a specific member of $\mathcal{R}$ as identified by a place on the line. Moreover, as we know from our original construction of the line in §19.4, these positions correspond to whole numbers. If our six field axioms had the property that the only field satisfying them corresponded to the geometric construction of the line, we would be done. But we still need the order axioms (see Appendix B) to capture $\mathcal{R}$.

There are many different structures that satisfy these six axioms and each of them has a 0 and a 1. Thus, using the assignment scheme above, we have a complete system of notation for the members of such a field corresponding to integers.

Directing our attention back to $\mathcal{R}$, as discussed in §19.4, we know that the integers can be identified as unique positions on the line and further that addition can be interpreted in terms of length. Using this relationship, every individual computation involving addition could be physically modelled using a finite portion of the real line and directed line segments in exactly the same way that every sum of counting numbers can be physically modelled by constructing collections of the appropriate size and combining them to find the sums. Thus, through the naming process, we can think of every positive integer as being a member of $\mathcal{R}$. Once we have identified a particular positive integer n, we know its additive inverse must also be in $\mathcal{R}$. Since there are only three types of integers, counting numbers, zero and additive inverses of counting numbers, it follows that all integers belong to $\mathcal{R}$. Thus, consistent with previous claims, we can write:

$$\mathcal{N} \subseteq \mathcal{I} \subseteq \mathcal{R}.$$

There is an addition table in $\mathcal{R}$, attached to this system of naming which is identical to the addition table we previously constructed. Nevertheless, we perform a calculation using the Commutative and Associative properties for addition, and the definitions of our symbols, to illustrate why the entries in the addition table are what we expect.

$$
\begin{aligned}
4 + 3 &= 4 + (2 + 1) = 4 + (1 + 2) = (4 + 1) + 2 \\
&= 5 + 2 = 5 + (1 + 1) = (5 + 1) + 1 \\
&= 6 + 1 = 7
\end{aligned}
$$

which is exactly the result in the addition table in §8.6.1. We stress that the successor idea has been incorporated into the assignment of names and that fact is essential to the process of finding the sum. The computation on each line simply subtracts 1

from the current value on the right, and adds it to the number on the left. So $4 + 3$ on the first line becomes $5 + 2$ on the second. The process continues until a single integer remains, in this case, 7. In terms of our real-world button analogy, each line corresponds to moving one button from the jar on the right to the jar on the left and stops when the jar on the right is empty. It's as simple as that.

A.7.1 Multiplication by Integers is Repetitive Addition

In §11.1 multiplication of counting numbers was defined to be repetitive addition. The reader may wonder whether this is still the case. That multiplication of two integers in $\mathcal{R}$ must come down to repetitive addition is enforced by the Distributive Law and the definitions of the symbols, as the following calculation shows:[7]

$$5 \times 3 = 5 \times (2 + 1) = 5 \times 2 + 5 \times 1 = (5 + 5) + 5.$$

Equally well, since the Distributive Law is two-sided, we can write

$$
\begin{aligned}
5 \times 3 &= (1 + 1 + 1 + 1 + 1) \times 3 \\
&= 1 \times 3 + 1 \times 3 + 1 \times 3 + 1 \times 3 + 1 \times 3 = 3 + 3 + 3 + 3 + 3
\end{aligned}
$$

so that in either direction, multiplication of counting numbers is repetitive addition.

The Distributive Law also ensures that when an arbitrary real is multiplied by a counting number on either the right or the left, the result is repetitive addition. To see this, let a denote any real number. Then

$$2 \times a = (1 + 1) \times a = 1 \times a + 1 \times a = a + a$$

and

$$a \times 2 = a \times (1 + 1) = a \times 1 + a \times 1 = a + a.$$

This relation, $2 \times a = a + a = a \times 2$, shows how to verify that multiplication of an arbitrary real number by a **counting number** is repetitive addition of a added to itself the number of times determined by the counting number.

A.8 Finding Common Fractions as a Subset of $\mathcal{R}$

As discussed above, **common fractions** are numbers that can be written in the form $\frac{m}{n}$ where $m, n \in \mathcal{I}$ and $n \neq 0$. In the last section we explained how the integers were

---

[7]Remember, multiplication takes precedence over addition.

identified within $\mathcal{R}$. In respect to the arithmetic of $\mathcal{R}$, we showed $\frac{1}{n} = n^{-1} \in \mathcal{R}$, whence

$$\frac{m}{n} = m \times \frac{1}{n} = m \times n^{-1} \equiv m \div n \in \mathcal{R}$$

by closure. In other words, common fractions are **ratios** of whole numbers where the divisor is not zero. Thus every such ratio of whole numbers corresponds to a particular real number, and as discussed above, a place on the line. We therefore define the **rational numbers**, $\mathcal{Q}$, by:

$$\mathcal{Q} = \left\{ \frac{n}{m} : m, \ n \in \mathcal{I} \ \ and \ \ n \neq 0 \right\}.$$

We could describe $\mathcal{Q}$ in words by:

> Q is the set of all fractions having an integer in the numerator and a non-zero integer in the denominator.

In still other words, the rational numbers are comprised of all numbers that can be obtained as a ratio of two integers, where the divisor must be non-zero. The ambitious reader can verify that $\mathcal{Q}$ under addition and multiplication satisfy the six field axioms. Indeed, $\mathcal{Q}$ is the smallest field that sits inside $\mathcal{R}$.

We stress that every common fraction is a real number and as will be shown below,

$$\mathcal{N} \subseteq \mathcal{I} \subseteq \mathcal{Q} \subseteq \mathcal{R}.$$

A.9   Rules of Arithmetic for Real Numbers

Our purpose in this section is to develop the remaining facts about the arithmetic of real numbers not already contained in Theorems A.1-A.9. These facts end up as computational rules taught to children. What makes these rules so useful is their universal application; they apply to all the numerical quantities we use to describe the world.

A+ curricula contemplate that children will learn how things work in a step-by-step manner that starts with making the connection between addition and counting, then multiplication as repetitive addition. In elementary, what qualifies as justification is being able to make these kinds of connections. By the end of the elementary curriculum children should be comfortable applying the rules of arithmetic in their algebraic form and have a fair idea of how they can be justified by tracing things back to basic principles: e.g., addition is counting, multiplication generates area, etc. We will derive the theorems from the axioms. In doing so the reader should focus on the Associative, Commutative and Distributive Laws as drivers of the development.

These axioms are based on our experience with counting and **CP** and mentors should be able to explain the connections to children in a clear manner.

Throughout this section x, y, $z \in \mathcal{R}$ are arbitrary.

Theorem A.10. $1 = 1^{-1}$ and $-1 = (-1)^{-1}$.

Proof. To establish $1 = 1^{-1}$, observe:

$$1 = 1 \times 1^{-1} = 1^{-1}.$$

The LH equality is due to the fact that 1^{-1} is the multiplicative inverse of 1. The RH equality is due to the fact that 1 is the multiplicative identity. Transitivity of equality gives $1 = 1^{-1}$.

To show $-1 = (-1)^{-1}$, by Theorem A.8 we have $(-1) \times (-1) = 1$, so that -1 is a multiplicative inverse for -1. Thus, uniqueness of multiplicative inverses gives $-1 = (-1)^{-1}$. $\boxed{\text{qed}}$

Theorem A.11. Given $x \neq 0 \in \mathcal{R}$,

$$(x^{-1})^{-1} = x \quad \text{and} \quad x = \frac{1}{\frac{1}{x}}.$$

Proof. Given $x \neq 0$, x^{-1} exists. To show $(x^{-1})^{-1} = x$, we first write

$$x^{-1} \times x = 1.$$

This equality tells us that x is a multiplicative inverse for x^{-1}. Since multiplicative inverses are unique, using the standard notation for the multiplicative inverse of x^{-1} on the RHS below, we can write:

$$x = (x^{-1})^{-1}$$

and the first equality in A.11 is proved.

To complete the proof, we must show:

$$x = \frac{1}{\frac{1}{x}}.$$

By definition, for any non-zero real number z, $\frac{1}{z} \equiv z^{-1}$. Substituting x^{-1} for z in the definition gives:

$$\frac{1}{x^{-1}} \equiv (x^{-1})^{-1}.$$

Substituting $\frac{1}{x}$ for x^{-1} in the denominator of the LHS of the last equation while replacing $(x^{-1})^{-1}$ by x on the RHS by using the first equality in A.11, we have

$$\frac{1}{\frac{1}{x}} = x.$$

Symmetry of equality completes the proof. $\boxed{\text{qed}}$

We can recast the last equation using only fractional forms as

$$x = \left(\frac{1}{x}\right)^{-1} = \frac{1}{\frac{1}{x}}.$$

Using the definition of division as a product, we have

$$y \div \frac{1}{x} \equiv y \times \left(\frac{1}{x}\right)^{-1} = y \times x$$

where the last step uses the previous equation. This is where the rule **invert and multiply** for division of fractions comes from.

Theorem A.12. If x, $y \in \mathcal{R}$ and $x \times y = 0$, then either, $x = 0$ or, $y = 0$.
  Proof. Assume $x \times y = 0$ but that $y \neq 0$. Then, the reciprocal of y exists, and

$$x = x \times 1 = x \times (y \times y^{-1})$$

where the 1 in $x \times 1$ is replaced by $y \times y^{-1}$ to obtain the RHS. Applying the Associative Law to the last expression above produces the LH equality below. Recalling the assumption that $x \times y = 0$, we can substitute 0 for $x \times y$ and obtain the expression on the RHS below

$$x = (x \times y) \times y^{-1} = 0 \times y^{-1}.$$

Since the product of any real number and 0 is 0 (Theorem A.5), the RHS is 0, whence

$$x = 0.$$

Thus, one of x and y, in this case x, must be 0. $\boxed{\text{qed}}$

We apply this fact to fractional forms by noting that $\frac{x}{y} = x \times y^{-1}$, where $y \neq 0$, whence $y^{-1} \neq 0$ and

$$\frac{x}{y} = 0 \text{ exactly if } x = 0.$$

In short, a **fraction is zero only if the numerator is zero**.
  Recall Theorem A.6

$$(-1) \times x = -x$$

and its consequence:

$$-(x \times y) = (-1) \times (x \times y) = (-x) \times y = x \times (-y)$$

which lets you move the centered dash anywhere in a product without changing the value. We apply A.6 and this equation to fractional forms.

Theorem A.13. Let $x, y \in \mathcal{R}$ with $y \neq 0$. Then

$$-\left(\frac{x}{y}\right) = \frac{-x}{y} = \frac{x}{-y}, \quad \text{and} \quad -\left(\frac{1}{y}\right) = \frac{-1}{y} = \frac{1}{-y}.$$

Proof. By definition, $\frac{x}{y} = x \times y^{-1}$ so that

$$\begin{aligned} -\left(\frac{x}{y}\right) &= -(x \times y^{-1}) = (-x) \times y^{-1} \\ &= x \times (-y^{-1}) \end{aligned}$$

where the two equalities on the right use the consequence of Theorem A.6. Applying the definition of fractional form to $(-x) \times y^{-1}$, we have

$$-\left(\frac{x}{y}\right) = (-x) \times y^{-1} = \frac{-x}{y}.$$

Setting $x = 1$ throughout the last equation gives $-\left(\frac{1}{y}\right) = \frac{-1}{y}$ and completes the proof of the first equality in both parts of A.13.

To obtain the second equality, we apply the rule of precedence stating that the exponent gets applied first to resolve $-y^{-1}$, hence

$$-y^{-1} = -\frac{1}{y}.$$

Using this equality to replace $-y^{-1}$ in the RHS of $-\left(\frac{x}{y}\right) = x \times -y^{-1}$ gives

$$-\left(\frac{x}{y}\right) = x \times \left(-\frac{1}{y}\right).$$

To complete the proof, we need to show $-\frac{1}{y} = \frac{1}{-y}$. We evaluate the RHS as follows

$$\frac{1}{-y} = \frac{1}{(-1) \times y} = \frac{1}{(-1)} \times \frac{1}{y}$$

where the product on the RHS is obtained using Theorem A.14 (proved independently of Theorem A.13 below). Since $\frac{1}{-1} = -1$ by Theorem A.10,

$$\frac{1}{(-1)} \times \frac{1}{y} = (-1) \times \frac{1}{y} = -\frac{1}{y}.$$

347

Transitivity of equality gives $\frac{1}{-y} = -\frac{1}{y}$ which is exactly what we need to conclude

$$-\left(\frac{x}{y}\right) = x \times (-y)^{-1} = \frac{x}{-y}.$$

qed

Theorem A.14. Suppose both x and y are not 0. Then

$$x^{-1} \times y^{-1} = (x \times y)^{-1} \quad \text{and} \quad \frac{1}{x} \times \frac{1}{y} = \frac{1}{x \times y}.$$

Proof. To obtain the first equation, we show $x^{-1} \times y^{-1}$ is a multiplicative inverse for $x \times y$. Since both x^{-1} and y^{-1} exist, we have

$$
\begin{aligned}
(x \times y) \times (x^{-1} \times y^{-1}) &= (x \times y) \times (y^{-1} \times x^{-1}) \\
&= ((x \times y) \times y^{-1}) \times x^{-1} \\
&= (x \times (y \times y^{-1})) \times x^{-1} \\
&= (x \times 1) \times x^{-1} \\
&= x \times x^{-1} = 1
\end{aligned}
$$

where each step applies the Commutative or Associative Law. Since multiplicative inverses are unique, we conclude:

$$(x \times y)^{-1} = x^{-1} \times y^{-1}.$$

The second equation we need to prove is obtained from the first by substitution using:

$$\frac{1}{x} \equiv x^{-1}, \quad \frac{1}{y} \equiv y^{-1}, \quad \text{and} \quad \frac{1}{x \times y} \equiv (x \times y)^{-1}.$$

qed

This theorem is simply stated in words as:

> **the multiplicative inverse of a product is the product of the multiplicative inverses.**

The computation above illustrates the utility of two facts. The first is that the definition of fractional form converts a fraction into a product. The second is that the Associative and Commutative Laws taken together permit one to move the individual factors around in a product however we please. We repeat this in the next theorem which provides a simple rule for multiplying any fractional forms.

Theorem A.15. Let x, y, z, $w \in \mathcal{R}$ with y, $z \neq 0$. Then

$$\frac{x}{y} \times \frac{w}{z} = \frac{x \times w}{y \times z}.$$

Proof. Establishing this equality begins by writing the fractions as products:

$$\frac{x}{y} \times \frac{w}{z} = (x \times y^{-1}) \times (w \times z^{-1}).$$

The rest amounts to moving the factors into the desired configuration using the Associative and Commutative Laws after which we convert back to fractional forms in the last line:

$$
\begin{aligned}
(x \times y^{-1}) \times (w \times z^{-1}) &= ((x \times y^{-1}) \times w) \times z^{-1}) \\
&= (x \times (y^{-1} \times w)) \times z^{-1}) \\
&= (x \times (w \times y^{-1})) \times z^{-1}) \\
&= ((x \times w) \times y^{-1}) \times z^{-1}) \\
&= (x \times w) \times (y^{-1} \times z^{-1}) \\
&= (x \times w) \times (y \times z)^{-1} = \frac{x \times w}{y \times z}.
\end{aligned}
$$

Transitivity of equality now yields the required equality:

$$\frac{x}{y} \times \frac{w}{z} = \frac{x \times w}{y \times z}.$$

$\boxed{\text{qed}}$

Theorem A.16. Let x, y, $z \in \mathcal{R}$. If y and z are both not 0, then

$$\frac{z}{z} = 1 \quad \text{and} \quad \frac{x \times z}{y \times z} = \frac{x}{y}.$$

Proof. Assume both y and z are non-zero. The LH equality follows from

$$\frac{z}{z} = z \times z^{-1} = 1.$$

The RH equality is obtained by applying the preceding theorem as follows:

$$
\begin{aligned}
\frac{x \times z}{y \times z} &= \frac{x}{y} \times \frac{z}{z} \\
&= \frac{x}{y} \times 1 \\
&= \frac{x}{y}.
\end{aligned}
$$

349

$\boxed{\text{qed}}$

The last result is most useful for simplifying fractional forms when the numerator and denominator contain a **common factor** (z)

$$\frac{x \times z}{y \times z} = \frac{x \times \not{z}}{y \times \not{z}} = \frac{x}{y}.$$

Theorem A.17. Let x, y, $w \in \mathcal{R}$ with $y \neq 0$. Then

$$\frac{x}{y} + \frac{w}{y} = \frac{x + w}{y}.$$

Proof. Using the definition of fractional form and the Distributive Law, we have

$$\begin{aligned}
\frac{x}{y} + \frac{w}{y} &= x \times y^{-1} + w \times y^{-1} \\
&= (x + w) \times y^{-1} = \frac{x + w}{y}
\end{aligned}$$

for two fractions having the same denominator. $\boxed{\text{qed}}$

Theorem A.18. Let x, y, z, $w \in \mathcal{R}$ with y, $z \neq 0$. Then

$$\frac{x}{y} + \frac{w}{z} = \frac{x \times z + w \times y}{y \times z}.$$

Proof. To see why this rule is valid, we produce equivalent fractional forms that have the same denominator, and then apply Theorem A.17. The computation is as follows:

$$\begin{aligned}
\frac{x}{y} + \frac{w}{z} &= \frac{x}{y} \times 1 + \frac{w}{z} \times 1 \\
&= \frac{x}{y} \times \frac{z}{z} + \frac{w}{z} \times \frac{y}{y} \\
&= \frac{x \times z}{y \times z} + \frac{w \times y}{z \times y} \\
&= \frac{x \times z + w \times y}{z \times y}.
\end{aligned}$$

The transitive law for equality produces:

$$\frac{x}{y} + \frac{w}{z} = \frac{x \times z + w \times y}{y \times z}$$

350

which completes the calculation. $\boxed{\text{qed}}$

The following theorems are referred to as **Cancellation Laws**. They have great utility when working with equations.

Theorem A.19. Let x, y, $z \in \mathcal{R}$. Then

1. if $x + y = x + z$, then $y = z$;

2. if $x \times y = x \times z$ and $x \neq 0$, then $y = z$.

Proof. The first law is obtained by adding $-x$ on both sides of $x + y = x + z$ to eliminate x. The second law is obtained by multiplying both sides of $x \times y = x \times z$ by x^{-1} to again eliminate x. $\boxed{\text{qed}}$

A.9.1 Summary of Arithmetic Rules for Real Numbers

Field Axioms

Since the field axioms are the foundation of everything, we begin with them.

Let x, y, $z \in \mathcal{R}$. Then the following statements are axioms about $+$ and $\times$ on $\mathcal{R}$:

1. Closure: $x + y \in \mathcal{R}$, and $x \times y \in \mathcal{R}$;

2. the Commutative, Associative and Distributive properties hold as shown:

$$x + y = y + x \quad \text{and} \quad x \times y = y \times x$$
$$(x + y) + z = x + (y + z) \quad \text{and} \quad (x \times y) \times z = x \times (y \times z)$$
$$(x + y) \times z = x \times z + y \times z \quad \text{and} \quad z \times x + z \times y = z \times (x + y);$$

3. $\mathcal{R}$ has an **additive identity**, denoted by 0, with the property that for every $x \in \mathcal{R}$
$$0 + x = x + 0 = x;$$

4. every element, x in $\mathcal{R}$ has an **additive inverse**, y, that satisfies
$$x + y = y + x = 0;$$

5. $\mathcal{R}$ has an **multiplicative identity**, denoted by 1 and $\neq 0$, with the property that for every $x \in \mathcal{R}$
$$1 \times x = x \times 1 = x;$$

6. for every $x \neq 0$, we can find a **multiplicative inverse** $y \in \mathcal{R}$ such that
$$x \times y = y \times x = 1.$$

Rules

The following rules of arithmetic apply to all real numbers and indeed to the elements of any field. Each is a theorem, a direct consequence of a theorem, or a definition. All denominators are assumed to be non-zero. These rules will are the basis for all algebraic manipulations with real numbers. All children should become comfortable with their use by the end of Grade 7.

1. the additive inverse, denoted $-x$, of any number is unique, and $-(-x) = x$;

2. $0 = -0$;

3. **subtraction** is defined in terms of addition by:
$$x - y \equiv x + (-y);$$

4. the multiplicative inverse of any non-zero number x is unique; it is referred to as the **reciprocal of** x and denoted x^{-1}; the reciprocal of x satisfies the equations
$$x^{-1} = \frac{1}{x}, \quad \text{and} \quad (x^{-1})^{-1} = x = \frac{1}{\frac{1}{x}};$$

5. $1 = 1^{-1}$ and $-1 = (-1)^{-1}$;

6. if $y \neq 0$, **division** of x by y is defined in terms of multiplication by the reciprocal of the divisor:
$$x \div y \equiv x \times y^{-1};$$

consistent with this definition and notation for the reciprocal, the **fractional form** $\frac{x}{y}$ is defined by
$$\frac{x}{y} \equiv x \times y^{-1} = x \times \frac{1}{y};$$

7. for every x, $0 \times x = 0$, hence 0 has no reciprocal and we cannot divide by 0;

8. $x \times y = 0$ exactly if at least one of x and y is 0;

9. if $\frac{x}{y} = 0$, then $x = 0$;

10. $(-1) \times x = -x$, whence
$$-(x \times y) = (-x) \times y = x \times (-y)$$

and,
$$-\frac{x}{y} = \frac{-x}{y} = \frac{x}{-y};$$

11. $(-1) \times (-1) = 1$ and $(-x) \times (-y) = x \times y$;

12. for x and $y \neq 0$

$$x^{-1} \times y^{-1} = (x \times y)^{-1} \quad \text{and} \quad \frac{1}{x} \times \frac{1}{y} = \frac{1}{x \times y};$$

13. for y and $z \neq 0$

$$\frac{x}{y} \times \frac{w}{z} = \frac{x \times w}{y \times z};$$

14. for y and $z \neq 0$

$$\frac{z}{z} = 1 \quad \text{and} \quad \frac{x \times z}{y \times z} = \frac{x}{y};$$

15. for $y \neq 0$

$$\frac{x}{y} + \frac{w}{y} = \frac{x + w}{y};$$

16. for y and $z \neq 0$

$$\frac{x}{y} + \frac{w}{z} = \frac{x \times z + w \times y}{y \times z};$$

17. the Cancellation laws:

$$\text{if} \quad x + z = y + z \quad \text{then} \quad x = y,$$

and,

$$\text{if} \quad x \times z = y \times z \quad \text{and} \quad z \neq 0 \quad \text{then} \quad x = y.$$

A.10   Notations for Arbitrary Real Numbers

Consider again the types of real numbers we know exist. We have integers, we have fractions that are not integers, but can be expressed as ratios of integers, and we have still other numbers like $\sqrt{2}$, or π, that arise in geometry, or elsewhere, and which have been shown to be neither integers, nor fractions. If we are to do the things required in the modern world, we need to have notations for all of these numbers.

Consider the rational numbers. As we have seen, they comprise the integers and ratios of integers. The decimal system provides a notation for each rational number. In some cases, the division procedure terminates in a finite number of steps[8] leaving a remainder of 0. In such a case, we know the decimal notation is exact. But repeating

---

[8]The definition of **finite** is that we can, in principle, count the number of digits.

decimals, such as $1.\overline{3}$, and irrational numbers, such as $\sqrt{2}$ and π do not. How do we know this?

In the case of a repeating decimal, we know it represents a fraction $\frac{m}{n}$. The procedure for finding the fraction from the repeating decimal was discussed in §18.6. We also know that every exact decimal number is representable by a **decimal fraction** having a denominator that is a power of ten. Thus, if a fraction is equal to an exact decimal number, as in the case of

$$\frac{1}{2} = 0.5 = \frac{5}{10},$$

all we have to do is subtract the fraction from the equivalent decimal fraction and we must get 0. For a fraction whose representation is a repeating decimal, for example $\frac{1}{3}$, this test can never be satisfied because whatever exact decimal number we choose, when we subtract it from a fraction having a repeating decimal, the difference will not be 0. We can make the result as small as we please, but we can never make it zero.

Consider irrational numbers. Such numbers are defined by some property of their behavior. For example, $\sqrt{2}$ is defined by the equation:

$$(\sqrt{2})^2 = 2.$$

So the question is: Is there a decimal number whose square is 2? Since every decimal number is also a fraction, the question would be answered by showing that there either is, or is not, a fraction whose square is 2. As indicated earlier, it was known to the Greeks that no fraction had this property. Thus, if we take any fraction $\frac{n}{m}$ whatsoever, we find

$$\left(\frac{n}{m}\right)^2 - 2 \neq 0.$$

Again, we can make the residual as small as we please. But we cannot make it zero! Thus, 1.414213562 may be what your calculator asserts is the square root of 2, but it is only an approximation. Although your calculator may tell you that the square of this number is 2, it is not and you can check this yourself by doing the multiplication long-hand using the standard procedure. The reason your calculator may "think" the square is 2 is because your calculator is limited in accuracy.

But there is one other important point about irrationals: their decimal notation is an **infinite non-repeating decimal**. Because of this, to identify a particular irrational, we have to use a property. For example, we can specify the square root of 3 by saying that

when this number is squared, we get 3,

or that when we multiply the length of the diameter of a circle by this number, we would get the circumference provides a property that allows us to identify $\sqrt{3}$ and π, respectively. We have to do this because we do not know what digits are very far out in the decimal expansion. Indeed, it is only since the advent of modern computers that the millionth digit in the decimal expansion for π became known. Since all the digits in the decimal expansion for π, and the like, will never be known, the only way to identify these numbers is by a property of their behavior. And the only way we can give them a name is by assigning a symbol like π, or an operational form like $\sqrt{3}$.

A.11   Teaching the Facts of Arithmetic

Section A.9 presented a summary list of rules that should be regarded as **factual content**. The items listed reflect the mathematical skills needed to complete high school and succeed in a post-secondary program that has a math component. Such programs run the gamut from accounting to kinesiology to zoology and are offered by community colleges and universities throughout North America.

The purpose in creating this list is to identify and highlight arithmetic facts that will have great application in a student's future learning experience so that, when and where appropriate, teachers and mentors may emphasize them to the benefit of students.

These facts fall into categories: **Axioms**, **Definitions**, **Theorems**, and **Conventions**. We consider the categories in relation to what students need to know.

A.11.1   Axioms

In the chapters preceding this one, we stressed that the axioms were intended to capture what we know about real-world processes such as counting collections and measuring lengths and areas.[9] In this respect arithmetic is the original experimental science whose purpose was to identify truths about the behavior of the world. Children need to know this. But while it is great for children to know what the axioms of arithmetic are designed to do, what is critical is that children are able to apply the axioms in computations. If you reflect on how the proofs were put together for the theorems, you will conclude that the Associative, Commutative and Distributive

---

[9]It is sometimes asserted by physicists that either the Special Relativity theory developed by Einstein or Quantum Mechanics is the best tested and most accurate theory we have that applies to the physical world. I would argue that this description properly applies to arithmetic, since every commercial computation, scientific computation, or engineering computation, verifies the truth of arithmetic as applied to real-world activities.

Laws drive the development. This assertion applies not only to the proofs of facts, but also to the development of computational procedures. The importance of these three axioms to the development of arithmetic is why A+ curricula want children to become comfortable with their use and not merely with their existence.

A.11.2 Definitions

Definitions play a critical role in the development of arithmetic. For example, definitions are of particular importance in respect to the notation for fractions and that is why the A+ curricula stresses ideas underlying what we have called the Notation Equation. Children need to understand that a definition provides a group of symbols, or a new word, with an exact meaning.

For example, x^{-1}, and its name, **reciprocal**, are given an exact arithmetical meaning by the following: if $x \neq 0$, then a number denoted by x^{-1} must exist and that number satisfies:

$$x \times x^{-1} = x^{-1} \times x = 1.$$

The arithmetical meaning, or property, is that for a non-zero quantity, the product of that quantity and its reciprocal will be 1. No computation has to be performed to know this fact about x and its reciprocal because the definition specifies this computational relation.

We have also used the symbol $\equiv$ to define new operations. For example, the operation of division, $\div$, is given meaning with: for x, $y \in \mathcal{R}$, $y \neq 0$,

$$x \div y \equiv x \times y^{-1}.$$

Notice that the utility and meaning of the definition of $\div$ is lost without knowledge of the definition of y^{-1}.

Children need to treat definitions as **rote** knowledge. They are simply facts to be learned exactly. In my experience one of the greatest mistakes that is made in teaching arithmetic to children in recent years is that learning mathematics is about **understanding** as opposed to acquiring factual content. There is a great deal of factual content in arithmetic. Understanding comes from assembling facts into a cohesive whole which can then be applied in a variety of contexts. This cannot happen without an exact knowledge of the facts, that is, rote learning.

A.11.3 Theorems

Theorems are arithmetic facts that are derived from the axioms. As such, theorems are statements that will be satisfied in any system that satisfies the axioms. For example, the equation

$$-(x \times y) = (-x) \times y = x \times (-y),$$

(Theorem A.7) has this property. Sometimes theorems assert that in prescribed circumstances we can draw certain conclusions. The Cancellation Law for addition (Rule 17) is like this:

$$\text{if} \quad x + z = y + z \quad \text{then} \quad x = y.$$

The prescribed condition is: if $x + z = y + z$. The conclusion is: $x = y$.

Again I emphasize that to be able to derive the theorem requires knowledge of the axioms as **factual content**. To see the relationships between the various rules requires knowledge of the rules and definitions as factual content.

A.11.4 Conventions

Conventions involve the **application of choice**. The most important conventions affecting arithmetic are the **order of precedence** rules. These rules tell us in what order computations are to be performed. They exist to remove ambiguity from ambiguous expressions. The simplest example is the rule that, unless otherwise indicated by parentheses, all multiplications should be performed before any additions. This means that

$$6 \times 5 + 7 = 37,$$

as opposed to computing $5 + 7$ first which leads to the incorrect calculation:

$$6 \times 5 + 7 \neq 72.$$

If we intend the latter, we have to introduce parentheses:

$$6 \times (5 + 7) = 72,$$

which require that $5 + 7$ be calculated first.

There are two rules of precedence not yet mentioned. The first states that the centered dash, $-$, applies to the **smallest quantity to its right in any expression**. Thus, in the expression

$$x - y + z,$$

the centered dash applies only to the y so that this expression properly interpreted is

$$x - y + z = x + (-y) + z.$$

Thus, in the following numerical example,

$$6 - 3 + 2 = 5, \quad \text{not} \quad 1.$$

To obtain 1, we have to use parentheses as follows

$$6 - (3 + 2) = 1.$$

The second (new) precedence rule states that the exponent -1 applies to the smallest quantity occurring to its left in any expression. To see what we mean, consider $x \times y^{-1}$. This expression produces

$$x \times y^{-1} = \frac{x}{y}$$

from which we see that the exponent -1 applies only to the y and not to the entire product which would yield an incorrect calculation

$$x \times y^{-1} \neq \frac{1}{x \times y}.$$

To force the exponent to apply to the entire product, we must use parentheses as follows:

$$(x \times y)^{-1} = \frac{1}{x \times y}.$$

Another convention is the choice of base for the numeration system. We have chosen base 10. However, computers choose to operate in base 2. Thus in base 2 for computers, $1 + 1 = 10$ which changes the notation for the successor of 1, but not the fact that this number is the successor of 1. What is critical to note is that in making these choices, the value of $1 + 1$ has not changed; it is still the successor of 1. Only the symbols we use to identify this number have changed. This is why these are conventions, not fundamental truths. That said, these **conventions must be carefully followed in order to correctly perform computations**.

A.11.5   Using Particular Rules

One way to view the summary list is as a set of instructions for performing all computations that a child will confront between Grade 3 and post-secondary. The truth of this assertion will become more evident as a child proceeds through the math curriculum. These rules will be applied again and again.

A question that every child should ask when confronted by a new computation is:

What rule or axiom tells me how to do this?

Every child needs to know that the rules and axioms are instructions for how to do all computations. Guessing is never involved and guessing how to perform a computation will certainly be wrong.[10] Children should also know that there is **nothing shameful** about not knowing and that the reasonable response to not knowing is: **Please tell me the procedure so I will know how to do it too**. The point is that as the rules are acquired, they can be treated as general instructions.

Let's take an example of how a rule is used as an instruction for computation. Consider **Rule 14** which asserts: given y and $z \neq 0$,

$$\frac{z}{z} = 1 \quad \text{and} \quad \frac{x \times z}{y \times z} = \frac{x}{y}.$$

Each equation has occurrences of various letters, as in

$$\frac{z}{z} = 1$$

which is simple because it only has one letter, namely z. To use this equation, we simply replace each occurrence of z by the same quantity, so long as it is not 0. The quantity being substituted for z can be a number, a combination of numbers, or an algebraic expression. Valid forms to be substituted would be:

$$5, \; -2.3, \; 5 \times w, \; \sqrt{w - 7},$$

and so forth. Again, there is one constraint in respect to z; we must know that the quantity is not zero. There may also be other constraints. For example in the present case, we need that $w - 7 \geq 0$, otherwise $\sqrt{w - 7}$ will not be a real number. Given the constraints are satisfied, we know with surety from Rule 14 that

$$\frac{5}{5} = \frac{-2.3}{-2.3} = \frac{5 \times w}{5 \times w} = \frac{\sqrt{w - 7}}{\sqrt{w - 7}} = 1.$$

No computation is necessary to replace anything on the LHS by 1.

Consider the second equation in Rule 14. Once again each letter can be replaced by any quantity. Here the first restriction is that if we put a quantity in for x, we must put that same quantity in for all other x's in the equation. The same is true for y and z. So the result of replacing both x's by $a + 2$ would be:

$$\frac{(a + 2) \times z}{y \times z} = \frac{a + 2}{y}.$$

---

[10]I had an *almost* in front of *certainly* because there is always the exception that proves the rule.

359

Another, correct substitution, which sets $x = \sqrt{w - 7}$, $y = 5$ and $z = (s + 4)^{-1}$ where $s + 4 \neq 0$, is:

$$\frac{\sqrt{w - 7} \times (s + 4)^{-1}}{5 \times (s + 4)^{-1}} = \frac{\sqrt{w - 7}}{5}.$$

Again, no calculation is required to get from the LHS to the RHS. This is why these rules are so powerful. But, in applying them one must be meticulous in verifying that the form you have for a LHS really can be obtained by substitution from the generic form in the rule and really does satisfy any restrictions, for example, in substituting for z in Rule 14, $s + 4 \neq 0$.

To summarize, the list of rules and axioms is short, certainly in comparison to the factual content of many other subjects. But each of these rules must be applied perfectly to have any utility at all.

Appendix B

Order Properties of $\mathcal{R}$

In Appendix A we studied the set of real numbers from the perspective of arithmetic. Our intuition about what we should take as basic principles (axioms) was based first on collections and counting numbers and second on the geometry of the line. The arithmetic rules that resulted from these considerations are summarized in §A.9.

In this Appendix we will develop the rules regarding order. In developing these rules we will again use what we learned from collections and geometry as a source of intuition.

B.1   Ordering Integers: A Brief Review

In Chapter 5 we developed the concepts of **more** and **less** based on pairing as applied to collections. Pairing led directly to counting and to counting numbers as stable numerical attributes of collections. Most important we discovered the counting numbers were ordered consecutively and that for counting numbers,

$$m \text{ is less than } n \quad \text{if and only if} \quad m + p = n \text{ for some } p.$$

In Chapter 19 we studied the integers which extend the counting numbers by creating zero and the additive inverses of counting numbers. Thus the integers consist of three kinds of numbers: **positive** (the counting numbers), zero (the additive identity) and **negative** (the additive inverses of counting numbers). The integers can also be represented geometrically as shown in §19.4, but a complete line that is unlimited in extent in both directions is required for this representation. In §19.1.3 we defined **less than** for all the integers using the standard symbol $<$:

$$m < n \quad \text{if and only if} \quad m + p = n \text{ for some counting number } p.$$

By adding $-m$ to both sides of the equation $m + p = n$ and doing the simplifying arithmetic, we see that this relation is completely equivalent to:

$$m < n \text{ if and only if } n + (-m) = p$$

where p is a counting number (positive integer).

We know that the real line also separates the real numbers into three types. As with integers, positive real numbers are on the right side of 0 and negative real numbers are to the left of 0 and addition of two real numbers can be simulated using directed line segments (see §A.5.4). We will take the algebraic relationship used to define the order relation applying to all integers as expressed above and apply it to all real numbers. The first step in the process requires us to identify some numbers as being positive. Based on geometry, we know which numbers should have this property. In what follows, we will see that our method identifies exactly this collection.

B.2   Axioms Governing Positive Real Numbers

The following four statements are axioms:

P1: some real numbers are **positive**;

P2: given $x \in \mathcal{R}$, exactly one of the following three assertions is true:

$$x = 0\,; \ x \text{ is positive}; \ -x \text{ is positive};$$

P3: if x and $y \in \mathcal{R}$ are positive, then $x + y$ is positive;

P4: if x and $y \in \mathcal{R}$ are positive, then $x \times y$ is positive.

P1-P4 together with the field axioms (see §A.9.1) define an **ordered field**.

As the reader can see, P1 and P2 are the two starting points. P1 merely asserts that there are some real numbers that are special in that they are positive. P2 then classifies each real number with respect to the property of being positive and is analogous to the three types of integers as either being a counting number, the additive inverse of a counting number, or zero. The relations between the three types of numbers are exactly like those for integers, but unlike the integers where the counting numbers were identified as being positive, P2 does not actually identify a single real number that is positive, not even 1! Instead, P3 and P4 say that the positive reals are closed under addition and multiplication. These two closure properties were automatic in respect to integers because the positive integers were known to be closed under both addition and multiplication. As we shall see, these two closure properties will force the counting numbers to be positive.

Our first theorem identifies 1 as being positive, which is what we need if ordered fields really are a model for real-world arithmetic.

Theorem B.1. 1 is positive.

Proof. To show 1 is positive, we apply P2. If we can eliminate the possibility that 1 is either zero or the additive inverse of a positive real number, then 1 will have to be positive.

We know that $1 \neq 0$, since this is a requirement of the field axioms (§A.6.2). The only remaining choices permitted by P2 are that 1 is positive, or -1 is positive, but both cannot be true. Suppose -1 is positive. Rule 11 of §A.9.1 states

$$(-1) \times (-1) = 1.$$

P4 tells us the product of positives is positive. Thus if -1 is positive, then its additive inverse 1 is also positive. But this is forbidden by P2. So 1 must be positive instead of -1, since this is the only choice left under P2. $\boxed{\text{qed}}$

Recall that the integers are a subset of the real numbers, and further that within $\mathcal{R}$, every counting number can be obtained by adding 1 to itself. So Theorem B.1 guarantees that every counting number is positive, and hence our axioms are identifying the correct subset of the integers as being positive. All the remaining integers can be obtained as additive inverses of counting numbers. Since such an integer, $-n$ where $n \in \mathcal{N}$, satisfies $-n + n = 0$, we have

$$-(-n) = n,$$

so that the additive inverse of $-n$ is positive. Thus, P2 classifies the integers, $\mathcal{I}$, into the same three groups as were identified in Chapter 11.

B.3   The Nomenclature of Order

We shall use the following nomenclature:

1. $x \in \mathcal{R}$ is **negative** provided $-x$ is positive;

2. for x and $y \in \mathcal{R}$, we write $x < y$, and say x **is less than** y, provided there is a positive $z \in \mathcal{R}$ such that $x + z = y$;

3. for x and $y \in \mathcal{R}$, we write $x \leq y$, and say x **is less than or equal to** y, provided $x < y$ or $x = y$;

4. for x and $y \in \mathcal{R}$, we write $y > x$, and say y **is greater than** x, provided $x < y$;

5. for x and $y \in \mathcal{R}$, we write $y \geq x$, and say y **is greater than or equal to** x, provided $x < y$ or $x = y$.

This usage is consistent with previous usage specified at the beginning of this chapter. We will appeal to these definitions in what follows.

B.4   Laws Respecting $<$ on $\mathcal{R}$

Let x, y and $z \in \mathcal{R}$ stand for arbitrary reals. Then the following four laws are satisfied by $<$:

1. the **Transitive Law**:

$$\text{if} \quad x < y \quad \text{and} \quad y < z, \quad \text{then} \quad x < z;$$

2. the **Trichotomy Law**, given any x and y, exactly one of the following holds:

$$x < y \quad \text{or,} \quad y < x \quad \text{or} \quad x = y;$$

3. the **Addition Law**:

$$\text{if} \quad x < y \quad \text{then} \quad x + z < y + z;$$

4. the **Multiplication Law**:

$$\text{if} \quad x < y \quad \text{and} \quad 0 < z, \quad \text{then} \quad x \times z < y \times z.$$

These four laws are identical to the laws in §11.4, except that they are made about real numbers and not integers. In §11.4 the laws were derived based on the definition

$$m < n \quad \text{if and only if} \quad m + p = n$$

for some counting number p. For real numbers, this definition is replaced by

$$x < y \quad \text{if and only if} \quad x + z = y$$

for some **positive** real number z. It is completely equivalent to:

$$x < y \quad \text{if and only if} \quad z = y + (-x) \text{ is a positive real number}$$

364

as shown in Theorem B.3.

The arguments underlying the truth of these four laws follow directly from the definition. We give one example.

Theorem B.2. Let x, y, $z \in \mathcal{R}$. Then the **Addition Law** states:

$$\text{if} \quad x < y \quad \text{then} \quad x + z < y + z;$$

Proof. Suppose $x < y$. By definition, for some positive $w \in \mathcal{R}$, $x + w = y$. Adding z to both sides of this equality gives

$$(x + w) + z = y + z.$$

Applying the Commutative and Associative Laws to the LHS gives

$$(x + z) + w = y + z.$$

Since w is a positive real number, by definition of $<$, we conclude that

$$x + z < y + z.$$

qed

The reader should compare this with Theorem 11.4 in §11.4. You will see the arguments are in essence the same.

B.5   Theorems Governing Order and $<$

The following theorems are valid in any ordered field and have many useful applications. In what follows, let x, y, $z \in \mathcal{R}$ stand for arbitrary reals.

Theorem B.3. $x < y$ if and only if $y - x$ is positive.

Proof. First we recall that $y - x \equiv y + (-x)$. From the definition of $<$, we know that $x < y$, means there is a positive z, such that

$$x + z = y.$$

Adding $-x$ on the left to both sides of this equation gives:

$$(-x) + (x + z) = (-x) + y.$$

The LHS reduces to z via

$$(-x) + (x + z) = (-x + x) + z = 0 + z = z.$$

365

The RHS to $y - x$ via

$$(-x) + y = y + (-x) \equiv y - x.$$

So $y - x$ is positive, since $z = y - x$, and z is positive.

Conversely, from $y - x$ is positive, we set $z = y - x$, and add x to both sides producing $x + z = x + (y - x)$. Reversing the steps leading to $y - x$ above, gives $x + z = y$, whence since z is positive, we have $x < y$. $\boxed{\text{qed}}$

Theorem B.4. Let $x \in \mathcal{R}$. Then x is positive if and only if the $0 < x$.

Proof. Let x be positive. Since 0 is the additive identity, $0 + x = x$ and by definition of $<$, $0 < x$. Conversely, if $0 < x$, then $x - 0 = x$ is positive by Theorem B.3. $\boxed{\text{qed}}$

Theorem B.5. x is negative if and only if $x < 0$.

Proof. If x is negative, then by definition $-x$ is positive, and $x + (-x) = 0$, so $x < 0$ by the definition of $<$.

Conversely, given $x < 0$, we know there exists positive z so that $x + z = 0$. But uniqueness of additive inverses tells us that $z = -x$, so $-x$ is positive and x is negative as claimed. $\boxed{\text{qed}}$

Theorem B.6. The product of two negative numbers is a positive number.

Proof. Recall Rule 11 of A.9.1 that $x \times y = (-x) \times (-y)$. Now if x and y are negative, then $-x$ and $-y$ are positive, whence their product is positive and by Rule 11, so is $x \times y$. $\boxed{\text{qed}}$

Theorem B.7. The product of a negative x and a positive y is negative.

Proof. Theorem B.5 asserts that x being negative is equivalent to $x < 0$. The Multiplication Law (§B.4 above) tells us that inequalities are preserved when we multiply through an inequality by a positive number like y. Thus, for $x < 0$,

$$x \times y < 0 \times y = 0,$$

so that $x \times y$ is negative by Theorem B.5, as claimed. $\boxed{\text{qed}}$

Theorem B.8. if $x \neq 0$, then $0 < x \times x$.

Proof. If $x \neq 0$, then either x is positive, or $-x$ is positive. Since Rule 11 §A.9.1 says $x \times x = (-x) \times (-x)$, in either case, $x \times x$ has to be positive. $\boxed{\text{qed}}$

Theorem B.9. If $0 < x$, then $0 < \frac{1}{x}$.

Proof. By P2, $\frac{1}{x}$ is either positive or negative. Suppose $\frac{1}{x}$ is negative. Then $\frac{1}{x} < 0$ and $1 = x \times \frac{1}{x} < 0$ by Theorem B.7 above. Since $0 < 1$ by Theorem B.1, this cannot be, so $0 < \frac{1}{x}$ as claimed. $\boxed{\text{qed}}$

Theorem B.10. If $x < 0$, then $\frac{1}{x} < 0$.

Proof. If $0 < \frac{1}{x}$, then $1 = x \times \frac{1}{x} < 0$ by Theorem B.7 above. As in the proof of 17.8, this cannot be, so $\frac{1}{x} < 0$ as claimed. $\boxed{\text{qed}}$

Theorem B.11. If $0 < x < y$, then $0 < \frac{1}{y} < \frac{1}{x}$.

Proof. Let $0 < x < y$. From Theorem B.9, the multiplicative inverses of x and y must also be positive. We also know the product of positive numbers is positive (P4), so $\frac{1}{x} \times \frac{1}{y}$ is positive, whence multiplying through $0 < x < y$ by $\frac{1}{x} \times \frac{1}{y}$ preserves the inequalities by the Multiplication Law so that

$$0 \times \left(\frac{1}{x} \times \frac{1}{y} \right) < x \times \left(\frac{1}{x} \times \frac{1}{y} \right) < y \times \left(\frac{1}{x} \times \frac{1}{y} \right).$$

Performing the indicated operations using the Associative and Commutative Laws leaves:

$$0 < \frac{1}{y} < \frac{1}{x},$$

which is what we wanted. $\boxed{\text{qed}}$

Theorem B.11 can be recast in the notation of common fractions and counting numbers:

$$0 < n < m \quad \text{if and only if} \quad 0 < \frac{1}{m} < \frac{1}{n}.$$

This result must be applied with care because the **inequalities reverse** when constructing reciprocals. One good way to think about this result is to remember

> **when you make the denominator of a fraction larger, you make the value of the fraction smaller**.

In the language of children: splitting a candy bar two ways is better than splitting the same bar three ways because each child gets more if there are only two. Getting children to think of these ideas in the most concrete ways possible builds understanding in the long run.

Theorem B.12. If $0 < x,\ y,\ w,\ z$, then

$$\frac{x}{y} < \frac{w}{z} \quad \text{if and only if} \quad x \times z < w \times y.$$

Proof. Theorem B.9 tells us $\frac{1}{y}$ and $\frac{1}{z}$ are both positive. Since the product of positives is positive, we have that the products $y \times z$ and $\frac{1}{y} \times \frac{1}{z}$ are both positive.

367

Thus, we know from the Multiplication Law that multiplying through any inequality by either one of these products will preserve the inequality. If we multiply the inequality on the LHS of Theorem B.12 by $y \times z$ we have

$$\frac{x}{y} \times (y \times z) < \frac{w}{z} \times (y \times z).$$

Since the y's cancel on the LHS and the z's on the RHS, we see this is just

$$x \times z < w \times y$$

which is the RHS of Theorem B.12. Conversely, if we multiply through this last inequality by the positive product $\frac{1}{y} \times \frac{1}{z}$, we obtain

$$(x \times z) \times \left(\frac{1}{y} \times \frac{1}{z}\right) < (w \times y) \times \left(\frac{1}{y} \times \frac{1}{z}\right).$$

Cancelling the z's on the LHS and the y's on the RHS and applying the Notation Equation to what is left produces $\frac{x}{y} < \frac{w}{z}$ and completes the equivalence. $\boxed{\text{qed}}$

Restating Theorem B.12 for positive common fractions, we have

$$\frac{q}{m} < \frac{p}{n} \quad \text{if and only if} \quad q \times n < p \times m.$$

The products being formed on the RHS are the **numerator of one fraction times the denominator of the other**. The process of forming these products is referred to as **cross multiplication**.

Theorem B.13. If $x < y$, then $x < \frac{x+y}{2} < y$.

Proof. Given $x < y$, the Addition Law tells us that $x + x < x + y$ and $x + y < y + y$. Since $x + x = x \times 2$ and $y + y = y \times 2$, we have

$$x \times 2 < x + y < y \times 2.$$

Since $0 < 2$ implies $0 < 2^{-1}$, we can multiply through the inequality by the multiplicative inverse of 2, to obtain

$$(x \times 2) \times 2^{-1} < (x + y) \times 2^{-1} < (y \times 2) \times 2^{-1}.$$

Converting these expressions to fractional forms and cancelling common factors leaves:

$$x < \frac{x + y}{2} < y.$$

$\boxed{\text{qed}}$

We note that Theorem B.13 merely asserts that

> **the arithmetic average of two unequal numbers lies strictly between them**.

B.5.1　Summary of Rules for Order

The following list contains the theorems of the last section in the same order that they were developed. These rules are so basic that they should become a matter of recall for students.

Let x, y, $z \in \mathcal{R}$ stand for arbitrary reals. Then:

OR 1　$x < y$ if and only if $y - x$ is positive.

OR 2　$0 < 1$, and if $n \in \mathcal{N}$, i.e., n is a counting number, then $1 \leq n$;

OR 3　x is negative if and only if $x < 0$;

OR 4　the product of two negative numbers is positive;

OR 5　the product of a negative number and a positive number is a negative number;

OR 6　if $x \neq 0$, then $0 < x \times x$;

OR 7　if $0 < x$, then $0 < x^{-1} = \frac{1}{x}$;

OR 8　if $x < 0$, then $x^{-1} < 0$;

OR 9　if $0 < x < y$, then $0 < y^{-1} < x^{-1}$, where now the order is **reversed**;

OR 10　if x, y, w, and z are positive, then

$$\frac{x}{y} < \frac{w}{z} \quad \text{if and only if} \quad x \times z < w \times y;$$

OR 11　if $x < y$, then $x < \frac{x+y}{2} < y$.

B.6　Intervals and Length

The first notions of the line are in the form of a measuring tape of unlimited length. This tape is essentially the positive half of the line and was considered first in Chapter 10 and next in Chapters 13-16 in respect to common fractions. The whole number line and its construction was described in §19.4 and finding all real numbers on this line was described in §A.2. How these ideas connect to measuring length as a numerical attribute were first discussed in Chapter 10 and again in §A.5.4 in the context of real numbers.

All the construction processes begin by specifying the positions of 0 and 1 on an otherwise empty line. As discussed in Chapter 13, there are many real numbers that fall between 0 and 1 on the line and we can consider the totality of these numbers as a set. Thus, the **unit interval** is the set of all real numbers x that satisfy $0 \leq x \leq 1$, or in standard set notation:

$$\{x : x \in \mathcal{R} \text{ and } 0 \leq x \leq 1\}$$

This interval defines the **unit length** (see §10.2, 19.4 and A.2). In particular, we know from the previous discussion that the position of any counting number, n, on the real line is exactly n units of distance to the right of 0. Further, as shown in §16.5 above, the position of any **proper fraction** $\frac{m}{n}$ is exactly $m \times \frac{1}{n}$ units of distance to the right of 0. From these two facts it follows that for any positive real number x, the position of x on the real line is x units of distance to the right of 0 which merely confirms what we discussed in §A.2. Negative real numbers y are identified by the property that $-y$ is **positive** and positioned a distance of $-y$ units to the left of 0 as discussed in §A 5.4.[1]

Since each real number is either positive, negative or 0, each has a position on the real line corresponding to its distance from 0. We want to generalize the these ideas.

Given two real numbers a and b that satisfy $a \leq b$, the set of real numbers x that satisfy

$a \leq x$ and $x \leq b$, that is, x is at least as large as a and x is no more than b

is called the **interval from** a **to** b. In set notation, the interval from a to b is written as:

$$\{x : a \leq x \leq b \text{ and } x \in \mathcal{R}\}.$$

The requirement that the left endpoint, a, is less than or equal to the right endpoint b ensures that an interval contains at least one real number, namely, a. In this notation the unit interval would be described as the interval from 0 to 1 and written in set notation as:

$$\{x : 0 \leq x \leq 1 \text{ and } x \in \mathcal{R}\}.$$

With any interval from a to b, we associate a **length**, defined to be $b - a$. The requirement that $a \leq b$ is important here because

$$a \leq b \quad \text{if and only if} \quad a + z = b$$

for some $z \geq 0$. Calculations leading to Theorem B.3 tell us $z = b - a$. Thus, the **length of an interval is a non-negative number**, which is exactly what we would

---

[1]Readers may wonder how we get from rationals to all real numbers. We will discuss this briefly at the end of this chapter when we introduce the **Completeness Axiom**.

expect from the real world. The unit interval has length $1 = 1 - 0$ which is also as it should be.

B.7   Size of a Real Number

There are lots of quantities in the real world that represent the **size** of something. Let's consider a few.

With respect to geometry, length provides a number that tells us the size of the distance between two points. Area provides a number that tells us the size of plots of land. Volume tells us the size of a cube, or a ball, or a quantity of liquid. In finance we have numbers that tell us the amount of money that a given bill represents. In physics we have mass as a numerical measure of the amount of resistance to motion, and so on. The one thing we can take from all these examples is that a numerical measure of size should be a non-negative number.

Mathematicians considered this question in respect to real numbers. Specifically, they asked

What is the size of a real number?

What is being looked for is an exact numerical measure that answers the question: How big is x?

For example, we might ask: How big is 50? Now if your immediate response is that the best answer to this question is 50, you'd be absolutely right! In respect to positive real numbers, to say the size of x is x exactly preserves the order relationships between positive real numbers developed in this chapter. To be clear, using this measure, that the size of x is x, for positive real numbers satisfies the following:

the size of x is less than the size of y if and only if $x < y$.

So we might consider saying the size of x was less than the size of y provided $x < y$ as a way of extending the notion of size to all real numbers. But consider,

$$-75 < 50.$$

Do we want to say the size of -75 is less than the size of 50? Or, consider

$$-75 < -10.$$

Do we want to say the size of -75 is less than the size of -10? Or worse yet,

$$-75 < 0.$$

Surely we want zero to have a smaller size than any other real number.

The solution to these questions is that we should take the size of a number, x, to be the length of the interval from 0 to x when x is non-negative, and the length of the interval from x to 0 when x is negative. In short, the distance from 0 to x.

This leads to the following rule for calculating the size of a real number, x:

if $0 \leq x$, the size of x is x; if $x < 0$, the size of x is $-x$.

The reader can verify that this rule gives exactly the length of the interval from 0 to x if x is positive, and the length of the interval from x to 0 if x is negative.

For the examples above, the size of 50 is 50; the size of -75 is $-(-75) = 75$ and the size of -10 is $-(-10) = 10$. So -10 has a smaller size than 50 which again has a smaller size than -75. As well, using this rule we have the size of 0 is less than or equal to the size of every other real number.

B.7.1 Absolute Value

The size of a number x is called its **absolute value** and is denoted by $|x|$. Using this notation, we rewrite the rule for calculation of size as:

if $0 \leq x$, then $|x| = x$, and if $x < 0$, then $|x| = -x$.

My personal view derived from my own teaching experience is that absolute value as a topic is best left to Grade 8. As can be seen, it is not a topic identified in the A+ Table in §1.5, nor in the full table in ACC which probably means it does not occur in A+ curricula before Grade 9. However, the CCSS-M identifies Grade 6 as the appropriate time to introduce absolute value to children.

Expressions involving absolute value are ubiquitous in higher mathematics. For example, a typical problem might be to find the solution set for $|x^3 - \frac{x}{2}| < \frac{1}{2}$. Students find working with such expressions intimidating because it is difficult to perform calculations with an expression like $|x^3 - \frac{x}{2}|$ using the rules for arithmetic. But the definition above gives instructions for eliminating the absolute value signs, namely, treat cases:

Case 1. $x^3 - \frac{x}{2} \geq 0$.

Case 2. $x^3 - \frac{x}{2} < 0$.

Under Case 1, you calculate with $x^3 - \frac{x}{2}$. Under Case 2 you calculate with $\frac{x}{2} - x^3$. Neither expression involves absolute value signs and is therefore subject to the ordinary rules of arithmetic. Children need to get used to the idea of using the definition to remove these signs from expressions so this becomes a **memory-based** problem solving tool.

It is important for children to remember the geometric interpretation of the absolute value of a number as its distance from 0. Thus, if x is positive, the number x will be found x units to the right of 0 on the line and if x is negative, then x will be found x units to the left of 0. This interpretation is illustrated below.

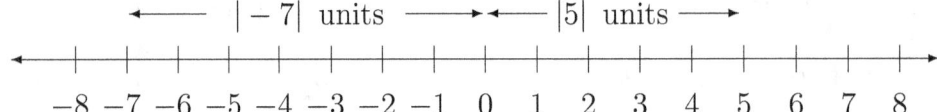

A graphical illustration of the positions of 5 and -7 in terms of their absolute values $|5|$ and $|-7|$. Geometrically, $|x|$ is the length of the interval from 0 to x.

B.8   Distance and Addition

In §A.5.4 we discussed representing the addition of two real numbers using the concept of directed distance. This idea incorporates both length, which is positive, and algebraic sign, $\pm$, in the form of direction.

The CCSS-M demand children have a thorough understanding of these ideas, but use of absolute value instead of directed distance. To see how, consider the problem of finding the quantity $3 + x$ on the real line. To find this sum we might think of starting at 3 on the line and then moving to $3+x$ in the manner described in §A.5.4 using directed line segments. Using the ideas concerning intervals and length, we know the distance between any two real numbers, $a \leq b$, is the length of the interval from a to b. So we need to determine the distance between 3 and $3+x$ and then, by starting at 3 and moving the required distance, either to the right or to the left, we will end up at $3 + x$.

How do we find the distance we need to move? We calculate the length of the interval from $3 + x$ to x. There are two possibilities concerning x, either it is positive, or it is negative. Again, applying the ideas in §A.5.4, if x is positive, then $3 \leq 3 + x$ so that $3 + x$ is to the right of 3 and the length of the interval from 3 to $3 + x$ is $(3 + x) - 3 = x$. On the other hand, if x is negative, then $3 + x \leq 3$ so that $3 + x$ is on the left of 3 and the length of the interval from 3 to $3 + x$ is $3 - (3 + x) = -x$. In other words, the length of the interval from 3 to $3+x$ is given by

$$|(3 + x) - 3| = |x|.$$

Thus, the absolute value of the difference between two numbers gives the distance between them **irrespective** of which is larger. This fact is one of the really useful properties of the absolute value and needs to become part of every student's tool kit. The only question is: When?

We give two numerical examples for $x = 3\frac{1}{2}$ and $x = -4$. If $x < 0$ we move to the left because we know $3 + x < 3$. If $0 < x$, we move to the right because we know $3 < 3 + x$. These ideas are illustrated below and replicate the notions presented in §A.5.4.

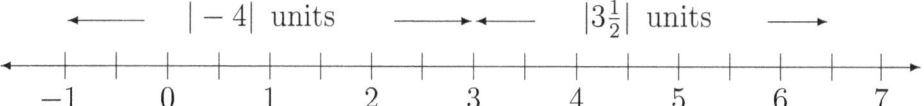

A geometric illustration of the result of adding $3\frac{1}{2}$ to 3 and -4 to 3. The first moves a distance equal to $|3\frac{1}{2}|$ units to the right of 3 and the second moves a distance $|-4|$ units to the left of 3.

The reader should be clear: there is nothing special about 3. If we fix a real number y on the line, then adding x simply moves one x units to the right or left of y depending on whether x is positive or negative as discussed above.

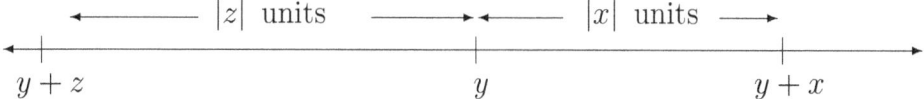

A geometric illustration of the result of adding $x > 0$ and $z < 0$ to a fixed y. The addition of x moves to a position $|x|$ units to the right from y. The addition of z moves to a position $|z|$ units to the left from y.

B.8.1 The Positive Integers in an Ordered Field

In Appendix A we remarked that the field axioms were insufficient to identify the real numbers and that we would need more axioms. As evidence supporting this claim, we note there are finite structures that satisfy all the field axioms. By adding the order axioms we exclude all **finite** structures as being candidates for $\mathcal{R}$ by the simple requirement that the sum of positive numbers be positive and the fact that we can prove the 1 is positive. In this section, we briefly explore the effect of the order axioms on the structure of $\mathcal{R}$. We begin with a careful definition of the positive integers.

Definition. A subset S of the real numbers is said to be **inductive** provided S has the following two properties:

(i) $1 \in S$;

(ii) if $x \in S$, then $x + 1 \in S$.

The **positive integers** are defined to be:[2]

$$\mathcal{N} = \bigcap \{S : S \subseteq \mathcal{R} \text{ such that } 1 \in S \text{ and if } x \in S \text{ then } x+1 \in S\}.$$

Because the operation $\bigcap$ requires $n \in \mathcal{N}$ to be a member of every $S \subseteq \mathcal{R}$ that is inductive, the positive integers are the smallest inductive subset of $\mathcal{R}$. What this means is that if S is any inductive subset of $\mathcal{R}$, then $\mathcal{N} \subseteq S$! We will use this fact in the proofs below.

Theorem B.14. Let

$$S = \{x : x \in \mathcal{R} \text{ and } 0 < x\}.$$

Then S is inductive.

Proof. Theorem B.2 tells us 1 is positive, while Theorem B.4 asserts x is positive if and only if $0 < x$. So $1 \in S$. Next, if $0 < x$, then $1 < x+1$ by the Addition Law. Since $<$ is transitive, $0 < x+1$, so $x+1 \in S$. Thus, S is inductive. $\boxed{\text{qed}}$

Notice that an immediate consequence of Theorems B.4 and B.14 is that all **positive integers** as defined by $\mathcal{N}$ above are, in fact, **positive in the sense of order**.

Our next theorem develops some key properties of $\mathcal{N}$ as defined above. These properties harken back to properties of the Natural numbers as counting numbers associated with collections.

Theorem B.15. Let $n \in \mathcal{N}$. Then the following are true about n:

(i) $1 \leq n$;

(ii) If $n \neq 1$, then for some $k \in \mathcal{N}$, $n = k+1$, i.e., n is a successor;

(iii) there is no $k \in \mathcal{N}$ such that $n < k < n+1$.

Proof. For (i), define the set K by

$$K = \{n : 1 \leq n \text{ and } n \in \mathcal{N}\}.$$

We show K is inductive.

Since $1 \leq 1 \in \mathcal{N}$, $1 \in K$. Further, if $n \in K$, then $1 \leq n$. Since $0 < 1$,

$$1 < 1+1 \leq n+1 \in \mathcal{N},$$

so that K is inductive. Thus, $\mathcal{N} \subseteq K$, which is what we need.

---

[2] The operation denoted by $\bigcap$ is set intersection as defined in IST. To get into such an intersection, an element must be a member of all the sets S over which the intersection is being computed.

For (ii), define

$$K = \{n : n \in \mathcal{N} \text{ and } (n = 1 \text{ or } n \text{ is a successor})\}.$$

Again, we show K is inductive.

Observe $1 \in K$ by definition of K and if $n \in K$, then $n \in \mathcal{N}$, whence $n+1 \in \mathcal{N}$. Since $n + 1$ is a successor, $n + 1 \in K$ and K is inductive. Thus, $\mathcal{N} \subseteq K$ follows and we are done.

For (iii), set

$$K = \{n : n \in \mathcal{N} \text{ such that there is no } k \in \mathcal{N} \text{ such that } n < k < n+1\}.$$

Again, we show K is inductive.

First assume that $1 \notin K$. Since $1 \in \mathcal{N}$, there must be a $k \in \mathcal{N}$ such that $1 < k < 2 = 1 + 1$. Such a k must be a successor by (ii) and by definition of successor, $k - 1 \in \mathcal{N}$ since $(k - 1) + 1 = k$. By our assumption $k < 1 + 1$. Adding -1 through this last inequality using the Addition Law tells us $k - 1 < 1$. This means $k - 1 \notin \mathcal{N}$, a contradiction. Thus, $1 \in K$ follows. A similar argument shows that $n \in K$ implies $n + 1 \in K$, so that K is inductive. Thus, $\mathcal{N} \subseteq K$ and (iii) is established. $\boxed{\text{qed}}$

As a consequence of these facts, we can write

$$0 < 1 < 2 < 3 < 4 < 5 < \cdots < n < n + 1 < \cdots$$

The fact that $n < n + 1$ forces all the positive integers to be distinct.

Finally we show the set of positive integers, $\mathcal{N}$, as defined above satisfy Peano's fifth axiom, namely:

Theorem B.16. If K is any subset of $\mathcal{N}$ having the following two properties,

(i) $1 \in K$;

(ii) if $n \in K$, then $n + 1 \in K$;

then $K = \mathcal{N}$.

Proof. By hypothesis $K \subseteq \mathcal{N}$. Because (i) and (ii) above, K is inductive, so $\mathcal{N} \subseteq K$. It is immediate $K = \mathcal{N}$. $\boxed{\text{qed}}$

The conclusion of all this is that every ordered field contains a subset that looks and behaves in exactly the manner we expect from our study of the Natural numbers as derived from cardinal numbers and collections.

B.9   The Completeness Axiom

In §13.1 and again in §A.2 we stated the principle that every identifiable point on the real line was associated with a unique real number. This idea is geometric, but our axioms are algebraic. Our purpose here is to briefly explain how this idea is captured as an algebraic axiom about $\mathcal{R}$ and to show that the positive integers relate to the rest of the reals in exactly the way we think they do. This material is included simply to complete the axiomatization of $\mathcal{R}$ and is not required at any point in an A+ curricula.

To complete the specification of $\mathcal{R}$, we will need to introduce two new ideas: **upper bound** and **least upper bound**.

Definition. Let S be a set of real numbers and $b \in \mathcal{R}$. If for every $x \in S$,

$$x \leq b,$$

we say b is an **upper bound** for S. Further, suppose $c \in \mathcal{R}$ is an upper bound for S. If for every upper bound b of S, $c \leq b$, we say c is the **least upper bound** for S.

An upper bound for a set, S, is simply any real number that is at least as big as every number in S. If a set has an upper bound, we say it is **bounded above**. In terms of the line, for S to have an upper bound, b, means that no number in S lies to the right of b on the line. Notice that if $b < b_1$ and b is an upper bound for S, then since b_1 is to the right of b and no member of S can be to the right of b, it follows that no member of S can be to the right of b_1. Thus, b_1 is also an upper bound.

The least upper bound is the smallest upper bound. To find the least upper bound, think of starting at an upper bound b and moving to the left. You do this until moving further to the left moves you to the left of a member of S. Thus the least upper bound is the left-most upper bound in respect to the line. The least upper bound for a set S may, or may not, be a member of S.

> **Completeness Axiom.** Let S be a non-empty set of real numbers. If S has an upper bound, then S has a least upper bound.

This axiom guarantees that every identifiable point on the real line is associated with a unique real number. This axiom is **not** satisfied by the ordered field $\mathcal{Q}$ of rational numbers as the following example shows:

$$S = \{x : x \in \mathcal{Q} \text{ and } x \times x \leq 2\}.$$

The reason is the least upper bound for S is $\sqrt{2} \notin \mathcal{Q}$.

Taken together, the eleven axioms for a complete ordered field guarantee that there is a unique structure that satisfies the axioms, namely, $\mathcal{R}$. All of this has very precise mathematical meaning that is beyond the scope of this book. In respect to how this axiom is used in the ordinary course of things, see ERA and FoA. For a discussion of why the reals are unique you need to consult an advanced text in mathematical logic.[3]

The Completeness Axiom is a powerful tool. One of its most important consequence is the following fact about the relationship between counting numbers and real numbers greater than 1.

Theorem B.17. Let $1 < x \in \mathcal{R}$. Then we can find $n \in \mathcal{N}$ such that

$$n < x \le n+1.$$

Proof. Fix $x \in \mathcal{R}$ such that $1 < x$ and consider the set K defined by

$$K = \{k : k \in \mathcal{N} \text{ and } k < x\}.$$

By hypothesis, $1 < x$. Since $1 \in \mathcal{N}$, we have $1 \in K$, whence K is non-empty. By definition, x is and upper bound for K, whence by the Completeness Axiom, K has a least upper bound. Call it z. Now if the conclusion of the theorem is false, x is an upper bound for $\mathcal{N}$ and for every $n \in \mathcal{N}$, $n \le z < x$. But since $n+1 \in \mathcal{N}$ whenever $n \in \mathcal{N}$, so that

$$n < n+1 < n+2 \le z < z+1,$$

for every $n \in \mathcal{N}$. Subtracting 1 through this inequality gives:

$$n \le z - 1 < z$$

for all $n \in \mathcal{N}$, whence $z - 1$ is an upper bound for K. But $z - 1 < z$ which contradicts our choice of z as the least upper bound for K. Thus, the conclusion of the theorem must follow. $\boxed{\text{qed}}$

Stated a different way, this theorem simply says that $\mathcal{N}$ has no upper bounds in $\mathcal{R}$. A consequence of Theorem B.17 is that for every $x \in \mathcal{R}$ we can find a $k \in \mathcal{I}$ such that

$$k \le x < k+1.$$

If we think about this theorem in terms of the construction of the whole number line, it is telling us that the process will eventually bracket each real number between consecutive integers which is exactly how we conceive of the real line. Using this inequality, we can show that an arbitrary real number, x, is the sum of an integer, m, and a residual $r \in \mathcal{R}$ which is in the unit interval.

---

[3]See **Introduction to Mathematical Logic**, J.R. Shoenfield.

Appendix C

Exponentiation

In Chapters 17 and 18 on decimal notation and decimal arithmetic, we made use of powers of 10 to represent decimal fractions and decimal numerals. In addition, at various times we quoted rules that govern the arithmetic of exponentials. In this appendix we present the basic theory.

C.1   What is Exponentiation?

We have already seen the exponent -1 and exponential notation used in a very limited way to specify **multiplicative inverses** as in §A.6.3:

$$4^{-1}, \quad \text{and} \quad x^{-1}.$$

The computations discussed in this chapter will extend this notation to include arbitrary integers as exponents, not just -1.

The reader will recall our emphasis on the idea that when one factor of a product was an integer, multiplication can be thought of as repetitive addition. The intention with respect to exponentiation is that when the exponent is an integer, the result should be **repetitive multiplication of the base** times 1. Keeping this intuitive idea in mind should be helpful in understanding this material.

As discussed in §19.1.2, there are exactly three types of integers, positive, negative and 0 and all can occur as exponents. Thus, we must be able to interpret exponentiation as repetitive multiplication for each type of integer. Central to our implementation of exponentiation will be the fact that

$$x = 1 \times x,$$

and Rule 10 (§A.9.1)

$$-x = (-1) \times x = x \times (-1)$$

applied to integers, namely, that an integer k is negative exactly if it is the additive inverse of a positive integer, in other words exactly if for some positive integer n,

$$k = -n = n \times (-1).$$

Lastly, in providing a meaning to quantities like 10^{-4} and 10^{-6}, we will have to ensure that we retain the existing meaning of the negative exponent -1 as specifying the **multiplicative inverse**.

C.2 General Definition of Exponentiation

We define the meaning of a^n, where $a \in \mathcal{R}$ is non-zero and $n \in \mathcal{I}$. The restriction that $a \neq 0$ is due to the fact that it is impossible to assign a consistent meaning to 0^0.

C.2.1 Non-negative Integer Exponents

Let a be any real number other than 0. For $0 \leq n \in \mathcal{I}$, the rules E1 and E2 define the calculation of a^n:

E1: for $n = 0$, $a^0 \equiv 1$;

E2: for $n > 0$, $a^n \equiv a \times a^{n-1}$.

The real number a is referred to as the **base** and the integer n as an **exponent**.

It is an immediate consequence of Rule 8 (§A.9.1) that $a^n \neq 0$ for all $n \in \mathcal{N}$.

The reader may wonder, why should E1 and E2 be true. The answer is that E1 and E2 are definitions. Because they are definitions, they can be anything we want. So, if somewhere back in the dim and distant past someone had suggested that a^0 ought to be 17, for example, that would be permissible as a definition. However, it would fail a consistency test and would be rejected on that basis. The question that readers should keep in mind is:

Is this definition consistent with previous work?

The only way to get a feeling for what these definitions mean is to actually do some computations. For $a \neq 0$:

$$
\begin{aligned}
a^1 &\equiv a \times a^{1-1} = a \times a^0 = a \times 1 = a, \\
a^2 &\equiv a \times a^{2-1} = a \times a^1 = a \times a, \\
a^3 &\equiv a \times a^{3-1} = a \times a^2 = a \times a \times a, \\
a^4 &\equiv a \times a^{4-1} = a \times a^3 = a \times a \times a \times a, \\
a^5 &\equiv a \times a^{5-1} = a \times a^4 = a \times a \times a \times a \times a,
\end{aligned}
$$

and so forth. Each computation involves a positive integer as the exponent, so each must begin by applying E2 to get the first expression on the RHS of $\equiv$. The next step is to reduce the exponent to an single non-negative integer and either apply the result in the previous line, or, in the case of line one, apply E1.

Readers will observe that in each computation:

if you count the number of factors of a in each product on the far RHS, it is the same as the exponent.

This is consistent with our intention that we can think of exponentiation as repetitive multiplication. Notice that we have not used parentheses in these calculations because so long as the operations consist entirely of multiplication, there is no need to use them.

If we set $n = m + 1$ and substitute into E2, E2 is recast as E2':

$$a^{m+1} = a \times a^{(m+1)-1} = a \times a^m.$$

This is an alternative form of E2 that is in common use.

C.2.2   Multiplicative Inverses and Negative Exponents

Every non-zero real number has a multiplicative inverse. Indeed, since the criterion for determining whether two numbers are multiplicative inverses of one another is simply to check whether their product is 1, every non-zero real number can be viewed as a multiplicative inverse. This fact is recorded in Rule 4 of §A.9.1 as

$$\left(x^{-1}\right)^{-1} = x$$

and forces E1 and E2 to apply to multiplicative inverses in the same way as any other real number.

For example, consider, $a = \frac{1}{5}$. Then

$$\begin{aligned}
\left(\frac{1}{5}\right)^2 &= \frac{1}{5} \times \frac{1}{5} \\
&= \frac{1 \times 1}{5 \times 5} = \frac{1}{25} = \frac{1}{5^2}.
\end{aligned}$$

The fact that $\frac{1}{5}$ is a multiplicative inverse does not change the application of E1 and E2 in the first line. The second line merely follows the rule for multiplying unit fractions found in §14.2.1. Further, **no matter how we choose to represent $\frac{1}{5}$, we must get the same answer**. This means that all of the following must have the value $\frac{1}{25}$

$$5^{-1} \times 5^{-1} = (5^{-1})^2 = (5^2)^{-1} = 25^{-1}.$$

Considerations like these tell us how we should define the result of computations involving negative integer exponents.

Recall that since there are exactly three kinds of integers (see §19.1),

if $m \in \mathcal{I}$ and $m < 0$, then $0 < -m = n \in \mathcal{N}$, whence $-n = m$.

So, for $m \in \mathcal{I}$ with $m < 0$, we define a^m by:

E3: for $m < 0$, $a^m = a^{-n} \equiv (a^n)^{-1} = \frac{1}{a^n}$.

Since an arbitrary integer must be either positive, negative or zero, E1-E3 define a^k for all possible integers k. A simple numerical illustration of this is:

$$7^0 = 1, \quad 7^2 = 49, \quad 7^{-2} = \frac{1}{49}.$$

Again we reiterate that E3 is a definition, so the issue is not whether E3 is true, but rather whether E3 is consistent with past computations and preserves our intent. The expression on the far RHS results from an application of Rule 4 of §A.9.1. Defining the effect of negative exponents in this way preserves all the notation developed for multiplicative inverses as we will see in what follows.

C.3 Integer Powers of 10

The arithmetic of reals is based entirely on decimal numbers which involve integer powers of 10. As well, computations involving powers of 10 are ubiquitous in the modern world. For this reason, we give special attention to powers of 10 before developing the general rules governing the behavior of exponents. Fluidity with calculations involving exponents is expected in Grade 8. However the calculations of decimal arithmetic are studied in Grade 5 and it is useful for mentors to know how exponents work since they are the basis for many calculations.

C.3.1 Definitions for Powers of 10

To start, we define what we mean by 10^n, where $n \in \mathcal{I}$. The integer n is referred to as an **exponent** and the number 10 is called the **base**. This usage is consistent with describing the Arabic System as being a **base ten** system. The result, 10^n, is referred to as a **power of** 10 and we say

$$10^n \text{ is: } 10 \text{ raised to the } n^{\text{th}} \text{ power}.$$

To give an entirely numerical example, we would say

10^4 is: 10 **raised to the** 4th **power**.

Recasting E1-E3 for $a = 10$ gives: for $n \in \mathcal{I}$, the rules E1, E2 and E3 define the calculation of 10^n:

E1: for $n = 0$, $10^0 = 1$;

E2: for $n > 0$, $10^n = 10 \times 10^{n-1}$;

E3: for $n \geq 1$, $10^{-n} \equiv (10^n)^{-1} = \frac{1}{10^n}$.

The following detailed calculations show how these definitions are applied in the case of 10. Knowing the results of these calculations is critical to children so we go through them in explicit detail.[1]

C.3.2 Positive Powers of 10.

Suppose we want to know the value of 10^1. Using E2, we see that

$$\begin{aligned} 10^1 &= 10 \times 10^{(1-1)} = 10 \times 10^0 \\ &= 10 \times 1 = 10, \end{aligned}$$

where to get the second line, we use E1.

To find a value for 10^2:

$$\begin{aligned} 10^2 &= 10 \times 10^{(2-1)} = 10 \times 10^1 \\ &= 10 \times 10 = 100. \end{aligned}$$

Notice that in calculating 10^2 we can use the fact that we have already found $10^1 = 10$ to get the second line.

Similarly, we have

$$10^3 = 10 \times 10^{(3-1)} = 10 \times 100 = 1000,$$

and

$$10^4 = 10 \times 10^3 = 10 \times 1000 = 10000,$$

and so forth. So for positive n, the above simply translates into:

10^n **denotes the number written as a** 1 **followed by** n **zeros**.

---

[1]If the CCSS-M are successfully implemented, no lab instructor in post-secondary should ever again see a student use a calculator to multiply by 10, which has been a common observation about students since the advent of calculators.

Performing calculations like these should convince students that exponentiation should be nothing more than **repetitive multiplication**. Thus, 10^7 is merely 7 copies of 10 multiplied together with 1. Further, 10^{12} should be twelve copies of 10 multiplied together with 1, or in our usual Arabic notation, a 1 followed by 12 zeros. Moreover, to multiply an integer by 10, students should automatically write the digit 0 on the right.

C.3.3 Negative Integer Powers of 10

For convenience, we restate E3:

E3: for $n \geq 1$, $10^{-n} \equiv (10^n)^{-1} = \frac{1}{10^n}$.

The effective part of E3 is in the relation

$$10^{-n} \equiv (10^n)^{-1}$$

because on the RHS we now have an exact instruction for how to do a calculation. Since $n > 0$, we can use E2 to compute 10^n. We then use the fact that $(10^n)^{-1}$ is the multiplicative inverse of 10^n, and this leads directly to

$$(10^n)^{-1} = \frac{1}{10^n}.$$

If we apply E3 to 10^{-1} we have,

$$
\begin{aligned}
10^{-1} &= (10^1)^{-1} \\
&= 10^{-1} = \frac{1}{10},
\end{aligned}
$$

which is exactly what we needed. The actual use of E3 in the first line takes 10^{-1} and turns it into $(10^1)^{-1}$. Since $10^1 = 10$ by E2, we get what we want.

For higher powers, we have

$$
\begin{aligned}
10^{-2} &= (10^2)^{-1} \\
&= 100^{-1} = \frac{1}{100},
\end{aligned}
$$

$$
\begin{aligned}
10^{-3} &= (10^3)^{-1} \\
&= 1000^{-1} = \frac{1}{1000}
\end{aligned}
$$

and

$$10^{-6} = (10^6)^{-1}$$

$$= 1000000^{-1} = \frac{1}{1000000}.$$

$$\frac{1}{10} = 10^{-1}, \quad \frac{1}{100} = 10^{-2}, \quad \text{and} \quad \frac{1}{1000} = 10^{-3}$$

in understanding decimal arithmetic.

C.4 Additive Inverses and Exponents

Every non-zero real number a is the additive inverse of some **other** non-zero real number. The rules E1 and E2 do not ask anything about a other than whether it is zero. If a given number is not zero, then E1 and E2 apply. Any other property of the particular number in question has no effect on the discussion. However, as pointed out in §A.6.3, ambiguities can arise when other operations are involved, for example, when the centered dash which is used to denote the additive inverse. An **order of precedence** rule is used to eliminate this ambiguity.

Consider the expression -2^4. The centered dash instructs: *find the additive inverse of* $\boxed{?}$. The exponent 4, instructs: *apply E2 to* $\boxed{?}$. There are two questions. The first question is: Which operation is performed first? The second question is: To what does the operation apply, in other words, What's in the box?

The precedence rule tells us that

raising to a power always happens first.

The answer to the second question is that the operation of exponentiation is always applied to the **smallest expression possible**.

Let's see how this works for -2^4. The first rule tells us that we have to compute the power first, so that the computation will look like:

$$- \left(\boxed{?}^4 \right) = - \left(\boxed{?} \times \boxed{?} \times \boxed{?} \times \boxed{?} \right)$$

where we have used parentheses to enforce computing the fourth power first. The answer to the second question is that the smallest expression that can go in the box is the 2 sitting to the left of the power in -2^4. Thus, these two rules force the following computation

$$-2^4 = -(2 \times 2 \times 2 \times 2) = -16.$$

If we want to raise -2 to the power of 4, then we have to use parentheses as in

$$(-2)^4 = (-2) \times (-2) \times (-2) \times (-2) = 16.$$

Here, the parentheses force the exponent to apply to -2, instead of simply 2.

Another example of the application of the *smallest expression* rule is:

$$x \times y^2 = x \times y \times y.$$

The exponent applies only to the y and not to the x. If we want the exponent to be applied to both the x and the y we must use parentheses:

$$(x \times y)^2 = (x \times y) \times (x \times y) = x^2 \times y^2.$$

Ultimately it is expected that every child will apply this rule automatically as they read an expression.

C.5   Rules Governing Calculations

We want to develop the key rules for dealing with integer exponents. To accomplish this, we will consistently use our ideas regarding a^n as a guide. Specifically, because of E3, every calculation with integer exponents must come down to finding a^n for some positive integer n and some non-zero a.

C.5.1   The Addition Law for Exponents

We consider what happens when two non-negative powers of a are multiplied together, that is:

What is $a^n \times a^m$?

First note that the entire computation involves multiplication of factors of the **same** base, namely, a. There are n factors of a associated with a^n and m factors of a associated with a^m. Determining the total number of factors comes down to counting, or using the power of addition by simply computing $n+m$. Second, recall that because the only operation is multiplication, the Commutative and Associative Laws say we can compute the product of the factors in any order we want as long as we don't change the total number of factors of a which is fixed at $n+m$. Since the number of factors of a in a^{n+m} is exactly $n+m$, these facts convince us that

$$a^n \times a^m = a^{n+m}.$$

This equation is known as the **Addition Law for Exponents** and we prove it next. To do so, we will apply Theorem B.15 which asserts that if $K \subseteq \mathcal{N}$ and K is inductive, then $K = \mathcal{N}$. Indeed, we will apply Theorem B.15 in every proof in this section and all these proofs will have the **same** three-step pattern, so that once you understand the pattern, you simply need to check the details at specific points in the pattern to verify the proof in your own mind.

Theorem C.1. Let $a \neq 0$ and $n, m \in \mathcal{I}$. Then

$$a^n \times a^m = a^{n+m}.$$

Proof. Let $a \in \mathcal{R}$, $a \neq 0$ be fixed. Since $a^0 = 1$, when either m or n is 0, the required equality in the theorem follows by inspection.

To establish the required equality for positive integers, let $m \in \mathcal{N}$ be arbitrary but fixed, and set

$$K = \{k : k \in \mathcal{N} \text{ and } a^k \times a^m = a^{k+m}\}.$$

(Step 1: Define K to be the set of positive integers for which the required equality is true. Then by definition, $K \subseteq \mathcal{N}$.)

Next observe that since $a^1 = a$,

$$a^1 \times a^m = a \times a^m = a^{m+1}$$

by E2', so that $1 \in K$. (Step 2: Show that $1 \in K$.)

Let $n \in K$ so that $a^n \times a^m = a^{n+m}$. Now using the Commutative and Associative Laws for both $+$ and $\times$, we have:

$$\begin{aligned} a^{n+1} \times a^m &= (a \times a^n) \times a^m = a \times (a^n \times a^m) \\ &= a \times a^{n+m}) = a^{1+(n+m)} = a^{(n+1)+m}. \end{aligned}$$

Thus, $n \in K$ implies $n + 1 \in K$. (Step 3: Show K is closed under successor.)

Thus, K is inductive and by Theorem B.16 is equal to $\mathcal{N}$. Since $m \in \mathcal{N}$ was arbitrary, $a^n \times a^m = a^{n+m}$ holds for all pairs of positive integers m and n.

The argument above treats all pairs of non-negative integers. For two negative integers, $-n$ and $-m$ where $n, m \in \mathcal{N}$, by E3 and Rule 4 of §A.9.1 we have,

$$\begin{aligned} a^{-n} \times a^{-m} &= (a^n)^{-1} \times (a^m)^{-1} \\ &= \frac{1}{a^n} \times \frac{1}{a^m}. \end{aligned}$$

Applying Rule 12 to the bottom product on the RHS, followed by the equality just proved for positive integers, followed by E3 gives

$$
\begin{aligned}
a^{-n} \times a^{-m} &= \frac{1}{a^n} \times \frac{1}{a^m} \\
&= \frac{1}{a^n \times a^m} = \frac{1}{a^{n+m}} \\
&= a^{-(n+m)}.
\end{aligned}
$$

Thus the required equality holds for two negative integers.

All other cases involve one positive and one negative integer. Thus, we let $0 < n \leq m$ and then use a minus sign as required to make one of the integers negative. For the case of $n = m$ and $-n$, we have

$$
\begin{aligned}
a^n \times a^{-n} &= (a^n) \times (a^n)^{-1} \\
&= 1 = a^0 \\
&= a^{n+(-n)}
\end{aligned}
$$

so that equality holds. Recasting for clarity gives

$$
a^n \times a^{-n} = a^{n+(-n)} = a^0 = 1.
$$

In the last two cases $0 < n < m$ and we treat $-n$ and m, and n and $-m$. The required equality for these two cases can be established by applying the following fact:

$$
m = (m + (-n)) + n
$$

where $0 < m + (-n)$. Thus,

$$
a^m = a^{(m+(-n))+n} = a^{(m+(-n))} \times a^n.
$$

Since multiplication is associative, multiplication on both sides by a^{-n} gives:

$$
a^m \times a^{-n} = \left(a^{(m+(-n))} \times a^n\right) \times a^{-n} = a^{(m+(-n))}.
$$

By E3, for a^{-m} we have:

$$
\begin{aligned}
a^{-m} &= \frac{1}{a^m} = \frac{1}{a^{(m+(-n))+n}} \\
&= \frac{1}{a^{(m+(-n))} \times a^n} = \frac{1}{a^{(m+(-n))}} \times \frac{1}{a^n} \\
&= a^{-(m+(-n))} \times a^{-n}
\end{aligned}
$$

388

where we have used Rule 12 of §A.9.1 in the second line. Since $-(m+(-n)) = n-m$, we have

$$a^{-m} = a^{n-m} \times a^{-n}.$$

Multiplication by a^n gives

$$a^{-m} \times a^n = a^{n-m}$$

which is the last required equality. $\boxed{\text{qed}}$

It is essential to remember in applying the Additive Law for Exponents (Theorem C.1) that there is a common base which cannot be 0, as in

$$z^4 \times z^7, \quad 3^6 \times 3^5, \quad \text{or} \quad (x-5)^3 \times (x-5)^{-8}.$$

The Addition Law does not apply in a situation of $a^3 \times b^2$ because a and b are different bases.

Using the Addition Law for Exponents in the cases above we get

$$z^{11}, \quad 3^{11}, \quad \text{and} \quad (x-5)^{-5},$$

respectively.

As the following computation shows, E3 is completely consistent with the Addition Law for Exponents:

$$7^{-2} = 7^{((-1)+(-1))} = 7^{-1} \times 7^{-1}.$$

We stress there is no other way to assign a meaning to 7^{-2} that is consistent with E1 and E2 and their consequences.

C.5.2 Product Law for Exponents

The second law concerns products of exponents. Let m be a fixed positive integer. Observe:

$$(a^m)^2 = a^m \times a^m = a^{m+m} = a^{m \times 2}$$

and similarly,

$$(a^m)^3 = a^m \times a^m \times a^m = a^{m+m+m} = a^{m \times 3}.$$

We could repeat the above sequence to obtain

$$(a^m)^4 = a^{m \times 4} \quad \text{or} \quad (a^m)^5 = a^{m \times 5}$$

or indeed

$$(a^m)^n = a^{m \times n}$$

where n is any other positive integer.

Another way to think about this is to ask:

389

What is the base for the exponent n in $(a^m)^n$?

The base is a^m because the smallest thing the exponent n can apply to is a^m and by Rule 8 §A.9.1 and the definition of exponentiation, $a^m \neq 0$. So the computation has to be simply multiplying n copies of a^m together: the Addition Law for Exponents now tells us to add the exponents. Since all the exponents are the same, namely m, and there are n of them, the sum must be $m \times n$, which is exactly what the Product Law for Exponents asserts.

The equation for evaluating $(a^m)^n$ is known as the **Product Law for Exponents** and we establish it next.

Theorem C.2. Let $0 \neq a \in \mathcal{R}$. If $0 \leq m, \ n \in \mathcal{I}$, then

$$(a^m)^n = a^{m \times n}.$$

Proof. If either $m = 0$ or $n = 0$, the result follows by inspection. Let $m \in \mathcal{N}$ be arbitrary but fixed. Define K (Step 1) by:

$$K = \{k : k \in \mathcal{N} \ \text{and} \ (a^m)^k = a^{m \times k}\}.$$

Observe $1 \in K$ because E2 tells us that: $(a^m)^1 = a^m = a^{m \times 1}$ (Step 2). Next suppose $n \in K$ so that

$$(a^m)^n = a^{m \times n}.$$

Now multiply both sides of the last equation on the left by a^m to obtain:

$$a^m \times (a^m)^n = a^m \times a^{m \times n}.$$

Notice that on the LHS, we can treat the quantity a^m as the base. Viewing the expression this way leads to

$$a^m \times (a^m)^n = (a^m)^{n+1}.$$

The RHS is:

$$a^m \times a^{m \times n} = a^{(m \times n) + m} = a^{m \times (n+1)}$$

where the last equality is by virtue of the Addition Law for Exponents. Since equality is transitive, we have

$$(a^m)^{n+1} = a^{m \times (n+1)}$$

which means $n + 1 \in K$ and K is closed under successor (Step 3). Thus, $K = \mathcal{N}$, as required. $\boxed{\text{qed}}$

C.5.3   Product of Bases Rule for Exponents

The **Product of Bases Rule for Exponents** asserts:

$$(a \times b)^n = a^n \times b^n.$$

To see why this should be true, consider $(a \times b)^3$. Using E2 we quickly obtain

$$(a \times b)^3 = (a \times b) \times (a \times b) \times (a \times b)$$

an expression that only involves multiplication. So the Associative and Commutative Laws guarantee we can rewrite it however we want, specifically as

$$(a \times b)^3 = (a \times a \times a) \times (b \times b \times b) = a^3 \times b^3.$$

A similar computation can be performed for any integer value of n. The point is that in $(a \times b)^n$, there are n factors of a and n factors of b and multiplying these factors together generates a^n and b^n, respectively.

The reader should be warned: this rule for products raised to a power applies **only to products and never to sums**.

Theorem C.3. If a, $b \in \mathcal{R}$ such that $a \times b \neq 0$ and $0 \leq n \in \mathcal{I}$, then

$$(a \times b)^n = a^n \times b^n.$$

Proof. Let a, $b \in \mathcal{R}$ such that their product is not 0. If $n = 0$, then the required equation follows by inspection. Define K by (Step 1):

$$K = \{k : k \in \mathcal{N} \text{ and } (a \times b)^k = a^k \times b^k\}.$$

Recalling the computation of a^1 using E2:

$$(a \times b)^1 = a \times b = a^1 \times b^1$$

so that $1 \in K$ (Step 2).

Next, suppose $n \in K$ so that

$$(a \times b)^n = a^n \times b^n.$$

Multiply both sides of this equality on the left by $a \times b$ to obtain:

$$(a \times b) \times (a \times b)^n = (a \times b) \times (a^n \times b^n).$$

By using the Commutative and Associative Laws and E2, we can obtain

$$(a \times b)^{n+1} = a^{n+1} \times b^{n+1}$$

so that $n+1 \in K$ and K is closed under successor (Step 3), whence $K = \mathcal{N}$. $\boxed{\text{qed}}$

C.5.4 Rule for Fractional Forms and Exponents

There is one other extremely useful fact which we state as our last theorem.

Theorem C.4. Let $a \neq 0$ and $n \in \mathcal{I}$, then

$$a^{-n} = \left(\frac{1}{a}\right)^n.$$

Proof. Let $a \in \mathcal{R}$, $a \neq 0$. We first establish the required equality for $n \in \mathcal{N}$.

E3 asserts: for $n \in \mathcal{N}$, $a^{-n} \equiv \frac{1}{a^n}$. By Rule 4 of §A.9.1, $\frac{1}{a^n}$ is the unique multiplicative inverse of a^n. So to obtain the required equality, all we have to do is show $\left(\frac{1}{a}\right)^n$ is also a multiplicative inverse for a^n. Thus, define K by (Step 1):

$$K = \{k : k \in \mathcal{N} \text{ and } a^k \times \left(\frac{1}{a}\right)^k = 1\}.$$

Since for $a \neq 0$, $a^1 = a$ by E2, $1 \in K$ (Step 2).

Assume that for some n, $n \in K$ so that

$$a^n \times \left(\frac{1}{a}\right)^n = 1.$$

Further, $a \times \frac{1}{a} = 1$. By multiplying these last two equations together and using the full power of the Associative and Commutative Laws, we have:

$$(a^n \times a) \times \left[\left(\frac{1}{a}\right)^n \times \frac{1}{a}\right] = \left[a^n \times \left(\frac{1}{a}\right)^n\right] \times \left[a \times \frac{1}{a}\right]$$
$$= 1 \times 1 = 1.$$

Applying E2 to each of the two products on the LHS above we obtain the LHS below:

$$a^{n+1} \times \left(\frac{1}{a}\right)^{n+1} = 1$$

whence $n + 1 \in K$ and K is closed under successor (Step 3). Thus $K = \mathcal{N}$ and $a^{-n} = \frac{1}{a^n}$ for $n \in \mathcal{N}$.

To complete the proof, let $m \in \mathcal{I}$ with $m < 0$. Then $m = -k$ for some $k \in \mathcal{N}$. We want to show that

$$a^{-m} = \left(\frac{1}{a}\right)^m.$$

By Rule 1 of §A.9.1, $-m = -(-k) = k$, so the the LHS satisfies $a^{-m} = a^k$. The RHS satisfies

$$\left(\frac{1}{a}\right)^m = \left(\frac{1}{a}\right)^{-k} = \left(\left(\frac{1}{a}\right)^k\right)^{-1}$$
$$= \left(\frac{1}{a^k}\right)^{-1} = a^k.$$

Since the LHS and the RHS of $a^{-m} = \left(\frac{1}{a}\right)^m$ are both equal to a^k, we conclude this is a valid equality completing the proof. $\boxed{\text{qed}}$

By using this theorem together with Theorem C.1, we can extend Theorems C.2-3 to all integers.

Our principal use of Theorem C.4 will be for powers of 10 in the context of decimal numerals. The key idea to remember is that for negative integer powers of 10:

if $-n$ is a negative integer, 10^{-n} is the unit fraction whose denominator is a 1 followed by n zeros, that is, $\frac{1}{10^n}$.

C.5.5   Summary of Rules for Exponents

Let a and b be fixed non-zero real numbers and n, m denote integers. The following two equations define exponential notation for non-negative integers:

E1: if $n = 0$, $a^0 = 1$;

E2: if $n > 0$, then $a^n = a \times a^{n-1}$.

The following equation governs calculations when the exponent is a negative integer:

E3: If $m = -n$ for some $n \in \mathcal{N}$, $a^m = (a^n)^{-1} = \frac{1}{a^n}$.

The Addition Law for Exponents:

$$a^n \times a^m = a^{n+m}.$$

The Product Law for Exponents:

$$(a^n)^m = a^{n \times m}.$$

The Product of Bases Law for Exponents:

$$(a \times b)^n = a^n \times b^n;$$

The Fractional Forms Law for Exponents:

$$a^{-n} = \frac{1}{a^n} = \left(\frac{1}{a}\right)^n.$$

Finally, there is one order of precedence rule that asserts:

in any expression, an exponent applies to the smallest part of the expression possible.

Bibliography

[1] S.K. Bleiler and D.R. Thompson, *Multidimentional Assessment of the CCSS-M*, **Teaching Children Mathematics**, Dec., 2012, 292-300.

[2] Charles M. Blow, *The Common Core and the Common Good*, NYT, 21 August, 2013.

[3] Leland Cogan, *et al.*, *Implementing the Common Core State Standards for Mathematics: A Comparison of Current District Content in 41 States*, Education Policy Center at Michigan State University, **WP32**, 2013, available online.

[4] Leland Cogan, *et al.*, *Implementing the Common Core State Standards for Mathematics: What We Know about Teachers of Mathematics in 41 States*, Education Policy Center at Michigan State University, **WP33**, 2013, available online.

[5] Leland Cogan, *et al.*, *Implementing the Common Core State Standards for Mathematics: What Parents Know and Support*, Education Policy Center at Michigan State University, **WP34**, 2013, available online.

[6] H.S. and C.M. Gaskill, *Parents' Guide to Common Core Arithmetic*, 2014, available from Amazon.

[7] H.S. Gaskill and P.P. Narayanaswami, *Elements of Real Analysis*, Prentice Hall, 1998, available from Amazon.

[8] Edmund Landau, **Foundations of Analysis**, available from Amazon.

[9] C. Lubienski and S. Lubienski, *The Public School Advantage*, 2014, available from Amazon.

[10] J.D. Monk, **Introduction to Set Theory**, available from Amazon.

[11] W. Schmidt and L. Cogan, *The Myth of Equal Content*, 2009, available online at: www.ascd.org/publications/educational-leadership/nov09/vol67/num03/The-Myth-of-Equal-Content.aspx

[12] W. Schmidt, R. Houang and L. Cogan, *A Coherent Curriculum: The Case of Mathematics*, **American Educator**, Summer 2002; available on-line.

[13] W.H. Schmidt, *et al.*, *A splintered vision: An investigation of U.S. science and mathematics education*, available from Amazon.

[14] J.R. Shoenfield, *Introduction to Mathematical Logic*, available from Amazon.

[15] Uri Treisman, *Iris M Carl Equity Address: Keeping Our Eyes on the Prize*, NCTM, Denver, April 19, 2013, available online at: www.nctm.org.

[16] Z. Usiskin, (2012) www.icme12.org/upload/submission/1881_F.pdf.

[17] www.medicaldaily.com/math-skills-childhood-can-permanently-affect-brain-formation-later-life-298516

[18] www.nature.com/neuro/journal/vaop/ncurrent/full/nn.3788.html.

[19] www.corestandards.org/about-the-standards/development-process/.

[20] *Shocking Number Of Canadian University Grads Don't Hit Basic Literacy Benchmark*, **The Huffington Post Canada**, Posted: 04/29/2014.

[21] http://nces.ed.gov/ the main site for educational data.

[22] http://nces.ed.gov/pubsearch/pubsinfo.asp?pubid=2014028 contains information on the most recent PISA.

[23] http://nces.ed.gov/nationsreportcard/ contains data from the National Assessment of Educational Progress.

[24] www.parcconline.org/parcc-assessment contains sample tests.

[25] www.wbez.org/news/education/chicago-teachers-union-votes-oppose-common-core-110152 Chicago teacher's union objection to CCSS.

[26] www.newsobserver.com/2014/09/22/4174322_common-core-review-begins.html?rh=1 North Carolina CCSS review.

[27] www.nctm.org/uploadedFiles/About_NCTM/Position_Statements/ contains **Formative Assessment**, a position of the National Council of Teachers of Mathematics.

Index

www.ingramcontent.com/pod-product-compliance
Lightning Source LLC
Chambersburg PA
CBHW080757180526
45168CB00006B/2235